Guide to Maritime Informatics

Alexander Artikis · Dimitris Zissis
Editors

Guide to Maritime Informatics

 Springer

Editors
Alexander Artikis
Department of Maritime Studies
University of Piraeus
Piraeus, Greece

Institute of Informatics &
Telecommunications
NCSR
Demokritos, Greece

Dimitris Zissis
Department of Product and Systems
Design Engineering
University of the Aegean
Syros, Greece

ISBN 978-3-030-61854-4   ISBN 978-3-030-61852-0   (eBook)
https://doi.org/10.1007/978-3-030-61852-0

This Springer imprint is published by the registered company Springer Nature Switzerland AG
The registered company address is: Gewerbestrasse 11, 6330 Cham, Switzerland

Preface

In the last 25 years, information systems have had a disruptive effect on society and business. Up until recently though, the majority of passengers and goods were transported by sea in many ways similar to the way they were at the turn of the previous century. Gradually, advanced information technologies are being introduced, in an attempt to make shipping safer, greener, more efficient and transparent. The emerging field of Maritime Informatics studies the application of Information Technology and Information Systems to maritime transportation. Maritime informatics can be considered as both a field of study and domain of application. As an application domain, it is the outlet of innovations originating from Data Science and Artificial Intelligence, while as a field of study, it sits on the fence between Computer Science and Marine Engineering. Its complexity lies within this duality, as it is faced with disciplinary barriers while demanding a systemic transdisciplinary approach.

At present, there is a growing body of knowledge developing, which remains undocumented in a single source or textbook designed to ease students and practitioners into this new field. The objective of this book is to collect the material required for an undergraduate or postgraduate student to develop the core knowledge of this domain, in an analytical approach through real-world examples and case studies. The aim is to present our audience with an overview of the main technological innovations which are having a disruptive effect on the maritime industry, describe their principal ideas, methods of operation and applications, and discuss future developments. The book is designed in such a way as to first introduce required knowledge, algorithmic approaches and technical details, before presenting real-world applications.

The book is structured around four main pillars. First, we focus on Maritime Data. Bereta et al. present a comparative analysis of the so-called Maritime Reporting Systems, such as the well-known Automatic Identification System (AIS). Moreover, Tzouramanis draws up a compilation of reliable, freely available maritime datasets. This way, readers are set on the path to successfully finding and selecting the data that fit their needs, and to navigating effortlessly the sea of on-line information.

Second, we present key techniques for Off-line Maritime Data Processing. Etienne et al. provide a step-by-step guide for building a relational maritime database, thus supporting the investigation of maritime traffic and vessel behaviour. Tampakis et al. present pre-processing techniques for cleaning, transforming and partitioning long GPS traces into meaningful portions of vessel movement, and overview the representative maritime knowledge discovery techniques, such as trajectory clustering, group behaviour identification, hot-spot analysis, frequent route and network discovery and data-driven predictive analytics. Subsequently, Andrienko and Andrienko show how Visual Analytics approaches can help in exploring properties of the maritime data, detecting problems and finding ways to clean and improve the data.

Third, we present three key steps for On-line Maritime Data Processing. Patroumpas presents techniques for maintaining summarised representations of vessel trajectories in an on-line fashion, based on AIS data streams. Santipantakis et al. present algorithms for on-line link discovery, i.e. identifying spatio-temporal relations between vessels, and areas of interest, such as when two vessels are close to each other, or when a vessel sails within a protected area. Then, Pitsikalis and Artikis present a formal computational model for combing the compressed trajectories and the spatio-temporal relations, in order to detect, in real time, composite maritime activities, such as ship-to-ship transfer and (illegal) fishing.

Fourth, we focus on Applications. Jousselme et al. present a comparative analysis of uncertainty representation and reasoning techniques with respect to maritime route deviation. Ducruet et al. apply graph theory and complex network methods to AIS data, in order to analyse the key properties of maritime networks. Finally, Adland shows how AIS data may guide data-driven market analysis in shipping. He outlines how AIS data may be used to track commodity flows and analyse the key variables for fleet efficiency, supply and demand.

To facilitate understanding, and allow for a consistent presentation, all chapters illustrate the presented techniques using AIS data. In particular, Chaps. 3–9 use the open, heterogeneous, integrated dataset for maritime intelligence that concerns the area of Brest, France.[1]

Moreover, the book is accompanied by a dedicated web page with on-line educational material, including datasets, SQL queries, algorithm implementations and visualisations:

http://maritime-informatics.com

The web page will be constantly updated, offering an up-to-date source of information for students and instructors.

---

[1]https://doi.org/10.5281/zenodo.1167595.

We hope that you will enjoy reading this book and that you will find answers to many related questions within. The overall aim has been to provide the audience with an informative book which will help introduce students to a range of related topics, while supporting practitioners in staying up to date with this fast-changing field.

Piraeus, Greece   Alexander Artikis
Syros, Greece   Dimitris Zissis

Acknowledgements

We are indebted to the authors of all chapters for making this book possible. We would also like to thank Manolis Pitsikalis and Alexandros Troupiotis-Kapeliaris for their invaluable, constant support in producing the book and its on-line material.

We have been supported by the INFORE project, which received funding from the European Union's Horizon 2020 research and innovation programme, under grant agreement No 825070.

Athens, Greece	Alexander Artikis
July 2020	Dimitris Zissis
	Editors

Contents

Part IV   Applications

Acronyms

ABQS	Amnesic Bounded Quadrant System
ACID	Atomicity, Consistency, Isolation and Durability
AIS	Automatic Identification System
API	Application Programming Interface
ASPL	Average Shortest Path Length
ATR	Automatic Target Recognition
AUC	Area Under Curve
BBA	Basic Belief Assignment
BPA	Basic Probability Assignment
BQS	Bounded Quadrant System
BTREE	Binary Tree
CEP	Composite Event Processing
CER	Composite Event Recognition
COG	Course over Ground
CRS	Coordinate Reference System
CSTDMA	Carrier Sense Time Division Multiple Access
CSV	Comma Separated Values
DBMS	Database Management System
DBSCAN	Density-Based Spatial Clustering of Applications with Noise
DC	Data Centre
DDP	Data Distribution Plan
DGPS	Differential GPS
DP	Douglas-Peucker algorithm
DWT	Deadweight
EMODnet	European Marine Observation and Data Network
eNav	electronic Navigation
EO	Earth Observation
EPIRB	Emergency Position-Indicating Radio Beacon
EPSG	European Petroleum Survey Group
ERC	European Research Council

ESA	European Space Agency
ESRI	Environmental Systems Research Institute
ETA	Estimated Time of Arrival
FBQS	Fast Bounded Quadrant System
FFA	Forward Freight Agreements
FMC	Fishery Monitoring Centre
GBT	Generalised Bayes Theorem
GCMP	General Co-Movement Pattern
GIS	Geographic Information System
GiST	Generalised Search Tree
GPS	Global Positioning System
GT	Gross Tonnage
IMO	International Maritime Organisation
ITU	International Telecommunication Union
IUCN	International Union for Conservation of Nature
KDE	Kernel Density Estimation
KML	Keyhole Markup Language
LD	Link Discovery
LNG	Liquified Natural Gas
LRIT	Long Range Identification and Tracking system
MAP	Maximum A Posteriori
MBR	Minimum Bounding Rectangle
MDL	Minimum Description Length
MER	Maximum Enclosed Rectangle
MMSI	Maritime Mobile Service Identity
MOB	Man Overboard
MoG	Mixture-of-Gaussian
MSA	Maritime Situational Awareness
MST	Minimum Spanning Tree
NAS	Navigational Assistance Service
NaTS	Neighbourhood-aware Trajectory Segmentation
NOAA	National Oceanic and Atmospheric Administration
NoSQL	Not-only-SQL
ODMG	Object Data Management Group
OGC	Open Geospatial Consortium
OODBMS	Object-Oriented DBMS
OPERB	One-Pass Error-Bounded algorithm
OQL	Object Query Language (of OODBMS)
RCC8	8 Region Connection Calculus
RMSE	Root Mean Square Error
ROC	Receiver Operating Characteristic
RoT	Rate of Turn
RTEC	Event Calculus for Run-Time reasoning
SaCO	Sampling, Clustering and Outlier
SAR	Search and Rescue

SED	Synchronous Euclidean Distance
SHP	Shapefile of Geometric Objects
SOG	Speed over Ground
SOLAS	Safety of Life at Sea
SOTDMA	Self-Organising Time Division Multiple Access
SQL	Structured Query Language
SQUISH	Spatial QUalIty Simplification Heuristic
SRID	Spatial Reference Identifier
STC	Space-Time Cube
TBM	Transferable Belief Model
THS	Trajectory Hot Spot
TREAD	Traffic Route Extraction and Anomaly Detection
URREF	Uncertainty Representation and Reasoning Framework
URRTs	Uncertainty Representation and Reasoning Techniques
VDES	VHF Data Exchange System
VMS	Vessel Monitoring System
VTS	Vessel Traffic System
WGS84	World Geodetic System 1984

Part I
Maritime Data

Chapter 1
Maritime Reporting Systems

Konstantina Bereta, Konstantinos Chatzikokolakis, and Dimitris Zissis

Abstract In recent years, numerous maritime systems track vessels while travelling across the oceans. Ship reporting systems are used to provide, gather or exchange information through radio reports. This information is used to provide data for multiple purposes including search and rescue, vessel traffic services, prevention of marine pollution and many more. In reality though researchers and scientists are finding out that these data sets provide a new set of possibilities for improving our understanding of what is happening or might be happening at sea. This chapter provides an introduction to the main vessel reporting systems available today, while discussing some of their shortcomings and strong points. In this context, several applications and potential uses are described.

1.1 Introduction

Unlike in the past, when tracking ships during their long voyages at sea was hampered by the lack of robust systems and data, nowadays numerous reporting systems are constantly reporting vessel positions. Today there are at least 23 mandatory commercial ship reporting systems, adopted by the International Maritime Organization (IMO) in accordance with the Safety of Life at Sea (SOLAS) regulation V/11 in the world. The SOLAS convention is a maritime treaty aiming at the establishment of safety specifications in the construction of vessels and the installation of equipment. Flag states are the states in which vessels are registered under their flag. Flag states should ensure that the vessels under their flag comply with the standards of the SOLAS convention.

K. Bereta (✉) · K. Chatzikokolakis
MarineTraffic, Athens, Greece
e-mail: konstantina.bereta@marinetraffic.com

K. Chatzikokolakis
e-mail: konstantinos.chatzikokolakis@marinetraffic.com

D. Zissis
University of the Aegean, Syros, Greece
e-mail: dzissis@aegean.gr

© Springer Nature Switzerland AG 2021
A. Artikis and D. Zissis (eds.), *Guide to Maritime Informatics*,
https://doi.org/10.1007/978-3-030-61852-0_1

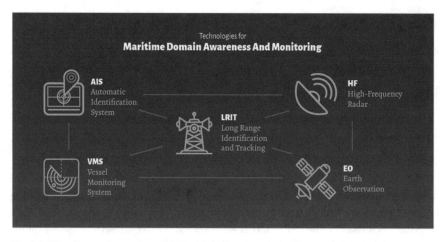

Fig. 1.1 Use of reporting systems for Maritime Situational Awareness

Figure 1.1 shows some of the vessel monitoring systems that are currently used to track vessels. In general, systems that can track vessels can be divided into two broad categories: cooperative and non-cooperative systems. Cooperative systems rely on the vessels crews' collaboration to identify and report the information about a vessel, while non-cooperative systems are designed to detect and track vessels that do not rely on the vessels crews' collaboration. Cooperative systems include the Automatic Identification System (AIS), the Vessel Monitoring System (VMS), the Long Range Identification and Tracking (LRIT) system, and others. Non-cooperative systems on the other hand, include coastal and high-frequency (HF) radar, active and passive sonar, ground- or vessel-based cameras (e.g., thermal), and satellite and airborne Earth Observation (EO) systems. EO systems can be divided into optical (generally visual and near-infrared) and Synthetic Aperture Radar (SAR) systems. Yet, despite the variety of tracking systems available and the massive data flows they produce, the "views" obtained by such data remain largely partial. In this context, data fusion is critical as it combines multi-origin information to determine relationships among the data; thus improving the understanding of a current complex environment. The advantage in fusing data from multiple sensors and sources is that the final estimated vessel trajectories are more accurate and with better confidence, extending to features that are impossible to perceive with individual sensors and sources, in less time, and at a lower cost. Also, it contributes towards better coverage and robustness to failure, thus improving the reliability and quality of the situational picture.

In this chapter we describe the most popular reporting systems, including those that are currently being developed and are expected to be used in the future. This chapter is intended for readers who wish to begin research in the field of maritime information systems. As such, emphasis is placed on a description of the available reporting systems, their technical details, produced datasets, while references are provided for further information. Some examples are provided as to how these datasets can offer insights into shipping patterns.

This chapter is organised as follows. Section 1.2 describes the Automatic Identification System (AIS). Section 1.3 describes VHF Data Exchange System (VDES) and Sect. 1.4 describes the Long Track Identification and Tracking system. Section 1.5 describes the Vessel Monitoring System for fisheries and Sect. 1.6 describes advanced applications that rely on data that derive from vessel reporting systems. Finally, Sect. 1.7 concludes the chapter and Sect. 1.8 presents some bibliographical notes.

1.2   The Automatic Identification System

The Automatic Identification System (AIS) was originally conceived as a navigational safety system to support vessel traffic services (VTS) in ports and harbours, but soon after its adoption, and especially after the International Maritime Organization (IMO) mandated AIS transceivers to be installed on-board a significant number of commercial vessels, it became the most popular vessel tracking system.

The Automatic Identification System [13] allows for efficient exchange of navigational data between ships and shore stations, thereby improving safety of navigation. Although this system was intended to be used primarily for safety of navigation purposes in ship-to-ship use (e.g., prevention of collisions), it may also be used for other maritime safety related communications, provided that the primary functions are not impaired; the system is autonomous, automatic, continuous and operates primarily in a broadcast. The system automatically broadcasts navigational information about vessels along with vessel characteristics (e.g., name) to all other installations (e.g., AIS stations, on-board transceivers) in a self-organized manner. Data transmissions are made in the VHF maritime mobile band. Since its introduction, AIS data has proven useful for monitoring vessels and extracting valuable information regarding vessel behaviour, operational patterns and performance statistics.

In the rest of this section, we briefly describe the AIS communication protocol, provide detailed analysis of the information broadcasted through the AIS including sample datasets and argue on the common issues and shortcomings of AIS.

1.2.1   AIS Equipment

AIS is based on the use of dedicated equipment that should be installed aboard (vessel stations), ashore (base stations) or on dedicated satellites (AIS-SAT). Vessels are equipped with transponders, e.g., stations that send and receive AIS messages. Transponders can either be class A or class B and have an integrated GPS that tracks the movement of the vessel it is installed on. The differences between class A and class B transponders will be presented in the next section. Base stations that are installed ashore are equipped with AIS receivers that receive AIS messages from vessels, but do not transmit. Dedicated satellites are also equipped with AIS receivers and this is very useful for areas with no or low coastal coverage (i.e., with no AIS receiver nearby).

Since 2004, the International Maritime Organization (IMO) requires that all commercial vessels over 300 Gross Tonnage (GT) travelling internationally to carry a Class A AIS transponder aboard. Vessels that do meet these requirements (e.g., smaller vessels, pleasure crafts, etc.) can be equipped with Class B AIS transponders. This requirement of IMO followed the 2002 SOLAS (Safety of Life at Sea) agreement's relative mandate.

AIS transponders use two dedicated VHF channels, AIS-1 (161.975 Mhz) and AIS-2 (162.025 Mhz). Class A transponders implement the Self Organizing Time Division Multiple Access (SOTDMA) protocol. The SOTDMA protocol is based on the division of time in slots. More specifically, a second is divided into 2250 slots, which means that base stations can receive at most one transmission every 26.67 ms. Each vessel should reserve a dedicated time slot in order to transmit an AIS message so that no other vessel transmits at the same time.

Class B transponders use the CSTDMA (Carrier Sense Time Division Multiple Access) protocol which interweaves with Class A transmissions by giving priority to SOTDMA transmissions.

1.2.2 AIS Messages

The ITU 1371-4 standard defines 64 different types of AIS messages that can be broadcast by AIS transceivers. These include Types 1, 2, 3, 18, and 19 which are position reports, including latitude, longitude, speed-over-ground (SOG), course-over-ground (COG), and other fields related to ship movement; while type 5 messages contain static-and-voyage information, which includes the IMO identifier, radio call sign (e.g., a unique designation for each radio station), ship name, ship dimensions, ship and cargo types. Other types of messages include supplementary information about the vessel and are not needed for monitoring the vessels' mobility, thus are not further described in this chapter.

AIS messages are distinguished on the following two categories: (i) dynamic, and (ii) static. Dynamic messages contain positional data about voyages. Static messages contain information related to vessel characteristics. The information (e.g., flag) changes less frequently than the respective information included in static messages. The information contained in both types of AIS messages is described below.

1.2.3 Dynamic AIS Messages

The dynamic AIS messages contain the following attributes:

- *Maritime Mobile Service Identity Number (MMSI)*. The MMSI is an identification number for each vessel station. However, it is not a unique identifier, as we will explain later in this chapter.
- *Rate of Turn*. This field contains data regarding the angle that the vessel turns right or left per minute. The values of this field range from 0 to 720 degrees.
- *Speed over Ground*. Speed over ground is the speed of the ship with respect to the ground. The value range of this attribute is from 0 to 102 knots (0.1-knot resolution).
- *Position Coordinates*. This field contains the latitude and the longitude of the position of the vessel.
- *Course over Ground (COG)*. COG describes the direction of motion with respect to the ground that a vessel has moved relative to the magnetic north pole or geographic north pole. The values are degrees up to 0.1° relative to true north.
- *Heading*. Heading describes the direction that a vessel is pointed at any time relative to the magnetic north pole or geographic north pole. Heading takes values from 0 to 359 degrees.
- *UTC seconds*. This is the second part of the timestamp when the subject data-packet was generated (in UTC time).
- *AIS Navigational status*. This field represents the navigational status of the vessel and it is completed manually by the crew. The different types of navigational status that can be reported in an AIS message are the following:

  - *Under way using engine*: the vessel is travelling using its engine.
  - *Anchored*: The vessel is not travelling and has dropped an anchor.
  - *Not under command*: The vessel is uncommanded. This can be either due to a hardware malfunction or a problem of the crew (e.g., the commander is injured).
  - *Restricted maneuverability*: There are constraints regarding the motion of the vessel (e.g., tugging heavy load).
  - *Constrained by her draught*: The draught (or draft) of a vessel is the vertical distance between the waterline and the bottom of the hull (keel). The draught is an indicator that shows how much a vessel is loaded. So, a vessel carrying heavy load might have maneuverability restrictions (e.g., navigating in shallow waters).
  - *Moored*: The vessel is moored at a fixed point (e.g., dock).
  - *Aground*: The vessel is touching the ground, after navigating in shallow water.
  - *Engaged in fishing activity*: The vessel is currently fishing.
  - *Underway sailing*: The vessel is travelling using sails instead of engine (it applies to sailing vessels).
  - AIS-SART (active), MOB-AIS, EPIRB-AIS. The AIS-SART is a self-contained radio device used to locate a survival craft or distressed vessel by sending updated position reports. MOB-AIS are personal beacons with an integrated GPS that can be used by a shipwrecked person to transmit a "Man Overboard" alert in order to be tracked by rescue services. The Emergency Position Indicating Radio Beacon (EPIRB) is installed on vessels to facilitate the search and rescue operations on case of emergency. Every EPIRB is registered through the

national search and rescue organisation associated to the vessel is installed on. In case of emergency, it is manually or automatically (e.g., when it touches the water) activated sending distress signals (e.g., emergency alerts) through AIS. These signals are received by the search and rescue services that are closer to the area of the accident.

Apart from the navigational status values described above, there are also values that are reserved for future use. For example, there is a placeholder for future amendment of navigational status for ships carrying dangerous goods (DG), harmful substances (HS), marine pollutants (MP) or IMO hazard or pollutant categories, high-speed craft (HSC), and wing in ground (WIG).

1.2.4 Static AIS Messages

Static information is provided by a subject vessel's crew and is transmitted every 6 min regardless of the vessel's movement status. The static AIS messages contain the following fields:

- *International Maritime Organisation number (IMO)*. This is a 9-digit number that uniquely identifies the vessel. Please note that this is not the same as the MMSI. The IMO number is assigned by IHS Maritime (Information Handling Services) when the vessel was constructed.[1] The MMSI can change, for example when the owner changes. Only propelled, seagoing vessels of 100 gross tons and above are assigned an IMO number.
- *Call Sign*. The international radio call sign assigned to the vessel by her country of registry.
- *Name*. The name of the vessel.
- *Type*. The type of the vessel (e.g., Tanker).
- *Dimensions*. Dimensions of ship in meters. More specifically, this field refers to: (a) dimension to bow, (b) dimension to stern, (c) dimension to port (left side of the vessel when facing the bow), and (d) dimension to starboard (i.e., right side of the vessel when facing the bow).
- *Location of the positioning system's antenna on-board the vessel.*
- *Type of positioning system (GPS, DGPS, Loran-C)*. Differencial GPS (DGPS) is a positioning system that performs positional corrections to GPS, providing more accurate positioning data. Loran-C (Long-range navigation) is a hyperbolic radio navigation system that allows a receiver to determine its position by listening to low frequency radio signals transmitted by fixed land-based radio beacons. Although Loran-C system is old, it can be used as backup system to the GPS, since GPS can be spoofed or jammed.

---

[1] https://ihsmarkit.com/products/imo-ship-company.html.

- *Draught*. The term draught (or draft) refers to the vertical distance between the waterline and the bottom of the hull (keel), with the thickness of the hull included. The value of this field is measured in meters.
- *Destination*. The destination as completed manually by the crew of the subject vessel (free text).
- *Estimated time of arrival (ETA)*. This is a UTC timestamp completed manually by the crew indicating the estimated time of arrival at destination.

There are some differences exchanged between vessels with class A and class B transponders, as shown in Table 1.1. For example, most of the vessels for which it is not mandatory to have class A transponders do not have an IMO, so this attribute is not used in messages sent from class B transponders. Rate of turn, navigational status, destination and ETA reports are also attributes that are not used in Class B AIS messages. Also, vessels with class B transponders do not transmit but are able to receive AIS messages related to safety.

Tables 1.2, 1.3, and 1.4 describe the reporting rates of vessels with class A and class B transponders. The AIS reporting rates of vessels depend on their navigational status (e.g., whether they are underway using engine or moored), their speed and course changes, and whether they are quipped with class A or class B transponders. Both class A and class B vessels that are anchored/moored or move very slow (up to 2 knots) send AIS messages every 3 min. Class B vessels with speed more than 2 knots send messages every 30 s, while class A have higher reporting rate that increases as the vessel accelerates and/or it is changing course. For example, a class A vessel that navigates with speed up to 14 knots needs to send AIS messages every 10 s.

1.2.5 Example Dataset

We provide example AIS messages through sample data of a dataset that is publicly available.[2] Table 1.5 shows a sample of static AIS messages (as the ones presented in Sect. 1.2.4 above) that contains the following attributes: MMSI, IMO, Call sign, Name, Type (the code corresponding to the vessel type), Dimension to bow, Dimension to stern, Dimension to port, Dimension to Starboard, Estimated arrival time (ETA), Draught, Destination, and the timestamp when the AIS message was received by an AIS receiver (e.g., terrestrial, satellite, or an AIS transceiver installed aboard the vessel). It is apparent from the table that some fields may be missing in some messages (e.g., destination in messages 2 and 4), or invalid in other (e.g., draught reported in messages 2 and 4). Table 1.6 depicts sample dynamic AIS messages from the same dataset. The dynamic messages contain the following attributes: MMSI, the code that corresponds to the navigational status of the vessel, the rate of turn, the speed over ground, the course over ground, the heading, the location of the vessel (longitude and latitude dimensions), and the timestamp.

---

[2]https://zenodo.org/record/1167595#.XQtPIY9RUuU.

Table 1.1 Attributes of AIS messages exchanged using Class A and Class B transponders [13]

Data	Class A (receive)	Class B (send)	Class B (receive)
Call sign	Yes	Yes	Yes
IMO	Yes	No	No
Length and beam	Yes	Yes	Yes
Antenna location	Yes	Yes	Yes
Draught	Yes	No	No
Cargo information	Yes	Yes	Yes
Destination	Yes	No	No
Estimated time of arrival	Yes	No	No
Time	Yes	Yes	Yes
Ship's position	Yes	Yes	Yes
Course over ground	Yes	Yes	Yes
Speed over ground	Yes	Yes	Yes
Gyro heading	Yes	Yes	Yes
Rate of turn	Yes	No	No
Navigational status	Yes	No	No
Safety message	Yes	No	Yes

Table 1.2 Class A systems

Ship's dynamic conditions	Reporting rate
Anchored/Moored	3 min
0-14 knots	10 s
0-14 knots and changing course	3.33 s
14-23 knots	6 s
14-23 knots and changing course	2 s
Faster than 23 knots	2 s
Faster than 23 knots and changing course	2 s

Table 1.3 Class B systems

Ship's dynamic conditions	Reporting rate
0-2 knots	3 min
Above 2 knots	30 s

Table 1.4 Other AIS sources

Special conditions	Reporting rate
Search and Rescue (SAR) aircraft	10 s
Aids to navigation	3 min
AIS base station	10 s or 3.33 s

Table 1.5 Example of static AIS messages

MMSI	304091000	228037600	228064900	227705102
IMO	9509255	0	8304816	262144
Call sign	V2GU5	FIHX	FITO	FGD5860
Ship name	HC JETTE-MARIT	AEROUANT BREIZH	VN SAPEUR	BINDY
Ship type	70	30	51	60
To bow	130	6	21	9
To stern	30	9	54	26
To starboard	18	5	10	5
To port	6	2	6	4
ETA	04-09 20:00	00-00 24:60	29-09 12:00	00-00 24:60
Draught	10.1	0	5.9	0
Destination	BREST		RADE DE BREST	
Time	1443650423	1443650457	1443650471	1443650474

Table 1.6 Example of dynamic AIS messages

MMSI	245257000	227705102	228131600	228051000	227574020
Navigational status	0	15	15	0	15
Rate of turn	0	−127	−127	−127	−127
Speed over ground	0.1	0	8.5	0	0.1
Course over ground	13.1	262.7	263.7	295	248.6
Heading	36	511	511	511	511
Longitude	−4.4657183	−4.4965715	−4.644325	−4.4851084	−4.4954414
Latitude	48.38249	48.38242	48.092247	48.38132	48.38366
Time	1443650402	1443650403	1443650404	1443650405	1443650406

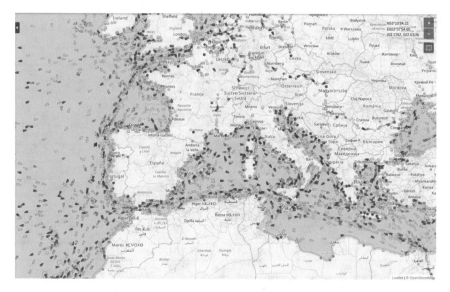

Fig. 1.2 Density maps layer of MarineTraffic's services

1.2.6 AIS Processing Difficulties and Challenges

The Automatic Identification System was initially designed to allow vessels to provide ship information automatically to other ships in the vicinity and to maritime authorities. The aim was to assist vessel's officers on the watch and coastal authorities to track maritime traffic and thus, reduce collision risk and improve the overall safety at sea. With the vast proliferation of vessel tracking systems, AIS has been used for vessel tracking services at a global scale. Such systems collect streams of AIS messages transmitted from the world's fleet and provide global ship tracking intelligence services such as those of MarineTraffic.[3] Figure 1.2 illustrates the density map of vessel traffic and highlights one of the capabilities the AIS tracking systems can offer.

However, since the initial purpose of the AIS communication system was not tracking vessels and their activities globally, some inherent characteristics of the communication protocol raise technical challenges that should be addressed to offer consistent and reliable information.

1. **Absence of unique ship identification**. Dynamic AIS messages (presented in Sect. 1.2.3) include the MMSI and IMO fields. The IMO number is a unique identifier for ships, registered ship owners and management companies that cannot be modified but is not mandatory for all the vessels. In fact, SOLAS regulation XI/3 made IMO number mandatory for all cargo vessels that are at least 300 Gross Tons (GT) and passenger vessels of at least 100 GT [13]. Vessels solely engaged

---

[3]https://www.marinetraffic.com/.

in fishing, ships without mechanical means of propulsion and pleasure yachts are just some examples of vessel types that are not obliged to have an IMO number. On the other hand, MMSI is a nine-digit number that is mandatory for all the vessels, but it is not a unique identifier (i.e., it can be modified under certain circumstances) [13]. According to ITU,[4] the first 3 digits of any MMSI number are called Maritime Identification Digits (MID) and indicate the respective vessel's flag. Thus, when a vessel owner decides to change the flag under which the vessel will sail, her MID will be updated and consequently her MMSI will change. The absence of a global unique ship id for all the vessels of the world fleet, dictates the necessity to parse the received AIS data, clean the stream from messages with invalid MMSI identifiers and assign a unique id to the rest of them before proceeding with any further processing.

2. **Prone to human errors**. Some of the information included in AIS messages are manually inserted by the vessel's crew. The reported destination and the Estimated Time of Arrival (ETA) are typical examples of inconsistent and unreliable information reported through AIS. For instance, a passenger vessel that is performing the same itinerary with multiple stops every day, may not change the destination for each stop and report only the final destination from the beginning of the voyage. Furthermore, "Piraeu", "Piraeus Port", "Piraeus Anchorage", "Pir", "P" are all acceptable values in the reported destination of a vessel travelling to the port of Piraeus. String similarity metrics can be used to deduce the correct destination. Similarly, the ETA field is also prone to errors as it may not correspond to the time needed to reach the next port but the time needed to reach the final destination. Manually inserted information is not reliable and data processing of AIS stream would provide more accurate results. For instance, calculating the ETA based on the vessel's speed or determining the vessel's destination based on its itinerary history or using pattern-of-life analysis [4] would provide more reliable information.

3. **Reporting Frequency**. According to the AIS communication protocol the reporting intervals are fluctuating and depend on the vessel's behaviour (e.g., speed, rate of turn). This decision was taken so as to avoid throttling the system with too many messages that would lead to packet collision and message re-transmission, but at the same time transmit frequent messages when moving with high speed or changing course so as to notify in time other vessels in the vicinity. From the data provider perspective the reception rate is the same as the vessel's transmission rate when the vessel is in range of a terrestrial station (which is approximately 50 km), but can be significantly lower when the vessel is sailing at open seas. In such case the satellites are used for monitoring and the update interval may range from few minutes up to several hours, depending on the satellite availability. All these lead to non-uniform distribution of collected data with significant communication gaps in some cases that may lead to inaccurate trajectory construction.

---

[4]https://www.itu.int/en/ITU-R/terrestrial/fmd/Pages/mid.aspx.

4. **Sensor malfunction**. It may occur that a vessel is transmitting erroneous information due to sensors' faulty operation. To discard such messages from the AIS
   data stream, feasibility analysis is essential to evaluate whether a vessel position
   is valid based on the vessel's past positions.

5. **Timestamping**. The only time-related information included in AIS messages is
   the seconds field of the UTC timestamp at which the AIS message was generated,
   as mentioned in Sect. 1.2.3. This is sufficient information for the vessels in the
   surrounding area, but when it comes to ship tracking data providers that store
   AIS messages for historical analysis, each message should be time referenced.
   This is usually done by assigning the UNIX epoch (i.e., seconds elapsed since
   01/01/1970) the moment each message is collected in base station. When the
   messages are aggregated in a central entity from the receivers, processing is needed
   to avoid duplicate messages (i.e., messages that were received from more than one
   station). Furthermore, it is likely that the messages arrive with a variable delay.
   This may be caused by network delay or due to collecting data from various
   sources (terrestrial stations or satellite-AIS stations) that may be out of synch.
   Message re-ordering or accepting messages with a delay in the stream system
   would tackle such issue.

1.2.7 AIS Applications & Use Cases

There is a growing body of literature on methods for exploiting AIS data for safety and
optimisation of seafaring, namely traffic analysis, anomaly detection, route extraction
and prediction, collision detection, path planning, weather routing and many more.

The work described in [20] introduces a method that identifies fishing activities
using AIS. The method uses AIS messages transmitted within an area of interest and
after some analysis tasks performed, such as the construction of the speed profile
of vessels, a map is produced that shows all fishing activities of the EU fleet in
high spatial and temporal resolution. In a similar direction, the work described in
[18] describes a big data approach that uses AIS data in large scale to identify port
operational areas. More specifically, the paper proposes an implementation of the
Kernel Density Estimation (KDE) algorithm using the MapReduce paradigm and
applies it on large volumes of AIS data for a specific area of interest (i.e., a busy
seaport), identifying activity areas suggesting port operations. The work described
in [27] presents another big data approach used for the extraction of global trade
patterns. In particular, AIS data are used for the extraction of routes, e.g., port-to-port
voyages categorized by ship type. Routes are constructed for each combination of
port of departure, port of arrival, and ship type from AIS messages using MapReduce.
Then, each set of voyages belonging to the same route per ship type are clustered
using K-Means. K-Means is a clustering algorithm that is based on the idea that n
elements are clustered into k groups, with each element belonging to the group with
the nearest mean.

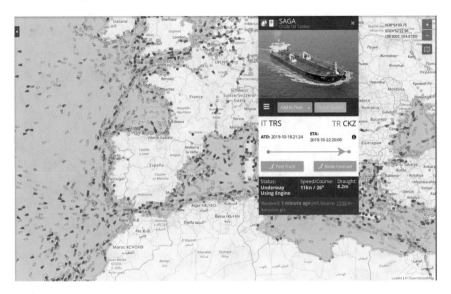

Fig. 1.3 Snapshot of the MarineTraffic Live Map showing the current positions of vessels and other data based on AIS messages

Since AIS was initially developed to assist in collision avoidance, soon after its establishment the first approaches for automatic detection of collisions emerged. A highlight of these approaches is for example the work described in [19], presenting an approach for collision avoidance in busy waterways by using AIS data.

The work described in [20] introduces the development of a framework that performs anomaly detection and route prediction based on AIS data. More specifically, it presents an unsupervised and incremental learning framework, which is called TREAD (Traffic Route Extraction and Anomaly Detection) for the extraction of movement patterns aiming at automatically detecting anomalies and projecting current trajectories and patterns into the future. In the context of anomaly detection, one common anomaly in the maritime domain is *spoofing*, which happens when a vessel attempts to camouflage her identity using the identification codes and/or the name of another vessel to hide her whereabouts. This might happen due to human error or on purpose, i.e., when the vessel is engaged in illegal activities and attempts to "hide" its identity. The work described in [17] introduces a big data approach that identifies spoofing events on AIS data streams.

Another interesting topic is the identification of events in the maritime domain. For example, the work documented in [23] proposes a rule-based system that uses Run-Time Event Calculus in order to perform recognition of complex events.

An interesting application that relies mainly on AIS data concerns MarineTraffic services. Figure 1.3 is an illustration of the MarineTraffic Live Map displaying the most recent positions of all vessels with active AIS transponders. Different colours correspond to different vessel types. Once a user clicks on a the position of a vessel

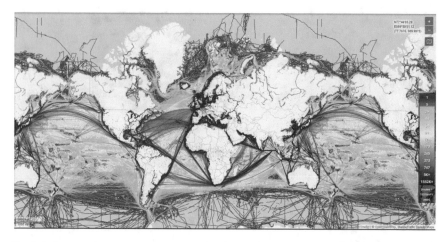

Fig. 1.4 Details of a vessel as derived from MarineTraffic's processing and analysis of AIS messages

on the Live Map, she is able to see some additional details about the vessel and its current voyage. For example, Fig. 1.4 shows the thumbnail of a photo showing a tanker named "SAGA", its type, its latest position and navigational status, destination, speed and draught. This information is made available after processing AIS messages transmitted by the vessel. The estimated time of arrival (ETA) to the reported destination is also displayed, but this derives from the analysis of AIS data related to the voyage (e.g., latest position, distance to the destination, speed, etc.).

1.2.8 Applications of AIS in the Book

The remaining chapters of this book present solutions regarding different aspects of processing, visualising, and analysing AIS data, as well as interesting applications that are based on AIS data.

The first part of the book describes maritime data. Tzouramanis [29] provides an overview of open maritime datasets, including a detailed list of the different sources of AIS data, and open datasets that could be used to enrich the information contained in vessel tracking data. For example, the chapter includes sources for bathymetry data, as well as sources for other maritime activities (e.g., locations of protected areas, platforms, etc.) that are useful for monitoring the maritime environment.

The second part of the book focuses on offline maritime data processing techniques. The work described in the chapter of Etienne et al. [7] presents an overview of geospatial relational databases and Geographical Information Systems that can be used in order to store and query data from the maritime domain. This chapter presents practical, hands-on scenarios for geospatial data processing using a modern geospatial relational database management system (PostGIS). The chapter of Tampakis et al. [28] introduces challenges that need to be addressed by existing data management frameworks, focusing on the geospatial and temporal nature of the data (e.g., spatial indices, etc.). This chapter presents techniques for pre-processing, cleaning and knowledge discovery for maritime data. The chapter of Andrienko et al. [3] presents an overview of visualisation techniques of maritime data combined with analytics tasks for data transformation, querying, filtering and data mining.

The third part of the book presents online maritime data processing techniques. The chapter of Patroumpas [21] provides an overview of online mobility tracking techniques against evolving maritime trajectories, focusing on trajectory simplification and compression. The chapter of Santipantakis et al. [26] presents link discovery techniques for maritime monitoring. The aim of link discovery is to enrich the information contained in a single dataset, such as trajectories of vessels, by associating it with other datasets, such as weather conditions. The chapter of Pitsikalis et al. [22] presents methodologies for identifying complex events using AIS data. This chapter first provides definitions of complex events and then describes a formal approach.

The last part of the book describes applications that are based on vessel tracking data. The chapter of Jousselme et al. [16] discusses approaches for performing anomaly detection under uncertainty. Subsequently, the chapter of Ducruet et al. [6] describes graph-theoretical network methods. These methods can be applied to an AIS dataset with port and inter-port data to derive new knowledge about: (i) the relative connectivity of ports, (ii) the impact of cargo diversity on the network structure, and (iii) the impact of structural or geographic patterns on the distribution of maritime flows. Finally, the chapter of Adland [1] describes the use of AIS data for shipping economics.

1.3   VHF Data Exchange System (VDES)

Since the establishment of the AIS protocol for vessel-to-vessel and vessel-to-coast radio communications, AIS became heavily used for safety of navigation and maritime situational awareness. In order to respond to the need of exchanging larger volumes of data through the AIS network which was getting overloaded, the International Telecommunications Union (ITU)[5] decided to revise the VHF marine band by adding designated channels for data transmission. In this direction, the development of the VHF Data Exchange System (VDES) [11] was proposed by ITU. VDES is also expected to increase data security by adding access control and authentication features for AIS radio traffic. This would potentially prevent data jamming, spoofing, etc. The foreseen contributions of VDES to the existing AIS protocol are summarised as follows:

- It supports faster data transfer rates than current VHF data link systems.

---

[5]https://www.itu.int/en/about/Pages/default.aspx.

- It supports both addressed (to a specific vessel or a fleet of vessels) and broadcast (to all units in the vicinity) transmissions.
- It offers increased reliability as it is optimised for data communication.
- It addresses the communication requirements for electronic Navigation, also known as eNAV. eNAV aims to improve berth-to-berth navigation and related services through data exchange in higher rates, improving the efficiency of maritime trade and transport. For example, the capability of VDES to transfer increased data volumes will allow the transmission of entire navigation plans, which is not possible currently in AIS. Electronic navigation is described in more detail in Sect. 1.6.1.

1.3.1   VDES Components

VDES is a system that consists of the following three sub-systems:

- AIS, which has been covered earlier.
- The Application Specific Messages system (ASM), which enables the exchange of application-specific messages. Such messages include information like meteorological conditions, collision possibility, danger region alert and route exchange.
- The VHF Data Exchange (VDE) component, that offers increased data transfer rates. VDE has also a satellite component, named VDE-SAT, that enables bidirectional ship-to-satellite communication.

As VDES is not expected to be fully operational until 2023, not all technical specifications about the implementation of VDES are defined at the time of the writing. Therefore, we provide below an overview of the requirements that have been foreseen for the implementation of a VDES network. First of all, as in the case of AIS, antennas will be used for transmitting and receiving data using a terrestrial and satellite link. VDES will build on AIS, so VDES transponders and receivers will be backwards compatible with AIS and ASM (Application Specific Messages system). This means that AIS and AIS-plus[6] equipment (e.g., transponders and receivers) will continue to be operational even after the establishment of VDES, but they will need to get upgraded in order to support the additional functionalities supported by VDES. VDES shore stations will either be upgraded AIS shore stations or new VDES stations at locations where no AIS stations exist.

Apart from the exchange of digital messages, the system should also be able to receive interrogating calls. For example, a vessel should report its position upon request. Other important issues that will be addressed concern the development of data integrity mechanisms and prioritisation of messages to ensure that critical messages (e.g., safety-related) will have higher priority than commercial services.

---

[6]AIS-plus is an extension of AIS that includes ASM.

1.3.2 VDES Applications

The new capabilities supported by VDES will be specifically useful for a number of use cases that heavily rely on data exchange between vessels, ship-to-shore and shore-to-ship communication. In this section we provide some examples.

Search and Rescue communications. Search and Rescue (SAR) communications will be improved using VDES and especially its satellite component, VDES-SAT. Since VDES-SAT will be able to receive and broadcast messages even when the coverage is low, it can be used in addition to VDES to relay distress alerts and locating signals.

Vessel traffic services (VTS). The high rate data services offered by VDES could significantly improve the Vessel Traffic Services, facilitating the navigation of vessels in traffic conditions and traffic organisation in general. The vessel traffic image can be created by the fusion of information from different sources, such as radar, AIS and multiple VTS centers (i.e., centers responsible for monitoring and controlling vessel traffic in facilities such as ports), as well as other arbitrary information sent from the vessels (e.g., navigational plan). Reliable and fast data exchange is crucial for VTS in order to provide up-to-date information about the traffic conditions in specific areas and monitor vessel routes and route deviations in an area.

Also, information needs to be exchanged between different parts: (i) vessels, (ii) port agencies, and (iii) VTS centers. In this direction, it might be requested by the VTS that vessels broadcast information regularly, at fixed time intervals, or on demand, i.e., then it is considered necessary. VDES compliant external systems installed ashore and on-board should support the exchange of information in this respect.

Another related service is the Navigational Assistance Service (NAS), which is defined by IMO as "a service to assist on-board navigational decision-making and to monitor its effects". Vessels can request to receive navigational assistance under specific circumstances, such as hardware malfunctions.

Route exchange. One of the most valuable pieces of information that a vessel can provide is its intended route at any given time, together with its current route, so that route deviations can be detected early. Route deviations may occur due to several reasons, such as weather conditions, hazardous situation, equipment failure, etc. In this way, route exchange could assist in the early detection of high risk situations, contributing to maritime safety. At the moment, the crew of a vessel reports changes in destination during a voyage. Using VDES and exploiting the high data exchange rates it offers, entire navigational plans could be transmitted.

Exchanging information about the route of vessels is important also for the shore authorities, in order to assess for example the possibility of congestion in the area of interest. Furthermore, route exchange could enable route and speed optimisation based on external conditions, such as weather and special conditions, hazardous situations, navigational limitations and traffic congestion.

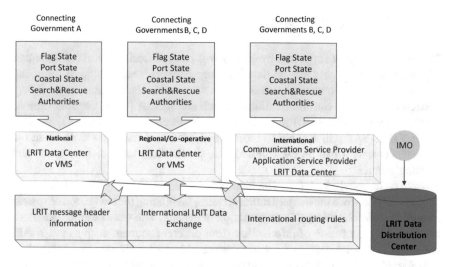

Fig. 1.5 LRIT system architecture (provided by IMO)

1.4 Long-Range Identification and Tracking

The Long-Range Identification and Tracking (LRIT) system is an international system that was adopted by IMO under its SOLAS convention for providing thorough identification and tracking of ships worldwide [14]. Figure 1.5 shows a high-level architecture of the LRIT system. The components of LRIT are described below.

- **LRIT transmission equipment**. This is the equipment that is installed on vessels in order to transmit and receive LRIT messages.
- **Communication service provider.** The communication service provider is responsible for providing the infrastructure and services required to support the communication path between the vessels and the application service providers. The application service providers are described below.
- **Application service providers (ASPs).** The information from the vessels is forwarded to the application service providers by the communication service provider. Once a message is received by the application service provider, it gets enriched with additional information. The complete message is then forwarded to the data centres. The application service provider also provides the implementation of the programming interface to the LRIT equipment that is installed on the vessels.
- **Data centres (DCs)**. Data centres are responsible for receiving, processing, and storing the information transmitted to and from the International LRIT Data Exchange (IDE), which will is described below. Vessels transmit LRIT messages to the data centres that correspond to the administrations that they belong to.
- **LRIT data distribution plan (DDP).** The data distribution plan contains the information regarding the distribution of data to the various Contracting Governments.

This information, accompanied by information about the coastal waters of each Contracting Government, should be available to the Data centres.

- **The International LRIT Data Exchange (IDE).** The IDE is responsible for routing messages to the appropriate data centres (DCs) based on the address and the URL fields of the message. IDE is not responsible for processing the actual content of the messages. At this stage, only the header content is read.
- **LRIT Data users.** LRIT data users may receive or request LRIT information in their capacity as a flag state, port state, coastal state or SAR service. Flag state is the state under the law of which a vessel is registered/licensed and it is considered as the "nationality" of the vessel. For example, a vessel carrying the Greek flag should operate under the Greek legislation.

1.4.1   Comparison with AIS

AIS and LRIT are independent and complementary. The work described in [5] provides a detailed comparison between AIS and LRIT. AIS messages have higher frequency, as each vessel obliged to have an AIS transponder should send messages every few seconds or minutes, depending on her navigational status and speed (see Tables 1.2, 1.3 and 1.4). On the other hand, LRIT covers areas of 1000 nautical miles off-shore, while terrestrial AIS stations cover approximately 50 nautical miles off-shore. This issue is addressed in AIS with the use of satellites, but communication in LRIT is more secure.

AIS messages transmitted by a vessel can be received by all vessels in the vincinity and ground stations, and are public. A satellite equipped with an AIS receiver collects AIS messages transmitted within the footprint of the satellite. On the other hand, LRIT messages are confidential and are only available to the LRIT data centres. Only contracting governmental organisations have access to this data, as well as Search and Rescue services, so the confidentiality and credibility of the data are ensured. LRIT data are not available for commercial purposes.

Regarding the content of the transmitted messages, AIS messages are richer as they contain more information than the current position at a given timestamp that is transmitted using LRIT. A significant advantage of LRIT, however, is that bi-directional communication is allowed, so for example, a vessel might be asked to report its position at any given time. In that sense, AIS is just a monitoring system, whereas LRIT is a communication system.

1.4.2   LRIT Applications

In [24], a comparison between AIS and LRIT was presented in terms of coverage, confidentiality and accuracy. It is shown how LRIT facilitates the prevention of maritime accidents, as its wide spatial coverage can be exploited to identify potential threats earlier, increasing reaction time and improving maritime safety. In [30], a statistical analysis was presented that showed that the number of piracy attacks declined in the Indian Ocean, an area prone to piracy events, due to the use of vessel tracking systems, including LRIT. Another interesting study that is based on LRIT and AIS data is the estimation of shipping emissions using tracking data, as described in [2]. In this work, the authors correlated vessel tracking information from LRIT and AIS with datasets containing information about the distributions of CO_2 and NO_x emissions from the EMEP[7] and EDGAR databases.[8] This way, density maps were produced showing the distribution of pollutant emissions along shipping lines.

1.5 The Vessel Monitoring System

The Vessel Monitoring System (VMS) is a system for monitoring fisheries [8]. It uses dedicated shipborne equipment that is installed on-board a fishing vessel and it is used to transmit data reports. Such reports contain information about the position of the vessel, speed, course and heading; VMS can also accommodate other auxiliary information (e.g., meteorological conditions). Such reports can either be sent automatically or manually by the operator. Data reports are sent to Fishery Monitoring Centres (FMCs), that store, validate and analyse the data. The transmission of messages from vessels to the FMCs is performed through the deployment of communication systems, both terrestrial and satellite. Some of these systems support bidirectional communication (i.e, not only vessel-to-FMC communication) and this enables the FMCs to poll or query vessels to send information (e.g., position, status of equipment, and so on) on demand.

The main components of VMS are the following:

- The shipborne VMS equipment, also known as the *VMS unit*, is installed permanently on the vessel and it mainly comprises a pair of transmitter and a receiver. The information that is transmitted by a vessel with VMS equipment installed contains the following fields: a unique identifier related to the equipment installed, the vessel's position, the date and time a data report is sent by the transmitter, and the speed, heading, and course of the vessel.
- The Fishery Monitoring centres (FMCs). The FMCs receive the data reports sent from the vessel's VMS equipment, extract the data, perform data validation, store the data and make it available for analysis. The data transmitted to the FMCs from the vessels are private and only accessible by the authorized personnel of the FMC.
- The VMS communication systems. These systems are responsible for securely transmitting data sent from the VMS shipborne equipment to the FMCs. These systems use both space and terrestrial sources. Depending on the communica-

---

[7]https://www.ceip.at/webdab_emepdatabase/.

[8]https://www.sec.gov/edgar.shtml.

tion service provider, a VMS communication system might support only one-direction communication (e.g., Argos (CLS)) or multi-directional communication (Inmarsat-C and Inmarsat-D+ or Orbcomm). In one-direction systems, the vessel automatically transmits information at a specified interval. Bi-directional systems allow vessels to be polled or queried by the FMCs. In this way, the FMCs are able to ask for information from vessels on demand.

Notably, VMS is not comparable with AIS, as VMS is an application for monitoring exclusively commercial fishing vessels, while AIS is a communication technology. Data coming from VMS can be used to increase navigational safety, to support Vessel Traffic Services (VTS) and can be further analysed to derive useful information, such as resources optimisation (e.g., fuel consumption). One of the most profound applications based on VMS data is the monitoring of fishing activities, as described in [9]. This work presents a case study that assesses the importance of VMS in fisheries management and focuses on the case of the Portuguese trawl fleet. Another study [25] performs an analysis using VMS data deriving trends in effort and yield of trawl fisheries in the Mediterranean Sea. A similar approach is also used in [31] that describes a method using VMS data to identify and annotate trips made by US fishing vessels in the North Pacific. In the work described in [32] an interesting study using VMS reveals an increase in fishing efficiency following regulatory changes in a demersal longline fishery.

1.6 Advanced Applications and Analytics Based on Maritime Reporting Systems and Data

In this section we describe applications that heavily rely on advanced analytics against data coming from reporting systems. First, we describe electronic navigation, which is an application that attempts to provide assistance to mariners automating navigational tasks as much as possible, in order to minimise human error, and increase safety of navigation. Then, we describe how data coming from reporting systems can be used to increase Maritime Situational Awareness.

1.6.1 Electronic Navigation (eNAV)

We describe E-Navigation (eNAV), which is not a reporting system per se, but an application that requires data fusion from different reporting systems and tools already installed aboard the vessels. eNAV is a broader concept, for which the technical implementations are not yet clearly defined. It is, however, one of the most important use-cases of the systems presented in the previous sections of this chapter, as it can contribute to the safety and security of navigation, automating navigation processing, providing assistance to the vessels crew, thus minimising the possibility of human error in the navigation process.

The definition of eNAV by IMO [15] is provided below.

> **E-Navigation (eNAV)** is the harmonized collection, integration, exchange, presentation and analysis of marine information on-board and ashore by electronic means, to enhance berth to berth navigation and related services for safety and security at sea, and protection of the marine environment.

eNAV systems aim to equip shipboard and ashore users with all the necessary decision support systems that will simplify maritime navigation and communication, making shipping safer and more efficient, providing timely warnings and preventing accidents such as collisions and groundings. It is foreseen though, that electronic navigation systems will not replace the human factor. Mariners will continue to have the same core role in decision making, but eNAV systems will be able to assist them in making more informed decisions. eNAV systems integrate information from various sources such as on-board sensors, hydrographic, meteorological and navigational information sent from ashore or from nearby vessels, alert management systems, etc. The information will be integrated transparently to the user and will be presented through a user-friendly interface. The development of systems that support eNAV should ensure both backward and forward compatibility so that compliance with existing and future systems is ensured, contributing to the interoperability of information exchange between vessels, and between vessels and the shore.

Using eNAV systems, shore-based operators and shipborne users will both benefit from the exchange of navigational information to improve services such as vessel traffic. eNAV systems involve the integration of multiple components that can be categorized as follows:

- *On-board.* Systems that provide assistance to the mariner using an interface, taking into consideration data coming from different sources, such as the vessels' in-house sensors and an alert management system.
- *Ashore.* Shore-based operators could be assisted through the exchange of navigational information to improve services such as management of vessel traffic.
- *Communications.* Navigation-related information could be shared between ships in an area of interest and be made available to third party applications.

eNAV would be particularly useful in the following use cases:

- **Safety of navigation**. The information exchanged through eNAV enables the involved officers (crew members, port authorities) to make decisions based on the navigational status of a specific vessel in relation to external conditions of the area of interest. Navigational information about nearby vessels could be integrated with other sources such as weather data, port congestion information, data describing facilities that could impact the navigational status of the vessel (e.g., wind farms), etc.

- **Environment protection**. Data acquired through eNAV (e.g., route plans), could be fused with other data and further analysed. For instance, a cost estimation analysis could be performed regarding the fuel consumption following different routes, as well as benchmarking of alternative routes with respect to cost and/or time. Different navigational plans could also be compared with respect to their impact on the environment (e.g., "green shipping").

1.6.2   Maritime Situational Awareness

Maritime Situational Awareness (MSA) is the effective understanding of activities, events and threats in the maritime environment that could impact global safety, security, economic activity or the environment. The primary goals of MSA include "enhancing transparency in the maritime domain to detect, deter and defeat threats" and "enable accurate, dynamic, and confident decisions and responses to the full spectrum of maritime threats", as stated in the US National Plan for Maritime Domain Awareness [10].

The most widely used collaborative tracking system is the Automatic Identification System (AIS). Collaborative tracking systems rely on the cooperation of the vessel's crew to provide information regarding the characteristics and the navigational status of the vessel. Therefore, a common criticism regarding AIS-based surveillance is that it says nothing about the presence of non-cooperative vessels, the "non-shiners" or so-called "dark targets", i.e., the vessels that intentionally do not transmit AIS messages despite being obliged to. Frequently, ships involved in illegal activities, attempt to hide their identification and positional information, either by counterfeiting, masking it or not transmitting at all. Thus, attempting to remain undetected by law-enforcement bodies throughout the whole duration of the illegal activity. On the other hand, non-cooperative systems do not rely on the cooperation with the vessel's crew to monitor vessels, therefore they are less sensitive to deception, but ship detection, identification and tracking, or speed and heading estimation are more complex, while often limited coverage and their reliability can vary depending heavily on environmental conditions.

The collection, fusion, analysis and dissemination of maritime intelligence and information are the fundamental building blocks of MSA. Data fusion combines multi-origin information to determine relationships among the data; thus improving the understanding of a current complex environment, but also attempting to predict its future state. The advantages in fusing data from multiple sensors and sources are that the final estimated vessel tracks are more accurate and with better confidence, extending to features that are impossible to perceive with individual sensors and sources, in less time, and at a lower cost. Also, it contributes towards better coverage and robustness to failure, thus improving the reliability and quality of the situational picture.

An essential pillar for this is building an accurate model of normalcy. The understanding of the complex maritime environment and a vessel behaviour though, can never be limited to simply adding up and connecting various vessel positions as they travel across the seas. Essentially, vessel-based maritime activity can be described in space and time, while classified to a number of known activities at sea (e.g., fishing). The spatial element describes recognised areas where maritime activity takes place; thus, including ports, fishing grounds, offshore energy infrastructure, dredging areas and others. The transit paths to and from these areas also describe the spatial element, e.g., commercial shipping and ferry routes, etc. While the temporal element often holds additional information for categorising these activities (e.g. fishing period and time of year).

At the core of this process is data mining, an essential step in the process consisting of applying data analysis and discovery algorithms that, under acceptable computational efficiency limitations, produce a particular enumeration of patterns over the data. Hence, extracting patterns also means fitting a model to the data, finding structures, or in general any high-level description of a set of data. Often vessels conducting illegal behaviour and trying to hide their behaviour follow a set of patterns depending on the activity they are engaged in: deviation from standard routes, unexpected AIS activity, unexpected port arrival, vessels in close distance (given a distance threshold), and vessels entering an area they were not expected to. **Anomaly detection**, thus, is considered as the detection of events that deviate from normalcy.

An example of such as deviation is the case of a vessel named Pluto. YM Pluto, departed from Ceuta, Spain on the 25th of April 2013 bound for Rotterdam. On the 27th of April 2013, the vessel was North of the Western coast of Portugal while the weather forecast was Northerly winds of Force 7 to 8 on the Beaufort scale, with severe gusts and very rough seas. During the early morning of 27 April 2013, the master of YM Pluto was on the forecastle deck in adverse weather conditions, attempting to stop a water leakage. Unexpectedly, the ship slammed into a very large wave. The master was exposed to the violent impact of the breaking wave and was severely injured. While arrangements were made to dispatch a helicopter to airlift the master, the vessel altered course heading towards the port of Averio. Figure 1.6 shows the commonly travelled route between Ceuta and Rotterdam in green colour and the actual trajectory of the vessel in red colour. The red star marks the position of the incident. The snapshot is from the Anomaly Detection platform of MarineTraffic,[9] i.e., a prototype system that detects maritime anomalies in real-time (e.g., collisions, vessels in proximity, route deviations, navigation in shallow waters).

---

[9]https://www.marinetraffic.com/anomaly-detection (service is accessible for guests after contacting MarineTraffic Research).

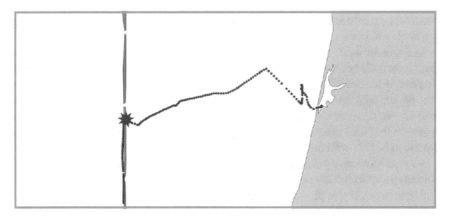

Fig. 1.6 Visualisation of the route deviation of the vessel "YM Pluto" during the time period of the incident while travelling from Ceuta to Rotterdam

1.7 Conclusions

In this chapter we described some of the main cooperative reporting systems that are widely used nowadays, such as the Automatic Identification System (AIS), the Long Range Identification and Tracking System (LRIT) and the Vessel Monitoring System (VMS) for fishing vessels. We also described the VHF Data Exchange System (VDES) that is now being developed and will be completed soon. Its adoption is expected to be highly impactful for maritime safety. Last we presented two applications that rely on data that derive from vessel reporting systems: Maritime Situational Awareness and eNAV.

1.8 Bibliographical Notes

The technical specifications for AIS, VDES, LRIT and VMS can be found in [8, 11, 13, 14]. The eNAV implementation strategy can be found in [12]. Further details about the comparison between Satellite-based AIS and LRIT can be found in [5]. Another comparison between AIS and LRIT is also documented in [24], in which it is showcased how LRIT can contribute to increase safety of navigation.

Watson and Haynie [31] and Watson et al. [32] describe two interesting applications of using VMS data. The former presents and approach that uses VMS to identify shipping routes made by fishing vessels in the US North Pacific. The latter describes how VMS was used in the context of a study that investigates the impact of regulatory changes in the efficiency of fishing activity. Another study described in [9] showcases an analysis based on VMS and AIS data for the Mediterranean Sea to identify trends regarding different types of fishing activities with a focus on trawlers.

Positional data were correlated with data about over-exploitation rates and primary production indices.

In a similar note, the work described in [20] introduces a method that identifies fishing activities of the EU fleet in high spatial and temporal resolution using AIS. Another work described in [18] presents an analysis against AIS data to identify port operational areas. The analysis is based on a distributed implementation of well-known data mining algorithms (e.g., KDE) to improve performance. A similar approach that uses distributed clustering for analysing large volumes of AIS data is described in [27] with the aim of extracting global trade patterns.

Regarding anomaly detection using AIS data, the methodology described in [19] presents an approach for collision avoidance in busy waterways. An anomaly detection framework is also presented in [20] that performs route prediction. An approach for detecting spoofing events against big AIS streams is introduced in [17]. The framework presented in [23] proposes a system based on Run-Time Event Calculus in order to perform recognition of complex events using a set of rules.

Regarding LRIT, the work described in [30] studies the impact of reporting systems on piracy events performing statistical analysis against LRIT data. Alessandrini et al. [2] correlates vessel tracking data with shipping emissions.

References

1. Adland, R.: Shipping economics and analytics. In: Artikis, A., Zissis, D. (eds.) Guide to Maritime Informatics, chap. 11. Springer, Berlin (2021)
2. Alessandrini, A., Guizzardi, D., Janssens-Maenhout, G., Pisoni, E., Trombetti, M., Vespe, M.: Estimation of shipping emissions using vessel long range identification and tracking data. J. Maps **13**, 946–954 (2017)
3. Andrienko, N., Andrienko, G.: Visual analytics of vessel movement. In: Artikis, A., Zissis, D. (eds.) Guide to Maritime Informatics, chap. 5. Springer, Berlin (2021)
4. Biltgen, P., Ryan, S.: Activity-based Intelligence: Principles and Applications. Artech House electronic warfare library. Artech House (2016). https://books.google.gr/books?id=4mcYjgEACAAJ
5. Chen, Y.: Satellite-based ais and its comparison with lrit. TransNav, Int. J. Marine Navig. Saf. Sea Transp. **8**(2), 183–187 (2014)
6. Ducruet, C., Berli, J., Spiliopoulos, G., Zissis, D.: Maritime network analysis: Connectivity and spatial distribution. In: Artikis, A., Zissis, D. (eds.) Guide to Maritime Informatics, chap. 10. Springer, Berlin (2021)
7. Etienne, L., Ray, C., Camossi, E., Iphar, C.: Maritime data processing in relational databases. In: Artikis, A., Zissis, D. (eds.) Guide to Maritime Informatics, chap. 3. Springer, Berlin (2021)
8. FAO.: VMS for fishery vessels. http://www.fao.org/fishery/topic/18103/en. Accessed 15 May 2019
9. Fonseca, T., Campos, A., Fonseca, P., Mendes, B., Henriques, V., Parente, J.: The importance of satellite-based vessel monitoring system (vms) for fisheries management: a case study in the portuguese trawl fleet. Maritime Engineering and Technology, pp. 19–24 (2012)
10. U.D. of Homeland Security.: National plan to achieve maritime domain awareness for the national strategy for maritime security. Technical report DHS (2005). https://www.dhs.gov/xlibrary/assets/HSPD_MDAPlan.pdf
11. IALA.: The technical specification of VDES. Technical report IALA (2018). https://www.iala-aism.org/product/g1139-technical-specification-vdes/

12. IMO.: Draft e-navigation strategy implementation plan. Technical report IMO (2014). http://www.imo.org/en/OurWork/Safety/Navigation/Documents/enavigation/SIP.pdf

13. IMO.: Technical characteristics for an automatic identification system using time division multiple access in the vhf maritime mobile frequency band. Technical report, ITU (2017). https://www.itu.int/dms_pubrec/itu-r/rec/m/R-REC-M.1371-5-201402-I!!PDF-E.pdf

14. IMO.: Long-range identification and tracking system. Technical report, IMO (2018). http://www.imo.org/en/OurWork/Safety/Navigation/Documents/LRIT/1259-Rev-7.pdf

15. Jonas, M., Oltmann, J.H.: Imo e-navigation implementation strategy - challenge for data modelling. TransNav, Int. J. Marnie Navig. Saf. Sea Transp. **7**(1), 45–49 (2013)

16. Jousselme, A.L., Iphar, C., Pallotta, G.: Uncertainty handling for maritime route deviation. In: Artikis, A., Zissis, D. (eds.) Guide to Maritime Informatics, chap. 9. Springer, Berlin (2021)

17. Kontopoulos, I., Spiliopoulos, G., Zissis, D., Chatzikokolakis, K., Artikis, A.: Countering real-time stream poisoning: An architecture for detecting vessel spoofing in streams of AIS data. In: 2018 IEEE 16th International Conference on Dependable, Autonomic and Secure Computing, 16th International Conference on Pervasive Intelligence and Computing, 4th International Conference on Big Data Intelligence and Computing and Cyber Science and Technology Congress 2018, Athens, Greece, August 12–15, 2018, pp. 981–986 (2018)

18. Millefiori, L.M., Zissis, D., Cazzanti, L., Arcieri, G.: A distributed approach to estimating sea port operational regions from lots of AIS data. In: 2016 IEEE International Conference on Big Data, BigData 2016, Washington DC, USA, December 5–8, 2016, pp. 1627–1632 (2016)

19. Min Mou, J., van der Tak, C., Ligteringen, H.: Study on collision avoidance in busy waterways by using AIS data. Ocean Eng. **37**, 483–490 (2010)

20. Pallotta, G., Vespe, M., Bryan, K.: Vessel pattern knowledge discovery from ais data: a framework for anomaly detection and route prediction. Entropy **15**, 2218–2245 (2013)

21. Patroumpas, K.: Online mobility tracking against evolving maritime trajectories. In: Artikis, A., Zissis, D. (eds.) Guide to Maritime Informatics, chap. 6. Springer, Berlin (2021)

22. Pitsikalis, M., Artikis, A.: Composite maritime event recognition. In: Artikis, A., Zissis, D. (eds.) Guide to Maritime Informatics, chap. 8. Springer, Berlin (2021)

23. Pitsikalis, M., Artikis, A., Dreo, R., Ray, C., Camossi, E., Jousselme, A.: Composite event recognition for maritime monitoring. In: Proceedings of the 13th ACM International Conference on Distributed and Event-based Systems, DEBS 2019, Darmstadt, Germany, June 24–28, 2019, pp. 163–174 (2019)

24. Qi, J., Guo, R., Wang, X., Zhang, H.: Research on risk long range identification for vessel traffic dynamic system. IOP Conf. Ser.: Mater. Sci. Eng. **231**, 012166 (2017)

25. Russo, T., Carpentieri, P., D'Andrea, L., de Angelis, P., Fiorentino, F., Franceschini, S., Garofalo, G., Labanchi, L., Parisi, A., Scardi, M., Cataudella, S.: Trends in effort and yield of trawl fisheries: a case study from the mediterranean sea. Front. Mar. Sci. **6**, 00153 (2019)

26. Santipantakis, G.M., Doulkeridis, C., Vouros, G.A.: Link discovery for maritime monitoring. In: Artikis, A., Zissis, D. (eds.) Guide to Maritime Informatics, chap. 7. Springer, Berlin (2021)

27. Spiliopoulos, G., Zissis, D., Chatzikokolakis, K.: A big data driven approach to extracting global trade patterns. In: Mobility Analytics for Spatio-Temporal and Social Data - First International Workshop, MATES 2017, Munich, Germany, September 1, 2017, Revised Selected Papers, pp. 109–121 (2017)

28. Tampakis, P., Sideridis, S., Nikitopoulos, P., Pelekis, N., Theodoridis, Y.: Maritime data analytics. In: Artikis, A., Zissis, D. (eds.) Guide to Maritime Informatics, chap. 4. Springer, Berlin (2021)

29. Tzouramanis, T.: Navigating the ocean of publicly available maritime data. In: Artikis, A., Zissis, D. (eds.) Guide to Maritime Informatics, chap. 2. Springer, Berlin (2021)

30. Vespe, M., Greidanus, H., Alvarez, M.: The declining impact of piracy on maritime transport in the indian ocean: statistical analysis of 5-year vessel tracking data. Marine Policy **59**, 9–15 (2015)

31. Watson, J., Haynie, A.: Using vessel monitoring system data to identify and characterize trips made by fishing vessels in the united states north pacific. PLoS ONE **11**, 0165173 (2016)

Chapter 2
Navigating the Ocean
of Publicly Available Maritime Data

Theodoros Tzouramanis

Abstract In the past, the data-capturing tasks involved in researching and developing maritime projects were often time-consuming and costly. It was not unusual that different project ventures, unaware that other projects were similarly engaged, were unknowingly simultaneously gathering data of a similar nature or even the very same data. As part of an initiative to prevent such a waste of resources, a number of international organizations decided to make high-quality maritime data, which had been painstakingly collected over many years, freely accessible to the public. The aim of this chapter is to draw up a compilation of maritime datasets that hundreds of core and strategic data sources across the world make freely available online at the time of writing. These datasets provide, in their large majority, and each one in its own field, a complete and detailed view of the entire aquatic environment of the earth. Ultimately, this chapter intends to contribute a useful working tool to all who operate as experts or as users of maritime geographic information systems. If this chapter manages in a helpful way to set readers on the path to successfully finding and selecting the data that fit their needs and to navigating effortlessly the sea of online information, then it will have achieved its goal.

2.1 Introduction

Maritime informatics is the field of information technology dedicated to the study of the oceans and coastal zones. The advances in respect of new technologies and equipment, such as research vessels, buoyancy sensors, orbital data from satellites and Global Positioning Systems (GPS) have broadened the scope and potential of research carried out in the maritime context and has turned into an interdisciplinary study continuously adapting to technological changes.

Data are the fuel that powers maritime infrastructures. This reliance has brought about a shift towards publicly available data establishing standards and formats that

T. Tzouramanis (✉)
Department of Computer Science and Biomedical Informatics,
University of Thessaly, Thessaly, Greece
e-mail: ttzouram@uth.gr

© Springer Nature Switzerland AG 2021
A. Artikis and D. Zissis (eds.), *Guide to Maritime Informatics*,
https://doi.org/10.1007/978-3-030-61852-0_2

are interchangeable between proprietary systems. As a result, academics and developers who would previously have had to start from the ground up and collect or buy expensive real-life maritime data from scratch to be able to do their work can now concentrate on the other important tasks and processes, since the hard and tedious groundwork of data collection has already been carried out by other specialised and fully reliable entities.

In the context of the increasing demand for specialized maritime data for experimental, operational and educational purposes, a number of studies (e.g. [1–6]) have published lists of online sources of maritime data, mainly in standard Geographic Information System (GIS) formats, for the purpose of building various applications for marine science, sustainable ocean governance, and the blue economy sector. In this same spirit, this chapter takes a fresh look at the available online data and gathers some of the most reliable up-to-date online sources offering various categories of free of charge data for maritime applications. These heterogeneous spatio-temporal naval, static or real-time, data include detailed vessels structural and identification data; ship-traffic data; data about maritime areas subject to limits and conditions or presenting risks for navigation; ports data; weather forecasting and climate data; cartographic data; historical marine data; etc.

The chapter unfolds as follows. Section 2.2 presents sources that provide vessel-related data, both in respect of the static characteristics of ships and of their historical and real-time positional data. Section 2.3 focuses on maritime data related to the aquatic domain both above and below the sea surface. Section 2.4 concentrates on data related to those areas of the globe which are constituted of land but which are at the same time related in a number of ways to the aquatic neighbouring or surrounding areas. Section 2.5 examines sources that provide maps and satellite images of the earth. Section 2.6 focuses on data related to marine climate and sea conservancy, and, finally, Sect. 2.7 summarises the study by drawing the chapter's conclusions.

2.2 Vessel-Related Data

Over the course of the last twenty years shipping traffic has increased by over 300% and it is still increasing at a rate of 3–4% per year. Ships that keep growing in size and speed now move 11 billion tons of goods around the globe per year. According to the 2019 annual review of maritime transport of the United Nations (UN) Conference on Trade and Development, the main UN body dealing with trade, investment and development issues, about 100,000 propelled seagoing merchant vessels of 100 or more gross tonnage are sailing the seas today (excluding military vessels, yachts, waterway vessels and fishing vessels, [7]). For all these reasons, several specialized online data sources have been developed to provide accurate historical and current publicly available data related to the ships that navigate the seas around the globe. Table 2.1 summarises some of the major data sources for such vessel-related data. These data sources offer information relating to the life-cycle of every ship, from the shipbuilding phase to its current state, including its seaworthiness or unseaworthiness,

Table 2.1 Major sources for vessel-related data

Data source	Data categories (kinds of data)	Data formats	Form of access	Number of vessels in the database*	Reference
Equasis	Official static vessel data and maritime transport companies data	Plain text, PDF	Free of charge with right-to-use restrictions	100,000	[8]
MarineTraffic	Historical and real-time vessel positional data; static vessel data; shipping companies and builders data, etc.	Multiple formats	Partly free of charge map data access with right-to-use restrictions	1,000,000	[9]
VesselTracker				160,000	[10]
VesselFinder		Plain text and map data		350,000	[11]
ShipFinder	Historical and real-time vessel positional data; static vessel data			Unknown	[12]
BalticShipping	Static vessel and shipping companies data	Plain text	Partly free of charge access with right-to-use restrictions	210,000	[13]
Maritime Vessels Directory				22,000	[14]
FleetMon	Historical and real-time vessel positional data; static vessel data; shipping companies and builders data, etc.	Plain text, map data, JSON	Subscription charges apply	500,000	[15]
Lloyd's List Intelligence	Historical (going back to 1734) and real-time vessel positional data; static vessel data; ship-ping companies and builders data, etc.	Multiple formats		More than 155,000 (it is difficult to find the precise number in the available documentation)	[16]
AISHub	Real-time vessel positional data	Multiple formats	Free of charge to anyone who offers AIS data in return for the service	Unknown/varies (only ships that are currently transmitting AIS messages are figured in the dataset)	[17]

*'Unknown': i.e. it is difficult to find the precise number in the available documentation

the history of its ownership, its structural, mechanical and operational characteristics and all the changes made to the ship overtime, the ship's more recent trajectories and voyages around the globe as well as its current position at sea or at anchor.

2.2.1 Static and Rarely Changing Characteristics

Going as far back as the year 1764, which marked the start of the data collection process by the Registry of Shipping ([18], later renamed as "Lloyd's Register"), a large variety of sources have developed offline and online databases to offer detailed static and historical data related to ships and shipping companies with a view to informing both the public and professionals about the condition of the vessels navigating around the globe.

Of these online databases, Equasis [8] is possibly the best-known today. It collates high-quality information from dozens of other existing official public and private information systems and data banks (public authorities and industrial organisations) and makes this information available in one single portal, with the aim of providing a comprehensive free of charge set of data for every ship listed therein. Since 2020, Equasis has been providing detailed information about ship characteristics: name; class/type; unique International Maritime Organisation (IMO) number and Maritime Mobile Service Identity (MMSI) number; flag; ownership; call sign; port of registry; tonnage; dimensions; survey dates; cargo; capacity; status; construction; engine details; machinery; photographs; and much more, for about 100,000 ships of 100 gross tonnage and more.

Several other—in most cases privately-funded—portals provide similar or supplementary information about ships, amongst which are MarineTraffic [9], Vessel-Tracker [10], the BalticShipping portal [13], the Maritime Vessels Directory [14], FleetMon [15], VesselFinder [19], etc.

The detailed facts and figures about ships provided by all the above directories put together, including characteristics, specifications and ownership, add up to more than 600 data fields for every vessel. However, none of these directories provides on its own all this information or provides it for the entire life cycle of every vessel. Thus, in order to build as comprehensive a profile as possible for every vessel, it might be necessary to compound the information contained in any number of these directories, using the common attributes of the IMO identification number and the MMSI vessel identification number, which are two unique and permanent identification numbers belonging to every ship.

In addition to the above, most of the these directories provide detailed information about business enterprises, services and other parties involved in the maritime domain, such as ship builders, managers and operators; ship suppliers and repairers; insurance and legal services companies; search and rescue teams; non-governmental organizations (NGOs), etc.

Some databases specialize in historical sources of information for research in maritime history and evolution, with details about ancient ships, including vessel types. These historical data, dating from 2000 B.C. onwards, are recorded in several studies [20], books [21–23] and online encyclopedias [24].

Another issue of fundamental importance in maritime databases concerns the flag of the nation under which a vessel is registered to sail. The term 'flag of convenience' (FOC) in merchant shipping refers to the registration of a ship in a country that is not necessarily that of the ship-owner, usually for reasons related to issues of compliance with regulations and of taxation. A number of these countries have been facing criticism, mainly from trade union organisations, for their sub-standard shipping regulations and for offering ship owners an opportunity to eschew their duty of responsibility to their crew and their duty of care for the environment. The Fair Practices Committee, a joint committee of the International Transport Workers' Federation, publishes every year a list of FOCs (which, at the time of writing, included 35 flags, [25]) and runs a campaign against flags of convenience to inform and to protect workers' rights.

2.2.2   Traffic Flow

Ship traffic data are produced using the Automatic Identification System (AIS) which was created to contribute to safe vessel navigation, and to provide support to port authorities in managing sea traffic more confidently. The AIS consists of transceivers that are mounted on ships and include a transmitter and a GPS receiver. The transmitter calculates the coordinates of the vessel's position, speed and course and transmits these data to two VHF AIS channels. Every vessel which has such an AIS device transmits 90–450 messages per 15 min, resulting in the daily transmission of millions of messages.

Base-stations or other vessels can receive these signals through their own transceiver or by means of a simple AIS receiver. Typically, base-stations and vessels with an external antenna located 15 m above sea level can receive AIS data from transmitters within a radius of 15–20 nautical miles. The range of base-stations with antennas at higher altitudes can extend to about 40–60 nautical miles. Finally, base-stations with antennas placed at a height of 700 m can receive a message from a distance of up to 200 nautical miles.

Several ship tracking systems exist on the web. Their main objective is to provide the general public and the owners of vessels with real-time information on ports and vessel movements. These information systems can collect millions of AIS messages per second from a rich network of cooperating base-stations.

MarineTraffic [9] is one of the best and most widely known online vessel tracking systems. It records at least 800 million vessel AIS messages monthly and it provides data for more than 1,000,000 marine assets such as vessels, ports, etc. In addition, it provides satellite AIS coverage, enhanced satellite tracking, advanced density maps, nautical charts, etc. The extensive AIS message coverage relies on a community of more than 20,000 volunteer base-stations from all over the world and also depends on local authorities being interested in joining this common effort and installing a receiver and sending the data they collect from their site to the MarineTraffic central server.

VesselFinder [11] and ShipFinder [12] are other widely used online ship tracking services. They offer efficient monitoring services, port calls, and a detailed voyage history of every ship. VesselFinder is provided by AISHub [17], which is the only dedicated online vessel monitoring service that freely distributes all of its data to anyone who also feeds AIS data into the service.

Besides the aforementioned information systems, FleetMon [26] and Lloyd's List Intelligence [16] of the Lloyd's List Group are two interactive online services offering detailed vessel movements, real-time AIS positioning, comprehensive information on ships, shipping companies, ports and casualties as well as credit reports, industry data and analysis including short-term market outlook reports. Access to their vessel tracking service is limited to those who are certified members.

Maritime routes represent a fundamental dimension of marine transportation. The IMO is responsible for the ships' routeing systems, including traffic separation schemes which are routes predetermined by the IMO in order to regulate the traffic at specific busy, confined waterways such as the English Channel, or where freedom of movement is inhibited by restricted sea-room, etc. Ships' Routeing [27] is a frequent IMO publication that incorporates the latest routeing rules that have been adopted. The IMO official ships' routeing rules are unfortunately not available yet in a digitized GIS form.

A comprehensive library of global ship routes, however, is provided in [28]. Here, in order to reconstruct the vessels' trajectories, a dataset of 8,000 ports and harbours, and 20,000 anchorage and waiting areas was used, as well as 100Gb of AIS data collected from roughly 19,000 unique ships, mainly bulkers and tankers, from the beginning of 2016 to the middle of 2018. The study then applied clustering methods [29] in order to find different clusters of vessel trajectories, which made it possible to compute the popularity of every maritime route. Figure 2.1 shows the popularity of some of these maritime routes. In this way, the existence of 43,847 distinct routes was established, reconstructed from one million trips. A number of similar methods can also be found in the literature [30].

Ships are advised to maintain anti-piracy watches while transiting high-risk areas and to report all piracy and armed robbery incidents to the local authorities, including any movements of boats and skiffs deemed suspicious. A free telephone service is provided by the Piracy Reporting Centre of the International Maritime Bureau (IMB) [31], which is an independent and non-governmental agency based in Kuala Lumpur. A live piracy and armed robbery map provided by the IMB can be found in [32]. A list of more than 8,000 reported incidents of piracy and armed robbery on a global scale since 1994 can be found in [33] and in [34].

Last but not least, in relation to data concerning to ship movement, are data about collisions between ships and whales or other marine species.[1] These are known as

---

[1] Please note that the major data sources providing vessel accidents data (involving a vessel sinking, foundering, grounding, or being otherwise lost) are reported in Sect. 2.6.3 because of the potential gravity of the impact on the environment of such incidents.

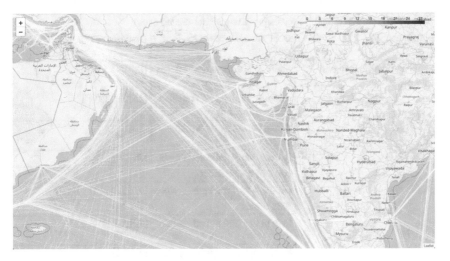

Fig. 2.1 Identifying sea routes from AIS data (*source* [28])

'ship strikes'. Ship strikes represent a major conservation and welfare issue affecting a large number of whale populations globally. Between 20% and 35% of all whales found dead show cuts and blunt trauma consistent with a ship strike. Jensen et al. [35] report 292 cases worldwide of ship strikes from 1885 to 2003. The database includes information about the whales (e.g., species, size, observed injuries), the type of the vessel involved and the location of the incident. A similar and more recent database that has compiled a list of about 150 cases of ship strikes in Australia can be found in [36].

Table 2.2 summarises this section's sources of reliable, publicly available vessel-related datasets.

2.3 Sea-Related Data

Sea-related data include data that describe the area of about the 70% of the globe that is covered by sea water. The data describe this area from the points of view of both above and below the surface of the water and include data for bathymetry, undersea cables, protected areas, wetlands, fish population, human-made structures, etc.

2.3.1 The Aquatic Environment Under the Surface of the Sea

Bathymetry studies the underwater depths. It is the undersea equivalent of hypsometry or topography. Bathymetric (or hydrographic) charts and maps are produced

Table 2.2 Summary of major sources providing vessel-related data

Data category	Data sub-categories (kinds of data)	Data sources
Static and rarely changing characteristics	Static & historical data related to ship characteristics,	[8–10, 13, 14],
	shipping companies, ship builders and repairers	[15, 18, 19]
	Personnel currently working on board	[13]
	Ancient ships and vessel types	[20–24]
	Flags of convenience	[25]
Traffic flow	Online vessel tracking services	[9, 11, 12, 16, 17, 26]
	Ship routes	[27, 28]
	Live piracy and armed robbery map	[32]
	Piracy and armed robbery activity	[33, 34]
	Ship strikes	[35, 36]

to support the safety of surface or of sub-surface navigation, and show sea floor relief or terrain as contour lines (i.e. lines drawn joining places of equal depth, called depth-contours or isobaths) and selected depths (soundings). An example of contour lines in the Southern Ocean at 500 m depth intervals is illustrated in Fig. 2.2a. The data produced by digital bathymetry are of crucial importance to practically every maritime information system. The best known high-quality bathymetry dataset is the one provided by the Seabed 2030 Project [37]. Its purpose is to amalgamate all the available bathymetric data from numerous sources of varying quality and coverage in order to produce the definitive map of the world's oceans' floors and to make it universally available and free of charge. However, despite the high quality of the dataset, the data provider strongly suggests that the dataset should not be used for navigation or for any other purpose depending on safety at sea.

In addition, several data sources offer multibeam deep-water bathymetric data (produced using multibeam echo sounders, a type of sonar that is used to map the seabed), the most comprehensive of which is the repository of the National Centers for Environmental Information (NCEI, [38]) of the United States of America (US) National Oceanic and Atmospheric Administration (NOAA): it hosts more than 9 million nautical miles of ship trackline data (magnetic, gravity and seismic reflection data) recorded from over 2,400 cruises and obtained from several sources worldwide. An example of multibeam surveys in the Southern Ocean is illustrated in Fig. 2.2b. Equally outstanding is the Bathymetry Portal [39] of the European Marine Observation and Data Network (EMODnet), as initiated by the European Commission, which offers a service for viewing and downloading free of charge a harmonised high-quality bathymetry dataset that covers all European marine waters.

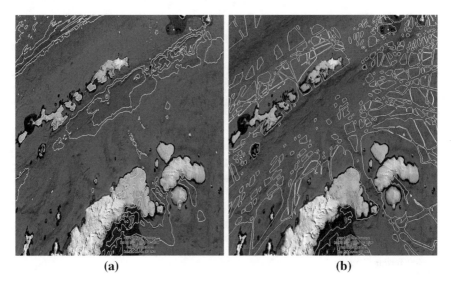

(a) (b)

Fig. 2.2 (a) Contour lines at 500 m intervals and (b) multibeam deep-water surveys of the same area in the Southern Ocean near the Antarctic Peninsula (data layer source: [40], background map-image source: [41])

Table 2.3 briefly compares the sources providing bathymetry data that are discussed in this subsection.

Data related to the location of petroleum fields under the seas and to the location of oil and gas drilling installations and platforms play an important role across a range of sectors, including the state and non-state sectors, the environmental sector, the social economy sector, etc. More than 2,000,000 wells have been installed onshore or offshore for the exploitation of over 65,000 oil and gas fields under the earth's surface. Lujala et al. [42] have compiled a free of charge global dataset that includes all known 383 offshore (and 890 onshore) oil and gas deposits throughout the world, providing both geographic coordinates and information on oil/gas discovery and production. The dataset is hosted at the Peace Research Institute Oslo.

The increased demand for subsea cables and pipelines has produced over two million kilometres of as-laid cables and pipelines worldwide (power, telecommunication and scientific, such as oceanographic or seismic cables; and pipelines that carry fluids such as oil or gas or water), which are sufficient to circle the globe 50 times. A publicly available interactive submarine cable map can be found in [43]. It shows most of the active and planned international submarine cable systems and their landing stations. Some other portals provide free of charge data for submarine cables and pipelines for specific countries and regions of the globe. For example, the EMODnet Human Activities portal [44] has launched a free of charge dataset of offshore pipelines that covers some parts of Europe.

All traces of human existence underwater which are one hundred years old or more, are protected by the United Nations Educational, Scientific and Cultural Orga-

Table 2.3 Major sources providing bathymetry and ocean topology data

Data source	Data categories (kinds of data)	Data formats	Covered area	Spatial resolution	Reference
The Seabed 2030 Project	Bathymetric contours	ESRI, ASCII, raster, GeoTIFF, NetCDF, Ocean Data View, etc.	Global coverage	15 * 15 arc seconds	[37]
The US NOAA NCEI's Repository	Multibeam bathymetric data	Raw multibeam files	Global coverage	Ranges from 1/9 arc seconds to 3 arc seconds	[38]
The EMODnet Digital Terrain Model	Bathymetric contours, bathymetric surveys (such as single and multibeam surveys), etc.	ESRI, ASCII, GeoTIFF, NetCDF, Fledermaus SD	European seas	3.75 * 3.75 arc seconds	[39]

nization (UNESCO) Convention on the Protection of the Underwater Cultural Heritage. There are several online catalogues of underwater cultural heritage sites and maritime archaeological parks with the remains of countless shipwrecks; aircraft wrecks and associated relics; submerged ancient buildings and settlements (including caves and wells), such as, for example, [45] which is hosted by the UNESCO or [46].

Shipwrecks alone are estimated to be in excess of three million worldwide [47]. The Wreck Site [48] is the world's largest online wreck database. It contains information about 186,900 wrecks but is only accessible for a fee. On the other hand [49] provides for free a database of shipwrecks, the remains of which have been located. Other portals contain detailed information about shipwrecks and aircraft wrecks. The Department of Culture of Ireland, for example, provides detailed information on about 1,200 shipwrecks [50]. Finally, the EMODnet Human Activities portal [51] provides information about wrecks, underwater sites and objects of historical and archaeological interest in the European seas. For reasons to do with their protection, this database does not give out the exact positions of wrecks.

Earthquakes are a common occurrence in the ocean. In terms of strength, they range from small tremors to tremors that can register as high as 9.2 on the richter scale. A sub-marine earthquake has the power to alter the seabed, which results in the creation of a series of waves. Depending on the duration and magnitude of the earthquake it has the power to cause a tsunami and spread destruction. Sub-marine earthquakes can also cause sub-marine damage, for example to pipelines that carry fluids or to cables, disrupting power and communication networks. The National Geophysical Data Centre of the US NOAA provides a database of earthquakes [52] which contains information about more than 10,000 devastating sub-marine and continental earthquakes from 2150 B.C. to the present day. Several other centres and portals provide real-time data on seismic events from all over the globe. One of these is the European-Mediterranean Seismological Centre [53].

Submarine volcanoes also represent a common feature of the ocean floor. The Global Volcanism Program [54] provides two global databases about volcanic activity, the Holocene volcano database [55], which includes 1,430 volcanoes that have been active approximately over the past 10,000 years (a complete list of current and past activity for every volcano is also available) and the Pleistocene volcano database [56], which currently lists 1,241 volcanoes thought to have been active throughout the past 2.5 million years (the list excludes the volcanoes that appear in the Holocene volcano database). Both under the sea surface and above the sea surface volcanoes are listed in the databases. It is worth mentioning that when the mid-oceanic ridge of a volcano is above sea-level volcanic islands are formed as was Iceland, for example.

2.3.2 The Aquatic Environment over the Surface of the Sea

Floating structures or platforms are artificial islands which are constructed to create floating airports, bridges, breakwaters, piers and docks, storage facilities (for oil and natural gas), wind and solar power plants. The geographical location of about 1,500 floating structures over the surface of the sea in specific locations around the globe, including offshore drilling rigs and meteorological stations, can be found in [57]. Several other sources provide detailed information including the location of offshore petroleum and gas related platforms for specific regions of the earth, such as in European seas [58].

Floating bridges and roads, while they are often temporary structures used in times of emergency, can also serve on a long-term basis in locations where water conditions are very sheltered. Two lists of some dozens of pontoon bridges located throughout the world can be found in [59, 60]. The detailed information that is available for every bridge includes its geographic coordinates, its length, photographic material, etc.

A wind farm or wind park is a group of wind turbines installed in the same location and used to produce electricity. The size of wind farms varies from a small number of turbines to several hundred covering an extensive area. The database of offshore wind farms in the European seas can be found in [61]. A comprehensive and regularly updated dataset of US onshore and offshore wind turbine locations and characteristics is freely available in [62].

At least 15% of the oceans is covered by ice for a part of a year. This means that, on average, sea ice covers almost 10 million square miles (about 25 million km^2) of the earth. Sea ice is closely monitored because changes in sea ice can be a reliable indicator of climate change. The Sea Ice Index [63] gives an insight into Arctic-wide and Antarctic-wide changes in sea ice. It represents a source of consistent, up-to-date sea ice extent and concentration data from November 1978 to the present.

2.3.3   Ecology, Marine Flora and Fauna

The global-scale approach to ecology and evolution has been attracting a growing amount of interest. It is an approach which focuses on patterns and determinants of species diversity and on the threats resulting from global change. Since these studies are bound to require global datasets of species distribution, they rely on databases such as FishBase [64], which is the largest global biodiversity information system on finfish. It provides a wide range of information on all of the 34,000 finfish species currently known, including taxonomy, biology, trophic ecology, life history and uses, as well as historical data going back 250 years. As for Global Fishing Watch [65], it provides ways of tracking and visualizing data about global fishing activity in near real-time (i.e. three days prior to present time) and the service is free of charge. With the cooperation of MarineTraffic, the platform makes it possible to monitor the global commercial fishing fleet via a public map. Anyone can download anonymized data about the past and present activities of fishing boats.

Aquaculture is the culture or the farming of fish, shellfish, and other forms of aquatic life in salt, brackish, or fresh water areas. The techniques used include hatching, seeding or planting, cultivating, feeding, raising, harvesting planted crops or natural crops, and processing. Several countries such as Chile in [66] and Spain in [67] provide online maps and repositories giving out information about aquaculture farms. The database on shellfish aquaculture production sites and locations in the European Union (EU) can be found in [68] and the database on marine finfish aquaculture in the EU can be found in [69].

The database that records the global distribution of the marine mammal families (including seals, sea lions and walrus, whales, dolphins and porpoises, manatees and dugongs) and chondrichthyes (sharks and rays) is provided in [70]. By attaching tags on marine mammals, such as southern elephant seals, the International MEOP (Marine Mammals Exploring the Oceans Pole to Pole) Consortium has been collecting since 2004, and distributes freely, a large dataset of trajectories of mammals together with other environmental data, such as the temperature and salinity of the oceans [71].

The International Union for Conservation of Nature (IUCN) Red List of Threatened Species, founded in 1964, has evolved to become the world's most comprehensive inventory of the global conservation status of biological species. With its strong scientific base, the IUCN Red List is recognized as the most authoritative guide on the status of biological diversity. To date, the list contains global assessments for more than 98,500 species (plants, animals and fungi), including several varieties and populations of species in the planet's marine environment. Of all the assessed species, 26,197 (i.e. 27%) are classified as vulnerable, critical or endangered. The

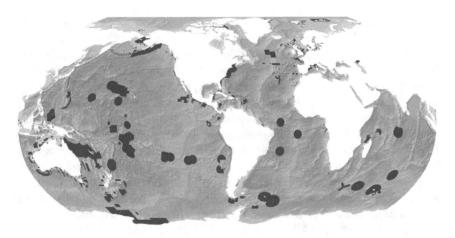

Fig. 2.3 The world map on maritime protected areas (*source* [72])

database can be found in [70] and is made freely available to help inform conservation planning and other non-commercial decision-making processes.

According to the `ProtectedPlanet.net` web portal, made available by the UN Environment World Conservation Monitoring Centre, the global coverage of marine protected areas is 7.6%. The World Database on Protected Areas (WDPA, [72]) is the most comprehensive global database of marine and terrestrial protected areas, updated on a monthly basis, and is one of the key global biodiversity datasets used by scientists, businesses, governments, international secretariats and other relevant authorities to inform planning, policy decisions and management. A map of the official marine protected areas included in the World Database on Protected Areas can be viewed in Fig. 2.3.

At a national level, countries define marine protected areas in their exclusive economic zone and it is incumbent upon them to legislate in order to protect these areas. At a regional level, countries with a marine estate may cooperate to define marine protected area networks [73], which are groups of marine protected areas that interact ecologically and/or socially. Some well-known networks are the Natura 2000 network (its dataset can be found in [74]) set up by the European Union, the MedPAN (Network of Marine Protected Areas Managers in the Mediterranean sea) established by the World Wildlife Fund (WWF) and other partner organisations (its dataset can be found in [75]), the NAMPAN (North American Marine Protected Areas Network, [76]), etc. The history of international agreements on global marine protection is provided by the Marine Conservation Institute in [77].

With regard to the earth's poles, areas in the Arctic under protection from all five Arctic nations can be found on the GeoNetwork catalog of spatially referenced resources [78], which is provided by the Conservation of Arctic Flora and Fauna and the Protection of the Arctic Marine Environments working groups of the Arctic Council. The same portal also provides the dataset of the marine areas of heightened ecological and cultural significance. Finally, the protected Antarctic areas can be found in [79].

A wetland is an area of land with water covering its surface or lying under its surface either all year round or at various times of the year. Wetlands therefore represent a distinct type of ecosystem, supporting both aquatic and terrestrial species. Wetlands can be found at the intersection of fresh water and salt water in coastal areas and inland. Wetlands are key regulators of climate change. They contribute positively to global warming through their greenhouse gas emissions, and negatively through the accumulation of organic material in histosols, particularly in peatlands. On February 2, 1971, the Convention on Wetlands was signed in Ramsar, Iran, as the intergovernmental treaty that provides the framework for the conservation and wise use of wetlands and their resources. The List of Wetlands of International Importance includes 2,341 designated sites from 170 countries, covering a total area of 252,489,973 hectares [80]. In addition, WWF and the Center for Environmental Systems Research at the University of Kassel in Germany created the Global Lakes and Wetlands Database [81]. This database includes lakes, reservoirs, and rivers that have a surface area larger or equal to $0.1 \, km^2$. The data are available for free downloading for non-commercial scientific, conservation and educational purposes.

Seagrasses are underwater flowering plants. Around the globe, seagrasses occupy shallow sedimentary shorelines where they operate as ecosystem engineers that, by virtue of their ability to cause physical state changes in their surrounding environment, can have a large impact on the availability of resources for other types of species, hence their important role in maintaining the health and stability of the places in which they live. The direct use by humans of the seagrass biomass and the benefits which human communities draw from them include carbon sequestration and storage; pharmaceutical developments; fish production, since they provide a habitat for fish species and other organisms which cannot live in unvegetated areas at the bottom of the sea; marine eco-tourism; education in environmental issues related to their significant role as a natural resources; etc. This means that the value of seagrasses surpasses the economic value of coral reefs and even the value of tropical rain forests and explains why seagrasses are ranked amongst the most valuable and fragile ecosystems on earth. In recent years, there have been reports about the significant decline of seagrass meadows (at a rate of about 29% in terms of area since the beginning of the twentieth century, at an annual rate of decline of about 1.5% and faster in recent years). Out of the 72 known species of seagrass, 10 are at risk of extinction and 3 are endangered. The dataset of the global distribution of seagrasses can be found in [82]. The seagrass species of Posidonia oceanica [83], which is exclusively endemic to the Mediterranean Sea and is on the IUCN Red List of Threatened Species, is also included in the dataset.

Table 2.4 summarizes the sources providing reliable, freely available sea-related datasets.

Table 2.4 Summary of major sources providing sea-related data

Data category	Data sub-categories (kinds of data)	Data sources
The aquatic environment under the surface of the sea	Bathymetric charts and maps	[37, 39, 40, 84]
	Multibeam deep-water bathymetric data	[38–40]
	Petroleum fields; oil and gas drilling installations	[42, 57, 58]
	Subsea cables and pipelines	[43, 44]
	Underwater cultural heritage sites and maritime archaeological parks	[45, 46, 51]
	Shipwrecks and aircraft wrecks	[48–51]
	Earthquakes	[52, 53]
	Volcanic activity	[55, 56]
The aquatic environment over the surface of the sea	Floating structures and platforms, including off-shore drilling rigs and meteorological stations	[57, 58]
	Pontoon bridges	[59, 60]
	Wind farms	[61, 62]
	Sea ice coverage	[63]
Ecology, marine flora and fauna	Finfish species distribution	[64]
	Fishing activity in near real-time	[65]
	Aquaculture farms	[66, 67]
	Shellfish and marine finfish aquaculture production	[68, 69]
	Distribution of marine mammals, chondrichthyes families, dolphins, seals, and turtle nesting sites	[70]
	Trajectories of marine mammals	[71]
	Marine protected areas	[72, 78, 79]
	Marine protected areas networks	[74–76, 78]
	International agreements on global marine protection	[77]
	Threatened species and their distribution range	[70]
	Wetlands	[80, 81]
	Distribution and biodiversity of seagrasses	[82]

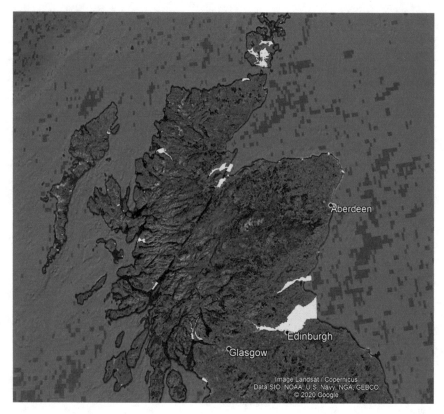

Fig. 2.4 A map designating in yellow the ports boundaries in Scotland (ports data layer source: [87], background map-image source: [41])

2.4 Land-Related Data

This section presents data that are related to the area amounting to about 30% of the globe which is land but which is in some way also related to its aquatic area. For example, the World Port Index [85] lists about 4,000 ports across the entire globe, describing their characteristics and the known facilities and available services which they provide. The geographic location of every port is given as a point (longitude and latitude) that most represents the central position of the port. However, since a port is not just a dot on a map but a series of terminals and anchorages, Wikimapia [86] and other sources provide datasets defining the port boundaries (polygons) on a global or on a more local scale. For example, the Scottish Government provides online the statutory limits of harbour authorities around Scotland [87]. These are illustrated in Fig. 2.4. Port databases with vessels live arrivals, schedules and weather information can also be found in MarineTraffic [88], FleetMon [89], etc. In addition, two lists of

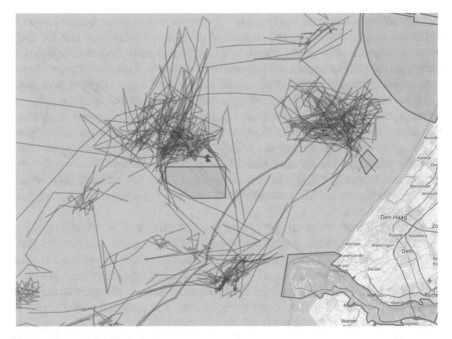

Fig. 2.5   Anchoring near Belgian and Dutch ports (*source* [28])

hundreds of marinas across a number of countries can be found in [90, 91], i.e. these are docks or basins providing moorings and supplies for yachts and small boats.

In addition, the Global Fishing Watch has used records of AIS vessel positions from 2012 to 2018 to develop an anchorages database [92] on the strength of the identification of locations where vessels congregate. In order to achieve this, they identified where individual vessels (specifically, individual MMSI) remained stationary (defined as when a vessel covers a distance under 0.5 km over a time span of at least 12 h: examples are provided in Fig. 2.5). If, in the same location within that 2012–2018 bracket, at least 20 unique MMSI had remained stationary, this location was identified as an anchorage point. Using this method, about 120,000 anchorages were identified and subsequently named using the World Port Index and the Geonames 1000 database [93], a version of the GeoNames database that contains over 25,000,000 geographical names [94]. This particular dataset does not include anchorages on rivers or lakes. Details relating to ways of classifying anchorage areas can be found in the chapter [95] by Andrienko and Andrienko.

Studies have also been undertaken to calculate the maritime distances between countries [96] and ports [97, 98]. Most of the distances represent the shortest navigable maritime routes. However, in some cases, longer routes that take advantage of favourable currents have been used. In other cases, longer distances result from the selection of routes on the basis of the need to avoid ice or other hazards to navigation, or of the need to follow required traffic separation schemes.

Lighthouses are towers, buildings, or other types of structures designed to emit light from a system of lamps and lenses and to serve as navigational aids for maritime pilots at sea or on inland waterways. Lighthouses mark dangerous coastlines, hazardous shoals, reefs, rocks and safe entries to harbours. A number of online directories provide extensive information about lighthouses. For every lighthouse on their list, these data sources provide photographs, maps, history, the colour of the light and its range, the weather, directions, postal addresses, and exact or approximate location coordinates. The Lighthouse Explorer Database [99] provides information about 8,259 lighthouses from all over the world, the ARLHS World List of Lights [100] provides information about more than 15,000 lighthouses across 233 countries, the Lighthouse Directory [101] provides information about more than 22,000 of the world's lighthouses and MarineTraffic [102] provides information about more than 13,500 lighthouses.

Although islands make up only 3% of the land surface of the planet, there are more than 175,000 islands around the globe that have a surface area over $0.1\,km^2$. These are home to more than 650 million inhabitants. In other terms, about 10% of the world population lives on islands. The UN Environmental Programme provides in [103] an alphabetical index of the 2,000 islands which they describe as the most significant ones on the planet. Basic environmental and geographic information is also included there. The dataset of all the small islands around the globe that are $2\,km^2$ or under in size, and the dataset of the coral reefs and atolls, can both be found in [104].

And last, the coastline is the line that forms the boundary between the coast and the shore, i.e. the line that forms the boundary between the land and the water. Several online sources provide versions of the coastline around the world in a variety of file formats and of levels of accuracy, such as the US NOAA [105], or Natural Earth [104]. The coastline of specific continents or countries are also available online,

Table 2.5 Summary of major sources providing land-related data

Data sub-categories (kinds of data)	Data sources
Ports, vessels live arrivals, schedules and weather information	[85–89]
Marinas	[90, 91]
Anchorages	[92]
Geographical names (place names)	[93, 94]
Maritime distances between countries	[96]
Maritime distances between ports	[97, 98]
Lighthouses	[99–102]
Significant islands	[103]
Small islands that are $2\,km^2$ or under in size, coral reefs and atolls	[104]
Coastline	[84, 104–107]

for example for Europe, in [106] and for the US, in [107]. The coastline of Antarctica and of the sub-Antarctic from the south pole to 40°S is provided by the Antarctic Digital Database [84].

Table 2.5 summarizes the sources providing reliable, freely-available land-related datasets.

2.5 Map and Imagery-Related Data

The exclusive economic zone of every country, which is the sea zone prescribed by the 1982 UN Convention on the Law of the Sea over which a country has special rights regarding the exploration and use of marine resources, including energy production from water and wind, can be found in [108].

Datasets of the political borders of almost every country on earth (i.e. about 247 countries at the time of writing) can be found in the Natural Earth portal [104] and in the World Borders database [109]. Another similar dataset showing different degrees of detail can be found in the University of Edinburgh DataShare portal, in [110] (coarse version) and in [111] (detailed version). The administrative areas of every country in the world, at all levels of sub-division (386,735 administrative areas in total) are provided in [112]. This high spatial resolution dataset is provided either country by country or for the entire globe in a single file.

Nautical charts are graphic representations of sea areas and their adjacent coastal regions. They are vital for passage planning (the procedure of developing a complete description of a vessel's voyage from start to finish), ocean crossings, coastal navigation and entering ports. Electronic nautical charts, more so than any other recent advancement, represent the way navigation and technology have sailed side-by-side into the 21st century. Several online databanks offer free electronic nautical charts for sailing the seas, such as the OpenSeaMap [2], which extends the coverage of OpenStreetMap [113] to the seas and freshwater bodies. The charts show lighthouses, lateral buoys, cardinal marks and other navigational aids. In the ports, facilities are mapped (port wall, pier, walkways, docks, fuelling stations, loading cranes, etc.). Similarly, public authorities, shipbuilders and repairers, as well as sanitation and utility facilities are displayed. Figure 2.6 shows an electronic nautical chart from the OpenSeaMap service. These charts are to be used along official charts and are not intended to replace these.

Since they are safe from the hazards which can befall modern navigation technologies and software tools, paper charts are still widely used by many navigators. A collection of more than 1,000 nautical charts for marine navigation in the oceans and coastal waters of the US and in the Great Lakes is provided by the US Office of Coast Survey in [114]. The charts may be printed at the true chart scale on plotters that accommodate 36" wide paper. However, only US NOAA paper nautical charts printed by a US NOAA certified chart agent, who will ensure that the charts are printed at the proper scale and quality, meet the US Coast Guard chart carriage requirements.

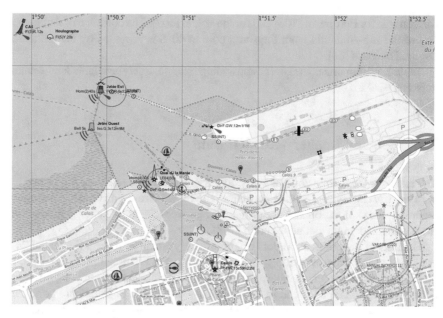

Fig. 2.6 An electronic nautical chart for the Port of Calais, France (*source* [2])

The Natural Earth portal features raster maps[2] of the globe at the 1:10 and at the 1:50 million-scales [104]. The embedded raster content includes land cover, ocean water, shaded relief (it illustrates the shape of the seabed in a realistic fashion by showing how its three-dimensional surface would be illuminated from a point light source), bathymetry, etc. All raster maps are accompanied by a text file (in TFW file format) so that it can be geo-referenced (geo-referencing is the process of relating a digitized map to its exact real-world location). In addition, several 500 m resolution global and by-region (colour and grayscale) maps of the earth at night are provided by the National Aeronautics and Space Administration (NASA) Earth Observatory in [115].

Several official sources provide free access to recent satellite images of the globe. The European Space Agency (ESA)'s Copernicus Open Access Hub [116] brings the latest free satellite images from all active Sentinels: radar imagery from Sentinel-1, optical multispectral Sentinel-2 imagery, etc. An imaging radar system provides its light to illuminate an area on the ground and take a clear radiowave picture regardless of the weather conditions and of any darkness. It is worth mentioning that imaging radar in high resolution is able to locate ships on the sea surface day and night, the reason being that when a ship is illuminated it returns a strong signal back to the

---

[2]These are maps in formats similar to digital photographs, the entire areas of which is subdivided into a grid of tiny cells or pixels, and the value of every one of these cells represents the colour or/and any other useful information (referring to the temperature, etc.) related to the corresponding location on the earth.

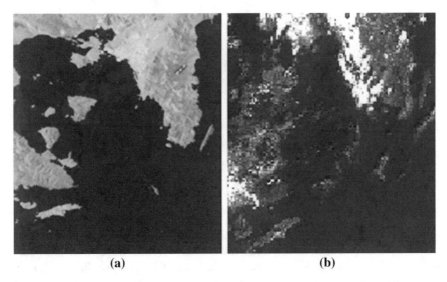

(a) (b)

Fig. 2.7 A radar Sentinel-1 image and a multispectral optical Sentinel-2 image for the port of Piraeus (Greece) and its surrounding area in June 2020 (*source* [116])

antenna and as a result a bright spot is captured. On the other hand, multi-spectral optical imaging may use a combination of the visible spectrum and the infrared spectrum, the ultraviolet spectrum and x-rays. The differences between these two types of satellite images stand out in Fig. 2.7, which illustrates a radar Sentinel-1 image and a multispectral optical Sentinel-2 image for the same area. The Earth Observatory (EO) Browser [117] stores medium and low-resolution satellite images, including the complete archives of all the Sentinel satellite missions, Landsat 5–7 and 8, MODIS, Envisat Meris, Proba-V, and GIBS satellite imagery, including an imagery mosaic of the globe derived from Sentinel-2, Landsat 8, MODIS satellites. Earthdata Search [118] provides a plethora of NASA's earth science imagery data. The amount of freely available current satellite images and historical imagery enlisted is impressive. In addition, Land Viewer [119] offers up-to-date miscellaneous free global satellite images from Landsat 7–8, Sentinel-1 and 2, CBERS-4, MODIS, aerial data from NAIP, and Landsat 4–5 historical satellite imagery. Finally, Remote Pixel [120] allows free access to and the download of Landsat 8 images and spectral bands.

Besides satellite images, several portals host broad collections of maritime-related high quality photographs, videos, vectors, drawings, illustrations and music files which are free to use, copy or adapt for both commercial and non-commercial purposes. Attribution crediting the creator (photographer, videographer, colorist, music source, etc.) or the media data source is not required although it is always appreciated. For example, the Pexels [121], the Pixabay [122] and the Burst [123] platforms allow the user to download and use (and even modify prior to publication) a sizeable collection of such multi-media data related to subjects including boats/ships (from

Table 2.6 Summary of major sources providing map and imagery-related data

Data sub-categories (kinds of data)	Data sources
Exclusive economic zones of all countries	[108]
Political borders of almost all countries	[104, 109–111]
Administrative areas of all countries	[112]
Nautical e-charts	[2]
Nautical paper charts	[114]
Maps of the globe (day and night)	[104, 115]
Satellite images	[116–120]
Images from non-satellite sources	[118, 121–123]

every possible angle, relating to piracy, cruises, cargos, containers, yachts, etc.), shipping, the oceans, the seas, waves, water, underwater, beaches, sailing, ports, harbours, marinas, diving, fishing, fish (sharks, dolphins, turtles, salmon, jellyfish, etc.), coral reefs, submarine life, the earth, oil rigs, pollution, etc. and relating to practically all the terms covered in this chapter. Similarly, no copyright restrictions apply to the uploading of multimedia data for sharing purposes.

Table 2.6 summarizes the sources providing reliable, publicly-available map and imagery-related datasets.

2.6 Climate and Sea Conservancy-Related Data

Due to their tremendous ability to store heat, the oceans and seas play a key role in the global redistribution of the energy absorbed by the sun and, hence, in the influence which they exercise upon the weather and climate system of the earth. Decades of research have also provided clear evidence of the central buffering role played by the oceans against the impact of climate change. The oceans have absorbed over 90% of the heat from climate change, and act as the natural sinks for roughly 30% of human-caused carbon dioxide emissions. Without the oceans acting as buffers, the heat resulting from climate change would already be intolerable for much of life on earth and this begs the question about protecting the oceans.

2.6.1 Weather Conditions

Techniques of marine forecasting, that of telling the future state of wind and wave conditions, have come a long way, bringing humanity into the modern era of marine weather forecasting using sophisticated computer programs, and marine observation via satellite and buoys. Today marine forecasting is a complex task of viewing data, both observed and model, and then of synthesizing these data through knowledge and experience in order to create forecasts in both graphic and text formats.

Several information services exist online that use thousands of onshore and off-shore observation stations around the world in order to provide historical and forecast weather data with global coverage. The OpenWeatherMap service [124], for example, provides weather data for every geographic location on earth (given a pair of known longitude/latitude values or the name of a place) and a weather Application Programming Interface (API) suitable for any cartographic application. Weather data are obtained from several global meteorological broadcasting services and from more than 40,000 weather stations around the globe. The data are provided in JSON, XML or HTML formats. At the time of writing, users can make up to 60 API calls per minute free of charge (i.e. queries to the service's database) to get current weather information for any geo-location; a 5-days weather forecast with a 3 h time step; weather maps including precipitation, clouds, pressure, temperature, wind, wave height and direction, etc.; historical ultraviolet (UV) data; weather alerts; etc.

Visual Crossing [125] is another global weather data provider. At the time of writing, it offers up to 250 calls per day for free access to weather forecast and weather history reports. This service, however, stands out in respect of its offer of free access to its full weather database to any user disposed to work as a weather data ambassador for the company (more details can be found on the service's website).

Weather Unlocked [126] delivers current live weather data and 3 hourly 7-days forecast data for worldwide locations. Its API returns responses in either JSON or XML format and it delivers a comprehensive set of weather elements including temperature, type and amount of precipitation, cloud cover, wind speed, gust speed, wind direction, humidity, etc. It offers 20,000 free calls per day with a limit of 100 calls per minute.

The Planet OS Datahub [127] provides a collection of open weather, climate and environmental data on a global scale. The current marine weather conditions and forecasting data include air pressure at sea level, surface temperature, surface and bottom salinity, surface and bottom water velocity, surface wave direction, visibility, wind speed and direction, direction of wind wave velocity, oxygen in the water, deviation of sea level from mean surface, ice fraction and thickness, water speed and turbidity, etc. Web links to the (mostly official) sources that are the providers of the data are also provided.

Table 2.7 briefly compares some of the major sources providing free global weather conditions and forecasting data in machine-readable formats.

The tropical cyclone stands out amongst extreme weather natural phenomena. Its rapidly rotating storm system is characterized by a closed low-level atmospheric circulation, strong winds, and a spiral arrangement of thunderstorms that produce heavy rain or squalls. The tracking and understanding of cyclones is important to the climatologists and other scientists who seek to uncover patterns and variability which could constitute clues to help us understand climate change. Understanding these patterns and being aware of potential cyclone strikes is of paramount importance to emergency management in cyclone strike areas. Historical data about over 6,000 global tropical cyclones are included in the IBTrACS database (International

Table 2.7 Some major sources providing weather conditions and forecasting data with global coverage

Data source	Data categories (kinds of data)	Data formats	Number of weather stations*	Number of free of charge calls offered	Reference
Open Weather Map	Current weather data, 5-days forecast with a 3 h time step; weather maps including precipitation, clouds, pressure, temperature, wind, wave height and direction, etc.	JSON, XML, HTML	40,000	60 calls/minute	[124]
Visual Crossing	15-days forecast and weather history reports since 1960.	JSON, CSV	Unknown (plus an archive of more than 100,000 stations)	250 calls/day (plus free access to its full database to weather data ambassadors)	[125]
Weather Unlocked	7-days forecast and past 7-days history, with elements including temperature, type and amount of precipitation, cloud cover, wind speed, gust speed, wind direction, humidity, etc.	JSON, JSONP, XML	Unknown	20,000 calls/day with a limit of 100 calls/minute	[126]
Weather-bit.io	Current conditions, 16-days forecast, severe weather alerts	JSON, CSV CSV	45,000 (plus an archive of more than 120,000 stations)	500 calls/day	[132]
Weather Stack	Current weather conditions	JSON	Unknown	1,000 calls/month	[133]

*'Unknown': i.e. it is difficult to find the precise number in the available documentation

Best Track Archive for Climate Stewardship, [128]), spanning the last 150 years. It provides for every cyclone, best track data in a centralized location to aid understanding of its distribution, frequency, and intensity. The database is updated on a weekly basis.

A tornado is a rapidly rotating column of air that is in contact with both the surface of the earth and a cumulonimbus cloud. The most extreme tornadoes can reach wind speeds of more than 300 miles per hour (480 km/h), their size can exceed 2 miles (3 km) in diameter, and they can cover a distance of dozens of miles (over 100 km) on the ground. The most frequent occurrences of tornadoes are in North America and in Europe. A database hosting tornado tracks (going as far back as 1950), along with hail and damaging convective winds (going as far back as 1955) in the US, Puerto Rico, and the US Virgin Islands is provided by the Storm Prediction Center of the US NOAA's National Weather Service [129]. In addition, the European Severe Weather Database (ESWD, [130]) has compiled several years of information on dust, sand or steam devils, tornado sightings, gustnados, large hail, heavy rain and snowfall, severe wind gusts, damaging lightning strikes and avalanches from all over Europe and around the Mediterranean. The database is provided by the European Severe Storms Laboratory (ESSL) and it is accessible through [131] for a fee; however, ESSL is open to discussing its terms when the use of the data is for academic or personal non-commercial purposes.

2.6.2 The Climate System

Climate is the long-term average of the weather, typically over a period of several years. The meteorological variables that are commonly measured are temperature, humidity, atmospheric pressure, wind, water salinity, etc. Climate, especially temperature, affects the health of all marine ecosystems. Knowledge of climate extremes, of their frequency and severity is also very important for human life which depends on those ecosystems. The World Ocean Database (WOD, [134]) is made up of a collection of quality-controlled ocean profile data that include historical (going back to the year 1800, where available) in-situ surface and subsurface oceanographic measurements of temperature, salinity, oxygen, phosphate, nitrate, silicate, chlorophyll, alkalinity, pH, plankton, etc. The WOD was first released in 1994 and it is updated approximately every four years.

The European Commission's Copernicus Marine Service [135] provides another collection of global ocean climate and forecast data updated daily, including top-to-bottom measurements and forecasts for temperature, salinity, currents, sea level, mixed layer depth, ice parameters, etc. It also includes hourly mean surface fields for sea level height, waves, temperature and currents; as well as daily and monthly mean files for several biogeochemical parameters (chlorophyll, nitrate, phosphate, silicate, dissolved oxygen, phytoplankton, PH, surface partial pressure of carbon dioxide, etc.) and more. All data are provided in high degree horizontal (sea-surface) resolution and in several vertical (below the sea-surface) depth levels.

The proximity of land areas to a sea or to an ocean affects the characteristics of the climate in those areas. These characteristics depend on the nearby water bodies and their wind-driven currents, the most notable one being (relatively to the latitude of these land areas) the moderately hot summers and the moderately cold winters. The climate of such regions, known as oceanic climate or maritime climate, is considered the converse of continental climate [136]. A dataset of spatially interpolated monthly weather and climate data for the global oceanic climate, and more generally for all the land areas around the globe, at a very high spatial resolution (approximately 1 km) is presented in [137]. The dataset includes monthly temperatures (minimum, maximum and average), precipitation, solar radiation, vapour pressure, wind speed and 19 bioclimatic variables, aggregated across a temporal range of years 1970–2000, and historical monthly weather data for the period 1960–2018, using data obtained from between 9,000 and 60,000 weather stations.

2.6.3   The Impact of Human Activity on the Health of the Ocean

The state of the ocean is vitally important to all living organisms on the planet. However, relentless human activity, both over and under the surface of the ocean, and particularly the predominance of maritime vessel traffic, represents a threat to the survival of the marine ecosystems and to oceanic life. The Third IMO Greenhouse Gas (GHG) Study [138] has estimated that, with the current growth rates of the world's maritime trade, international shipping could account for some 10% of global greenhouse gas emissions by 2050, undermining the objectives of the 2015 UN Framework Convention on Climate Change (i.e. the Paris Agreement, [139]).

For this reason, the IMO has recently initiated a data collection process [140] that requires the owners of large ships (of more than 5,000 gross tonnage) engaged in international shipping to report the fuel consumption of their ships to the IMO's Flag Member States where the ships are registered. The Flag Member States must in turn report to the IMO with aggregated data, using the IMO Ship Fuel Oil Consumption Database, which is a module of the Global Integrated Shipping Information System (GISIS) platform [141]. On the basis of the data received, the IMO compiles an annual emissions summary report. The above data collection process started in 2019 and the first data observations have not yet been announced at the time of writing.

As a result of the frequency of shipping activities (i.e. the shipping industry delivers 90% of all world trade), all ships are at risk of accidents with all kinds of destructive consequences, including casualties, financial loss and various kinds of environmental pollution. A comprehensive ship accidents dataset from the year 1990 onwards is provided by the Marine Casualties and Incidents Reports [142] issued by the IMO. Every accident in its report includes information about the event, such as the time when the accident occurred and the geographical coordinates of the location, the initial event, a summary, the type of casualty and the type of vessel involved (the database contains in their fullest form approximately 160 data fields, potentially providing a very detailed picture for every accident). The dataset included more than 11,000 accidents at the time of writing. Another comprehensive list of accidents involving a ship sinking, foundering, grounding, or being otherwise lost is available in [143]. The records go back to the year 1747.

Authorities around the world provide lists of maritime accidents and investigation reports drawn up about specific accidents which have occurred in waters under their responsibility. These case studies offer valuable details in respect of the location and the exact manner in which these accidents at sea happened and in respect of the advice they offer on how to prevent such accidents from happening again in the future. Catalogues of addresses of these governmental investigation bodies around the world can be found online, for example in [144, 145], the latter providing the online and postal addresses of all of the EU national accident investigation authorities.

Drilling for oil is another human activity putting the ocean's health at risk. Oil spills (accidental discharges) are events of a catastrophic nature entailing the destruction of the natural environment and long-lasting consequences for the ocean's ecosystem. A database listing more than 200 confirmed oil spills caused by ships, mainly tankers, by combined carriers and barges, since the 1900's, can be found in [146]. The database records the type of oil that was spilled, the amount that was spilt, the cause and location of the incident and the vessel involved. Details which include the geographical locations of the areas affected for the 50 worst oil spill disasters up to 2010 can be found in [147].

Plastic pollution, like all forms of pollution, is ubiquitous in the marine environment. The estimations of the global abundance and weight of floating plastics depend on the data that could be obtained, which can be insufficient, particularly from the southern hemisphere and remote regions. Although only 3% of the global plastic waste ends up in the ocean (bags, bottles, caps/lids, straws and stirrers, etc.), an estimation [148] of the total number of plastic particles floating in the world's oceans and of their weight has come up with a minimum number of 5.25 trillion particles, weighing 268,940 tons.

In addition, the NASA's Garbage Patch Visualization Experiment by Shirah and Mitchell [150] has used the trajectories from floating scientific buoys that the US NOAA had been disseminating in the oceans over the period 1979–2013 from different locations on earth (e.g. from research stations, ships or planes). The moving positions of these floating buoys having been recorded over time in the Global Drifter Database [151], their migration patterns were revealed. Their journey was traced up until the point when they reached their final destinations at the five major gyres of the global ocean (i.e. at what Ebbesmeyer and Scigliano in [152] coined "garbage patches", as shown by the five 'cycles' in Fig. 2.8), at which point the buoys subsequently changed the pattern of their journey and took a new course, adopting a wide and slow circular pattern of motion as shown by the arrows in Fig. 2.8. It was also observed that these patterns of the buoys' journeys match the patterns of the journeys followed by the oceans' currents around the globe.

Data from several other studies have also brought to light evidence that the floating oceanic debris (human-created waste that has deliberately or accidentally been

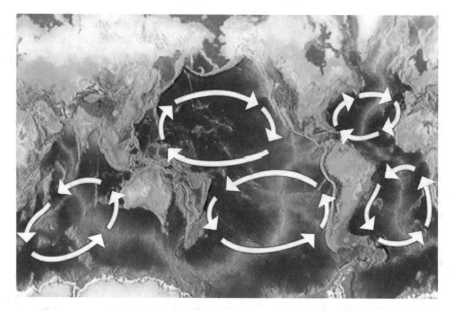

Fig. 2.8 The five main ocean gyres which have become garbage patches (*source* [149])

released in a sea or ocean) tends to build up at the center of these ocean gyres. For example, in [153] a total of 3,070 samples across the open ocean have been collected in order to identify hot spots of surface-level pollution. The frequency of occurrence of plastic debris in these samples was considerably high, 88%. The study clearly shows that the highest concentrations of plastic debris on the surface of the sea are in the five major ocean gyres.

In addition, the Sea Education Association Plastics Project 2010 Expedition [154] provides the geographic coordinates of more than 6,000 locations in the North Atlantic Subtropical Gyre and of 2,500 locations in the eastern Pacific Ocean in which the concentration of surface marine debris was measured (in pieces per square kilometre).

However, while about 60–80% of all marine debris is composed of plastic, fishing gear and equipment (nets, fishing line etc.), cigarettes, paper, wood, metal and other manufactured materials and industrial waste are also found along coastlines world-wide and at all the different depths of the ocean. The Deep-sea Debris Database [155] provides videos and photos of about 4,000 large pieces of marine debris (cans, bags, bottles, batteries, shoes, beach balls, fishing nets, barrels, etc.) sunk into deep-sea. Information about the locations and the depth at which the debris was found is also available. The database reports that the record depth at which pieces of waste (a flimsy shopping bag and a lumber) have been found was 10,900 m inside the Mariana Trench in the western Pacific Ocean.

The quality of the water in coastal zones is monitored at the national level with yearly reports that evaluate them on the basis of compliance with standards for microbiological parameters (total coliforms and faecal coliforms) and physicochemical parameters (mineral oils, surface-active substances and phenols). For example, Greece, a major European tourist destination, publishes detailed yearly assessments of bathing water quality for over 1,500 organized bathing water sites on the basis of the guidelines of EU directives. The geographic locations and all the relevant profile information for these sites is provided in [156]. At a higher regional level, the European Environmental Agency maintains the Waterbase [157], a detailed database providing information about the status and quality of Europe's water resources, including coastal and marine waters and about the emissions on surface waters from point and diffuse sources pollution.

Table 2.8 summarizes the sources providing reliable, publicly-available climate and sea conservancy-related datasets.

Table 2.8 Summary of major sources providing climate and sea conservancy-related data

Data category	Data sub-categories (kinds of data)	Data sources
Weather conditions	Weather conditions and forecasting	[124–127, 132, 133]
	Tropical cyclones	[128]
	Tornados	[129, 131]
	Severe storms	[131]
The climate system	Climate in areas covered by the ocean	[134, 135]
	Climate in areas in the proximity of the seas	[137]
The impact of human activity on the health of the ocean	Ship fuel consumption (greenhouse gas emissions)	[140]
	Accidents involving vessels	[142, 143]
	Governmental investigation bodies of maritime accidents	[144, 145]
	Oil spills	[146, 147]
	Estimation of the plastic particles floating in the oceans	[148]
	Trajectories of the oceans' currents and of the floating oceanic debris	[151]
	Hot spots of oceanic pollution	[153–155]
	Quality of sea water	[156, 157]

2.7 Discussion and Concluding Remarks

The value of maritime data collection is very far reaching. While it is a fundamental dimension of the study of the oceans and of coastal environments, it is at the same time the raw material which feeds, challenges and supports our ways of thinking about the seas of our planet. Maritime data give us an insight into the depths of the issues that have surrounded the relationship of the human race with the marine world throughout the ages. Since the dawn of time mankind has been confronted with the indomitable unknown that is the sea and every single art form has expressed this in its own way, offering interpretations using the knowledge available at the time or through myth and fantasy. All the literatures of the world have contributed a page to the timeless story of this relationship, from Homer's Iliad and Odyssey to the Tale of Sindbad the Sailor, Victor Hugo's Les Travailleurs de la Mer, Jules Vernes' Twenty Thousand Leagues Under the Seas, or Jack London's The Pirates of San Francisco, to mention but the best known out of a myriad examples. While the steady progress of science throughout the centuries has enabled us to harness the seas to our needs, the exponential technological advances which are shaping our lives today require that we adapt our ways and, as the issues confronting the future relationship of the human race with the oceans are becoming increasingly complex, we shall need increasing amounts of increasingly advanced forms and kinds of data to help us establish improved ways of managing this relationship. These data should enable us to carry out better informed evaluations of the ways in which the seas support the global economy and of the impact human activity has on the maritime environment; they should help us to understand the biological and ecological characteristics of the planet's environment; they should put us in a position to identify priority areas for management and conservation; guides spatial planning processes; guides the initiation of further quantitative and qualitative field-based research and analysis for the common benefit of both the worldwide community and of the planet which we inhabit.

Since the collection of maritime data on such a scale requires enormous commitments in terms of time, effort, financial investment and human and other resources, existing reliable datasets collected by trustworthy entities operating in the field represent an unparalleled advantage. Even more importantly, the use of these same readily available online datasets facilitates objective comparisons of different models, algorithms and systems, and gives us an insight into the merits of different proposals and methods. And last, the integration of many kinds of maritime data originating from different online sources around the world makes possible the more accurate and deeper understanding of global and complex issues which the unavailability of key data had not allowed to investigate properly until now. One such example might be offered by the SUMO system [158], which can detect unidentified maritime objects by comparing satellite Synthetic Aperture Radar (SAR) images, AIS ship self-reporting data, the coastline around the world, and raster maps. The proposed system has the appealing potential to provide a better picture of what is happening at sea, for example it can spot vessels that are not reporting their positioning data. The integration of different kinds of maritime data from several reliable sources has also been the focus of [3], the authors of which compiled a popular publicly available dataset bringing together, under the umbrella, four categories of data (navigation data, vessel-oriented data, geographic data, and environmental data). The dataset can be useful as a benchmark tool in a variety of maritime applications, including maritime traffic analysis, assessing the impact of human activities at sea, the prognostication and more comprehensive assessment of environmental changes, and assessment of multi-source information fusion algorithms.

The plethora of free of charge datasets that have been released online in the recent years will foster research, development and education in maritime informatics. These datasets may vary a lot in terms of domains and applications, of their resolution, their data format, or even in terms of the region of the globe which they cover. This chapter has endeavoured to explore the contents and specialised characteristics of a significant collection of close to 150 publicly available, useful and well-documented datasets across a wide and heterogeneous variety of data and fields related to the maritime domain, including nature, trade, transport, tourism, development, natural hazards, employment, security, energy, seabed mining, pollution, aquaculture, fishing, coastal erosion, climate, biogeography, conservation, and much more. Table 2.9 sets out the categories of maritime data following the taxonomy adopted earlier in the chapter and refers to the number of online data sources surveyed in this chapter. The provided datasets are presented, described and, whenever possible, briefly compared from different angles and perspectives, aiming to put the reader looking for a suitable dataset on the right track. A major challenge was to try and ensure that the vast majority of the datasets surveyed have global coverage, include every sea and as large an area as possible of the aquatic surface of the earth.

However, while the maritime data and the characteristics of the datasets covered here are comprehensive, they cannot be exhaustive and might not meet the reader's needs. Under the tidal wave of technological advances in the fields of monitoring and analysing the maritime world, an even deeper ocean of publicly available maritime data is waiting to surface and to flood the digital world It is hoped that the chapter will provide an informative springboard which readers will use to search for, identify,

Table 2.9 The number of online sources surveyed in this chapter per maritime data category

Data category	Number of data sub-categories	Number of data sources
Vessel-related data	9	28
Sea-related data	24	50
Land-related data	10	24
Map and imagery-related data	8	19
Climate and sea conservancy-related data	14	27
Total	**65**	**148**

evaluate and judiciously select other sources of information which they will come across in the future.

It is also hoped that this chapter will provide readers with a sense of direction in respect of the quality and standards requirements that will have to be met by future real-life maritime datasets to ensure that they are reliable and on a par with the existing high-quality datasets, and that they are compatible with them. In view of the expected exponential growth of online data, an additional challenge will be to ensure that resources are not wasted in the production of redundant data.

An ambition of the present data review is to inspire a drive to plan and initiate measurement strategies for the collection of critical data in domains presently lacking in highly accurate and essential maritime datasets on a global or on a more regional scale. Across almost every discipline there are areas of knowledge which could benefit from such datasets. To mention but three examples: opportunities could be created to fill in existing knowledge gaps leading to the compilation of databases of paleopathology data related to the marine flora and fauna. Palaeontologists have only made sporadic breakthroughs and contributed indirectly to this domain, with studies focusing mainly on dinosaurs and Cenozoic mammals (such as the species of Squalodon [159], an extinct genus of toothed whales that lived about 33–14 million years ago); other opportunities could lead to the monolithic recording of the worldwide maritime paths of undocumented human migration from poorer to richer countries around the globe, along with other important related information such as the number of people that successfully crossed and the numbers of those that died along the various routes channelling human migration; as a third and last example, opportunities should not be missed to record data relating to the impact on all aspect of marine life and of the maritime domain of extreme events, such as massive earthquakes, war, or human/animal pandemics on a global scale. Gathering and integrating such heterogeneous data and making the datasets available would ideally inform strategies and lead to quick response decisions when extreme events re-occur.

Disclaimer

All the data presented herewith are the property of their owners and are discussed for information and educational purposes only. The author of this chapter has made every effort to examine most of the data but cannot take responsibility for their quality and accuracy and cannot, therefore, be liable for any damage caused by inaccuracies in the data or resulting from the failure of the data to function on a particular system or setting. The reader is also advised to refer to the Conditions and Terms of Usage of the data before making any attempt to use them. Any use of the data is solely the reader's responsibility and made at the reader's own risk.

References

1. Coastal Wiki: Marine data portals operated by institutions in Europe (2020). Available at: http://www.coastalwiki.org/wiki/Marine_data_portals_and_tools
2. OpenSeaMap: The free nautical chart (2020). Available at: https://www.openseamap.org/index.php?L=1
3. Ray, C., Dréo, R., Camossi, E., Jousselme, A.L., Iphar, C.: Heterogeneous integrated dataset for maritime intelligence, surveillance, and reconnaissance. Data Brief **25**, 104141 (2019)
4. Kalyvas, C., Kokkos, A., Tzouramanis, T.: A survey of official online sources of high-quality free-of-charge geospatial data for maritime geographic information systems applications. Inf. Syst. **65**, 36–51 (2017)
5. US Geological Survey: GIS data (2020). Available at: https://www.usgs.gov/products/data-and-tools/gis-data
6. Free GIS Data: A categorised list of freely available geographic datasets (2020). Available at: https://freegisdata.rtwilson.com/
7. United Nations Conference on Trade and Development, Review of Maritime Transport 2019, Geneva (2020). Available at: https://unctad.org/en/PublicationsLibrary/rmt2019_en.pdf
8. Equasis: The Equasis portal (2020). Available at: http://www.equasis.org/
9. MarineTraffic: Vessels database (2020). Available at: https://www.marinetraffic.com/en/data/?asset_type=vessels
10. VesselTracker: Vessels (2020). Available at: https://www.vesseltracker.com/en/vessels.html
11. VesselFinder: AIS coverage map (2020). Available at: https://www.vesselfinder.com/
12. ShipFinder: AIS coverage map (2020). Available at: http://www.shipfinder.co
13. BalticShipping: Vessels (2020). Available at: https://www.balticshipping.com/vessels
14. Maritime Activity Reports: Maritime Vessels Directory (2020). Available at: https://directory.marinelink.com/ships/
15. FleetMon: Discover more than 500,000 vessels (2020). Available at: https://www.fleetmon.com/vessels/
16. Lloyd's List Group: Lloyd's list intelligence (2020). Available at: https://www.lloydslistintelligence.com/
17. AISHub: AIS coverage map (2020). Available at: http://www.aishub.net/
18. Lloyd's Register Foundation: Lloyd's register of ships online (2020). Available at: https://hec.lrfoundation.org.uk/archive-library/lloyds-register-of-ships-online
19. VesselFinder: Vessel database (2020). Available at: https://www.vesselfinder.com/vessels
20. Wood, A.K.: Warships of the Ancient World. Osprey Publishing, Oxford (2020). Available at: https://ospreypublishing.com/warships-of-the-ancient-world
21. Culver, H.B.: The Book of Old Ships: From Egyptian Galleys to Clipper Ships. Dover Publications, Mineola (1992)
22. Gibbons, T., Ford, R., Hewson, R., Jackson, R. (eds.): The Encyclopedia of Ships. Thunder Bay Press, San Diego (2001)
23. Batchelor, J., Chant, C.: The Complete Encyclopedia of Sailing Ships: 2000 BC - 2006 AD. Book Sales, Inc., New York (2008)
24. Wikipedia: Category: ancient ships (2020). Available at: https://en.wikipedia.org/wiki/Category:Ancient_ships
25. International Transport Workers' Federation (ITF): Flags of convenience (2020). Available at: https://www.itfglobal.org/en/sector/seafarers/flags-of-convenience
26. FleetMon: AIS coverage map (2020). Available at: https://www.fleetmon.com/
27. International Maritime Organization: Ships' Routeing (2019). ISBN 978-9280100495
28. Novikov, A.: Creating sea routes from the sea of AIS data (2020). Available at: https://towardsdatascience.com/creating-sea-routes-from-the-sea-of-ais-data-30bc68d8530e
29. Pan, S., Yin, J.: Extracting shipping route patterns by trajectory clustering model based on automatic identification system data. Sustainability **10**(7), 2327 (2018)
30. Silveira, P., Teixeira, A.P., Guedes-Soares, C.: AIS based shipping routes using the dijkstra algorithm. TransNav: Int. J. Mar. Navig. Saf. Sea Transp. **13** (2019)

31. International Maritime Bureau: Piracy Reporting Centre (2020). Available at: https://www.icc-ccs.org/index.php/piracy-reporting-centre

32. International Maritime Bureau: Piracy and armed robbery map (2020). Available at: https://www.icc-ccs.org/index.php/piracy-reporting-centre/live-piracy-map

33. International Maritime Bureau: Live piracy and armed robbery report (2020). Available at: https://www.icc-ccs.org/index.php/piracy-reporting-centre/live-piracy-report

34. International Maritime Organization: Reported incidents of piracy and armed robbery (2020). Available at: https://gisis.imo.org/Public/PAR/Search.aspx

35. Jensen, A.S., Silber, G.K.: Large whale ship strike database. US Department of Commerce, NOAA Technical Memorandum. NMFS-OPR-25, 37p. (2003)

36. Peel, D., Smith, J., Childerhouse, S.: Historic Vessel Strikes with Whales in Australian Waters. v1. CSIRO. Data Collection (2018)

37. The Nippon Foundation: The GEBCO seabed 2030 project (2020). Available at: https://seabed2030.gebco.net/

38. National Centers for Environmental Information: Multibeam bathymetry, US NOAA (2020). Available at: https://www.ngdc.noaa.gov/mgg/bathymetry/multibeam.html

39. European Marine Observation and Data Network (EMODnet): Bathymetry (2020). Available at: http://www.emodnet-bathymetry.eu/

40. The Scientific Committee on Antarctic Research: International bathymetric chart of the southern ocean (2020). Available at: https://www.scar.org/science/ibcso/resources/

41. Google Earth: The world's most detailed globe (2020). Available at: https://www.google.com/earth/

42. Lujala, P., Ketil Rod, J., Thieme, N.: Fighting over oil: introducing a new dataset. Confl. Manag. Peace Sci. **24**(3), 239–256 (2007). Data are available at: https://www.prio.org/Data/Geographical-and-Resource-Datasets/Petroleum-Dataset/

43. Telegeography: The submarine cable map (2020). Available at: https://www.submarinecablemap.com/

44. EMODnet Human Activities: Pipelines (2020). Available at: http://www.emodnet-humanactivities.eu/search-results.php?dataname=Pipelines

45. United Nations Educational, Scientific and Cultural Organization (UNESCO): Databases - underwater cultural heritage sites (2020). Available at: http://www.unesco.org/new/en/culture/themes/underwater-cultural-heritage/underwater-cultural-heritage/databases/

46. Pers sida: Wreck databases and lists (2020). Available at: https://www.abc.se/~m10354/uwa/wreckbas.htm

47. Croome, Angela: Sinking fast. New Sci. **161**(2169), 49 (1999)

48. The Wreck Site: The wreck database (2020). Available at: https://www.wrecksite.eu/

49. WikiPedia: Lists of shipwrecks (2020). Available at: https://en.wikipedia.org/wiki/Lists_of_shipwrecks

50. Department of Culture of Ireland, Heritage and the Gaeltacht: National monuments service: Wreck viewer (2020). Available at: http://dahg.maps.arcgis.com/apps/webappviewer/index.html?id=89e50518e5f4437abfa6284ff39fd640

51. EMODnet Human Activities: Ship wrecks (2020). Available at: http://www.emodnet-humanactivities.eu/search-results.php?dataname=Ship+Wrecks

52. National Geophysical Data Center/World Data Service (NGDC/WDS): Significant earthquake database. National Geophysical Data Center, US NOAA (2020). https://doi.org/10.7289/V5TD9V7K. Available at: https://www.ngdc.noaa.gov/nndc/struts/form?t=101650&s=1&d=1

53. European-Mediterranean Seismological Centre (EMSC): Real time seismicity (2020). Available at: https://www.emsc-csem.org/Earthquake/

54. Global Volcanism Program: Volcanoes of the world, v. 4.8.0. In: Venzke, E. (ed.). Smithsonian Institution (2013). https://doi.org/10.5479/si.GVP.VOTW4-2013

55. Global Volcanism Program: The Holocene volcano database. In: Venzke, E. (ed.). Smithsonian Institution (2013). Available at: http://volcano.si.edu/list_volcano_holocene.cfm

56. Global Volcanism Program: The Pleistocene volcano database. In: Venzke, E (ed.). Smithsonian Institution (2013). Available at: http://volcano.si.edu/list_volcano_pleistocene.cfm
57. National Data Buoy Center: Stations map, US NOAA (2020). Available at: https://www.ndbc.noaa.gov/
58. EMODnet Human Activities: Oil and gas (active licenses, boreholes, offshore installations) (2020). Available at: http://www.emodnet-humanactivities.eu/view-data.php
59. Wikipedia: List of pontoon bridges (2020). Available at: https://en.wikipedia.org/wiki/List_of_pontoon_bridges
60. Janberg, N.: Pontoon bridges list, Structurae international database and gallery of structures (2020). Available at: https://structurae.net/en/structures/bridges/pontoon-bridges/list
61. EMODnet Human Activities: Wind farms (polygons) (2020). Available at: https://www.emodnet-humanactivities.eu/search-results.php?dataname=Wind+Farms+%28Polygons%29
62. Hoen, B.D., Diffendorfer, J.E., Rand, J.T., Kramer, L.A., Garrity, C.P., Hunt, H.E.: United States wind turbine database. US Geological Survey, American Wind Energy Association, and Lawrence Berkeley National Laboratory data release: USWTDB V2.3 (2020). Available at: https://eerscmap.usgs.gov/uswtdb
63. Fetterer, F., Knowles, K., Meier, W.N., Savoie, M., Windnagel, A.K.: Sea ice index (updated daily), Version 3. Boulder, Colorado US NSIDC: National Snow and Ice Data Center (2020). Available at: https://nsidc.org/data/seaice_index/archives
64. Fishbase: A global information system on fishes (2020). Available at: https://www.fishbase.in/home.htm
65. Global Fishing Watch: Datasets and code (2020). Available at: https://globalfishingwatch.org/datasets-and-code/
66. Chile Infraestructura de Datos Espaciales Institucional: Visualizador de Mapas - Aplicación de Visualización de Mapas de la Subsecretaría de Pesca y Acuicultura (in Spanish) (2020). Available at: http://mapas.subpesca.cl/ideviewer/
67. Gobierno de Espana: Acuivisor (2020). Available at: https://servicio.pesca.mapama.es/acuivisor/
68. EMODnet Human Activities: Shellfish production (2020). Available at: https://www.emodnet-humanactivities.eu/search-results.php?dataname=Shellfish+Production
69. EMODnet Human Activities: Finfish production (2019). https://www.emodnet-humanactivities.eu/search-results.php?dataname=Finfish+Production
70. International Union for Conservation of Nature (IUCN): The red list of threatened species: spatial data download. Version 2019-3 (2020). Available at: https://www.iucnredlist.org/resources/spatial-data-download
71. MEOP Consortium: Databases (2020). Available at: http://www.meop.net/database/meop-databases/
72. Protected Planet: The world database on protected areas (WDPA), United Nations Environment Programme World Conservation Monitoring Centre (UNEP-WCMC) and International Union for Conservation of Nature (IUCN), Cambridge, U.K. (2020). Available at: https://www.protectedplanet.net/
73. Laffoley, d'A.D. (ed.): Towards Networks of marine protected areas. The MPA plan of action for IUCN's world commission on protected areas. IUCN WCPA, Gland, Switzerland, 28p. (2008)
74. EMODnet Human Activities: Natura 2000 (2020). Available at: https://www.emodnet-humanactivities.eu/search-results.php?dataname=Natura+2000
75. MedPAN Association: The marine protected areas in the Mediterranean (MAPAMED) database (2020). Available at: https://medpan.org/main_activities/mapamed/
76. The North American Marine Protected Areas Network (NAMPAN) (2020). Available at: https://nampan.openchannels.org/
77. Marine Conservation Institute: History of international agreements on global marine protection (2020). Available at: http://www.mpatlas.org/progress/targets/
78. The GeoNetwork Catalog: Arctic protected areas (2017). Available at:http://geo.abds.is/geonetwork/srv/api/records/2e56ee1f-50a9-4983-88f4-edaa8588950d

79. Terauds, A.: An update to the Antarctic specially protected areas (ASPAs), Australian Antarctic Data Centre (2016). Available at: https://data.aad.gov.au/metadata/records/AAS_4296_Antarctic_Specially_Protected_Areas_v2

80. The Convention on Wetlands: The list of wetlands of international importance, Ramsar, Iran (1971). Available at: https://www.ramsar.org/sites/default/files/documents/library/sitelist.pdf

81. Lehner, B., Doell, P.: Development and validation of a global database of lakes, reservoirs and wetlands. J. Hydrol. **296**(1–4), 1–22 (2004). Data are available at: https://www.worldwildlife.org/pages/global-lakes-and-wetlands-database

82. United Nations Environment World Conservation Monitoring Centre, Short FT: Global Distribution of Seagrasses (version 6). Sixth update to the data layer used in Green and Short (2003), Cambridge (UK): UN Environment World Conservation Monitoring Centre (2018). Available at: http://data.unep-wcmc.org/datasets/7

83. Pergent, G., Semroud, R., Djellouli, A., Langar, H., Duarte, C.: Posidonia oceanica. In: International Union for Conservation of Nature (IUCN) Red List of Threatened Species. Version 2012.2 (2012). Data are available at: http://www.iucnredlist.org

84. British Antarctic Survey Geodata Portal: Antarctic digital database (2020). Available at: https://add.data.bas.ac.uk/repository/entry/show?entryid=f477219b-9121-44d6-afa6-d8552762dc45

85. National Geospatial-Intelligence Agency (NGA) - Maritime Security Office: World port index (2020). Available at: https://data.humdata.org/dataset/world-port-index

86. Wikimapia: How to use Wikimapia API (2020). Available at: http://wikimapia.org/api/

87. The Scottish Government: Area Management - Ports and harbours - Statutory harbour limit boundaries (2020). Available at: https://data.gov.uk/dataset/area-management-ports-and-harbours-statutory-harbour-limit-boundaries

88. MarineTraffic: Ports database (2020). Available at: https://www.marinetraffic.com/en/data/?asset_type=ports

89. FleetMon: Port database with arrivals, schedules and weather information (2020). Available at: https://www.fleetmon.com/ports/

90. Marinas.com: Marinas, The Wanderlust Group (2020). Available at: https://marinas.com/browse/marina

91. Wikipedia: List of marinas (2020). Available at: https://en.wikipedia.org/wiki/List_of_marinas

92. Global Fishing Watch: Datasets and code: anchorages (2020). Available at: https://globalfishingwatch.org/datasets-and-code/anchorages/

93. Global Fishing Watch: the Geonames 1000 database (2020). Available at: https://github.com/GlobalFishingWatch/anchorages_pipeline/blob/master/pipe_anchorages/data/port_lists/geonames_1000.csv

94. The GeoNames Geographical Database (2020). Available at: https://www.geonames.org/

95. Andrienko, N., Andrienko, G.: Visual analytics of vessel movement. Chapter 5, In: Artikis, A., Zissis, D. (eds.) Guide to Maritime Informatics. Springer, Berlin (2021)

96. Bertoli, S., Goujon, M., Santoni, O.: The CERDI-seadistance database. Études et Documents, n° 7, Centre Éttudes et de Recherches sur le Développement International (CERDI) - Clermont Auvergne (2016). Available at: https://halshs.archives-ouvertes.fr/halshs-01288748

97. National Geospatial-Intelligence Agency: The distances between Ports (Pub 151), The United States Government (2001). Available at: https://maddenmaritime.files.wordpress.com/2015/01/pub151-distances-btw-ports.pdf

98. SeaRoutes: Route calculator (2020). Available at: https://www.searoutes.com/routing

99. Lighthouse Digest Magazine: The lighthouse explorer database (2020). Available at: http://www.lighthousedigest.com/Digest/database/searchdatabase.cfm

100. Amateur Radio Lighthouse Society: World list of lights (2020). Available at: http://wlol.arlhs.com/index.php

101. Rowlett, R.: The lighthouse directory. University of North Carolina at Chapel Hill (2020). Available at: http://www.ibiblio.org/lighthouse/

102. MarineTraffic: Lights (2020). Available at: https://www.marinetraffic.com/en/data/?asset_type=lights

103. Dahl, A.L.: Island Directory, United Nations Environment Programme (UNEP) Regional Seas Directories and Bibliographies No. 35, 573p. (1991). Available at: http://islands.unep.ch/isldir.htm

104. Natural Earth: Free vector and raster map data at 1:10m, 1:50m, and 1:110m scales (2020). Available at: https://www.naturalearthdata.com/downloads/

105. Wessel, P., Smith, W.H.F.: A global, self-consistent, hierarchical, high-resolution shoreline database. J. Geophys. Res. **101**(B4), 8741–8743 (1996). Available at: https://www.ngdc.noaa.gov/mgg/shorelines/

106. European Environment Agency: Coastline for analysis (2020). Available at: https://www.eea.europa.eu/data-and-maps/data/eea-coastline-for-analysis-2

107. US data.gov: TIGER/Line Shapefile, 2017, nation, US, Coastline National Shapefile (2020). Available at: https://catalog.data.gov/dataset/tiger-line-shapefile-2017-nation-u-s-coastline-national-shapefile

108. MarineRegions: Shapefiles (2020). Available at: http://www.marineregions.org/downloads.php

109. Sandvik, B.: World borders dataset, thematic mapping (2020). Available at: http://thematicmapping.org/downloads/world_borders.php

110. McGarva, G.: World political boundaries - coarse. University of Edinburgh (2017). https://doi.org/10.7488/ds/1789

111. Pope, A.: World political boundaries - detailed. University of Edinburgh (2017). https://doi.org/10.7488/ds/1934

112. Database of Global Administrative Areas: Maps and data (2020). Available at: https://gadm.org/index.html

113. OpenStreetMap: A map of the world. Available at: https://www.openstreetmap.org/

114. US Office of Coast Survey: Find nautical charts (2020). Available at: https://nauticalcharts.noaa.gov/

115. NASA Earth Observatory: Earth at night: flat maps (2020). Available at: https://earthobservatory.nasa.gov/features/NightLights/page3.php

116. European Space Agency: Copernicus open access hub (2020). Available at: https://scihub.copernicus.eu/

117. Sinergise: Earth observatory (EO) browser (2020). Available at: http://apps.sentinel-hub.com/eo-browser/

118. NASA: Earth data (2020). Available at: https://search.earthdata.nasa.gov/

119. EOS Data Analytics, Inc: Land viewer (2020). Available at: https://eos.com/landviewer/

120. MapBox: Remote pixel (2020). Available at: https://search.remotepixel.ca/

121. The best free stock photos & videos shared by talented creators (2020). Available at: https://www.pexels.com/

122. Pixabay: Stunning free images and royalty free stock (2020). Available at: https://pixabay.com/

123. Burst: Download free, high-resolution images (2020). Available at: https://burst.shopify.com/

124. OpenWeather (2020). Available at: https://openweathermap.org/

125. Visual Crossing: Weather forecast and weather history data (2020). Available at: https://www.visualcrossing.com/weather-data

126. Weather Unlocked (2020). Available at: https://developer.weatherunlocked.com/

127. Planet OS Datahub: Earth data at your fingertips (2020). Available at: https://planetos.com/

128. Knapp, K.R., Kruk, M.C., Levinson, D.H., Diamond, H.J., Neumann, C.J: The International Best Track Archive for Climate Stewardship (IBTrACS): Unifying tropical cyclone best track data. Bull. Am. Meteorol. Soc. **91**, 363–376 (2010). Data are available at: https://www.ncdc.noaa.gov/ibtracs/index.php?name=ib-v4-access

129. US NOAA National Weather Service: Storm prediction center severe weather GIS (2020). Available at: https://www.spc.noaa.gov/gis/svrgis/

130. Dotzek, N., Groenemeijer, P., Feuerstein, B., Holzer, A.M.: Overview of ESSL's severe convective storms research using the European severe weather database (ESWD). Atmos. Res. **93**(1–3), 575–586 (2009)

131. European Severe Weather Database (2020). Available at: https://eswd.eu/
132. Weatherbit.io: The high performance weather API (2020). Available at: https://www.weatherbit.io/
133. WeatherStack: Real-time and historical world weather data API (2020). Available at: https://weatherstack.com/
134. US NOAA National Centers for Environmental Information: The world ocean database (2020). Available at: https://www.nodc.noaa.gov/OC5/WOD/pr_wod.html
135. European Commission: Copernicus marine service (2020). Available at: https://marine.copernicus.eu/
136. Oliver, J.E.: Maritime climate. In: Oliver, J.E. (ed.) Encyclopedia of World Climatology. Encyclopedia of Earth Sciences Series, pp. 477–479. Springer, Dordrecht (2005)
137. Fick, S.E., Hijmans, R.J.: WorldClim 2: new 1-km spatial resolution climate surfaces for global land areas. Int. J. Climatol. 37(12), 4302–4315 (2017). Data are available at: https://www.worldclim.org/data/worldclim21.html
138. Smith, T.W.P., Jalkanen, J.P., Anderson, B.A., Corbett, J.J., Faber, J., Hanayama, S., Pandey, A.: Third IMO greenhouse gas study 2014. Retrieved from http://www.imo.org/en/OurWork/Environment/PollutionPrevention/AirPollution/Documents/Third%20Greenhouse%20Gas%20Study/GHG3%20Executive%20Summary%20and%20Report.pdf
139. United Nations Climate Change: Framework convention on climate change (Paris agreement) (2015). Available at: https://unfccc.int/process-and-meetings/the-paris-agreement/the-paris-agreement
140. IMO: Data collection system for fuel oil consumption of ships (2020). Available at: http://www.imo.org/en/OurWork/Environment/PollutionPrevention/AirPollution/Pages/Data-Collection-System.aspx
141. IMO: the global integrated shipping information system (GISIS) (2020). Available at: https://gisis.imo.org/
142. International Maritime Organization: Marine casualties and incidents (2020). Available at: https://gisis.imo.org/Public/MCI/Default.aspx
143. Wikipedia: Lists of shipwrecks by year (2020). Available at: https://en.wikipedia.org/wiki/Category:Lists_of_shipwrecks_by_year
144. Kuehmayer, J.R.: Marine accident and casualty investigation boards (2008). Available at: http://www.amem.at/pdf/AMEM_Marine_Accidents.pdf
145. European Maritime Safety Agency: EU accident investigation bodies (2020). Available at: http://www.emsa.europa.eu/contact-points.html
146. Wikipedia: List of oil spills (2020). Available at: https://en.wikipedia.org/wiki/List_of_oil_spills
147. Tryse, D., Tryse, E.: Oil spill database: the worst oil spills in history (2010). Available at: http://david.tryse.net/googleearth/oilspill.html
148. Eriksen, M., Lebreton, L.C., Carson, H.S., Thiel, M., Moore, C.J., Borerro, J.C., Reisser, J.: Plastic pollution in the world's oceans: more than 5 trillion plastic pieces weighing over 250,000 tons afloat at sea. PloS ONE 9(12), e111913 (2014)
149. Wikipedia: Ocean gyre (2020). Available at: https://en.wikipedia.org/wiki/Ocean_gyre
150. Shirah, G., Mitchell, H.: Garbage patch visualization experiment, NASA's Scientific Visualization Studio (2015). Available at: https://svs.gsfc.nasa.gov/cgi-bin/details.cgi?aid=4174
151. US NOAA Atlantic Oceanographic and Metrological Laboratory: GDP drifter data (2020). Available at: https://www.aoml.noaa.gov/phod/gdp/data.php
152. Ebbesmeyer, C., Scigliano, E.: Flotsametrics and the floating world: how one man's obsession with runaway sneakers and rubber ducks revolutionized ocean science. Smithsonian Books/HarperCollins Publishers (2009)
153. Cózar, A., Echevarría, F., González-Gordillo, J.I., Irigoien, X., Úbeda, B., Hernández-León, S., Palma, A.T., Navarro, S., García-de-Lomas, J., Ruiz, A., Fernández-de-Puelles, M.L.: Plastic debris in the open ocean. Proc. Natl. Acad. Sci. 111(28), 10239–10244 (2014). Data are available at: http://metamalaspina.imedea.uib-csic.es/geonetwork

154. Marine Geoscience Data System (MGDS): Floating plastics: a wide-reaching problem. One Shared Ocean (2020). Data are available at: http://onesharedocean.org/data#256

155. Japan Agency for Marine-Earth Science and Technology: The deep-sea debris database (2020). Available at: http://www.godac.jamstec.go.jp/catalog/dsdebris/e/

156. Greek Ministry of Environment: Bathing water profiles registry (2020). Available at: http://bathingwaterprofiles.gr/en/

157. European Environmental Agency: Waterbase - water quality (2020). Available at: https://www.eea.europa.eu/data-and-maps/data/waterbase-water-quality-2

158. Greidanus, H., Alvarez, M., Santamaria, C., Thoorens, F.X., Kourti, N., Argentieri, P.: The SUMO ship detector algorithm for satellite radar images. Remote Sens. **9**(3), 246 (2017)

159. Godfrey, S.J.: On the olfactory apparatus in the Miocene odontocete Squalodon sp. (Squalodontidae). Comptes Rendus Palevol **12**(7–8), 519–530 (2013)

Part II
Off-Line Maritime Data Processing

Chapter 3
Maritime Data Processing in Relational Databases

Laurent Etienne, Cyril Ray, Elena Camossi, and Clément Iphar

Abstract Maritime data processing research has long used spatio-temporal relational databases. This model suits well the requirements of off-line applications dealing with average-size and known in advance geographic data that can be represented in tabular form. This chapter explores off-line maritime data processing in such relational databases and provides a step-by-step guide to build a maritime database for investigating maritime traffic and vessel behaviour. Along the chapter, examples and exercises are proposed to build a maritime database using the data available in the open, *heterogeneous, integrated dataset for maritime intelligence, surveillance, and reconnaissance* that is described in [42]. The dataset exemplifies the variety of data that are nowadays available for monitoring the activities at sea, mainly the Automatic Identification System (AIS), which is openly broadcast and provides worldwide information on the maritime traffic. All the examples and the exercises refer to the syntax of the widespread relational database management system *PostgreSQL* and its spatial extension *PostGIS*, which are an established and standard-based combination for spatial data representation and querying. Along the chapter, the reader is guided to experience the spatio-temporal features offered by the database management system, including spatial and temporal data types, indexes, queries and functions, to incrementally investigate vessel behaviours and the resulting maritime traffic.

L. Etienne (✉)
LabISEN, Brest, France
e-mail: laurent.etienne@isen-ouest.yncrea.fr

C. Ray
Naval Academy Research Institute, Brest, France
e-mail: cyril.ray@ecole-navale.fr

E. Camossi · C. Iphar
NATO STO Centre for Maritime Research and Experimentation, La Spezia, Italy
e-mail: elena.camossi@cmre.nato.int

C. Iphar
e-mail: clement.iphar@cmre.nato.int

© Springer Nature Switzerland AG 2021
A. Artikis and D. Zissis (eds.), *Guide to Maritime Informatics*,
https://doi.org/10.1007/978-3-030-61852-0_3

3.1 Introduction

The analysis and the processing of data from automated surveillance technologies
like the Automatic Identification System (AIS), Synthetic Aperture Radar (SAR),
satellite images and coastal radars (*cf.* the chapter of Bereta et al. [7]) is an emerging
topic. Once fused with contextual information like coastlines, nautical charts, fishing
areas, maritime protected areas, sea state and weather conditions, these positioning
data can be analysed in order to understand vessel mobility, monitor maritime traffic,
uncover activities or risks for the environment, living resources and navigation, and
so on [42]. The research is expeditiously progressing, and many analysis techniques
are applied to attain the above major goals using maritime navigational data. Some
of the leading maritime informatics research topics include:

- *Vessel trajectory analysis and prediction*, recently focusing mainly on AIS and
  complementary sources to predict and analyse vessel trajectories [8].
- *Traffic forecasting*, like port volume handling and cargo throughput forecast [24].
- *Collision prevention*, analysing risk of ships collisions through the analysis of their
  navigational behaviour [22].
- *Ship detection, classification and identification*, analysing videos from cameras in
  delimited areas, for instance to detect small vessels [25], or remote sensing images
  in larger areas [14].
- *Anomaly detection*, mostly adopting data-driven methods to model normal traffic
  against which any irregular behaviour is associated to potential threats [44].
- *Analysis of human activities at sea*, such as fishing, illegal traffic (human beings,
  narcotics, goods), piracy [39].
- *Multi-source information fusion*, to reduce uncertainty in the data available and
  achieve better accuracy in analysis [33].
- *Dynamics of maritime transportation networks*, where trajectory data are considered at an aggregated level through network abstractions designed to analyse
  behavioral patterns [60].

Research work processing maritime data requires modelling and analysing moving objects associated to maritime navigation and traffic. Considering an existing
batch of data, two main processing approaches exist. Some works consider direct processing of data files, using general-purpose data science environments (e.g. *Python, R,
Matlab*) or developing customised tools for spatio-temporal data processing (e.g. [3]).
Other works use database management systems (DBMS), which nowadays offer a
native support for modelling spatial information, optimised functionalities for spatial
indexing and analysis, and efficient integration with the query language.

The latter approach, which is the one illustrated in this chapter, enables a quick
development of moving object applications. Thanks to built-in spatial data types
and extended DBMS support, developers can exploit the spatial dimension of the
application entities. For instance, they can explicitly represent spatial objects and
execute spatial queries to compare their topological properties or to calculate their
geographical distance. The available spatial types of such DBMS can be extended

with user-defined data types and functions, tailored to the specific application or task. The soundness and the efficiency of the resulting model are guaranteed by the DBMS.

This chapter focuses on the support offered by relational DBMS, which are proposed for maritime data storage and querying. While time-consuming for handling very large datasets, and while being mainly used for off-line analysis, relational DBMS suit well the requirements of applications dealing with average-size heterogeneous data that can be described according to a known structured and fixed schema, i.e. where data properties are known in advance and can be represented in tabular form.

Along the chapter, many SQL (Structured Query Language) [35] examples are presented to the reader, who is asked to solve some exercises.[1] A basic knowledge of relational databases and SQL is necessary to understand the examples and solve the exercises. All together, examples and exercises illustrate the spatio-temporal capabilities of DBMSs that can be progressively combined to analyse maritime data for investigating vessel traffic. The examples in this chapter refer to *PostgreSQL*,[2] a widespread relational DBMS. *PostgreSQL* is a very popular DBMS, fully open source, very robust, and its spatial support, which is provided by *PostGIS* [53], is compliant to well established standards for representing spatial features. *PostgreSQL* also integrates a native temporal support, which is necessary to analyse vessel movements, that corresponds to Allen's algebra operators [1].

The rest of the chapter is organised as follows. In Sect. 3.2 we overview the existing approaches for representing and managing spatial data and moving objects. Section 3.3 introduces the software the reader needs in order to create, query and visualise a maritime database. Section 3.4 explains the steps required to create a maritime database, given the *heterogeneous, integrated dataset for maritime intelligence, surveillance, and reconnaissance* that is described in [42]. Using this maritime dataset, Sect. 3.5 illustrates typical queries and functions, useful to derive additional information and reason on vessel movements. Finally, Sect. 3.6 concludes the chapter.

3.2 Systems for Spatial Data and Moving Objects

In this section, we overview how spatial, specifically geographic data, and moving objects are handled, either using Geographic Information Systems (GIS), or DBMS. The approaches and software products supporting only spatial data are introduced first, followed by some of the existing systems for spatio-temporal data.

---

[1]The solutions of the exercises are available, with additional material (e.g. resulting geographic data), online, together with the reference dataset: https://zenodo.org/record/3930660.

[2]https://www.postgresql.org.

3.2.1 Handling Spatial Data

For decades, the representation and analysis of geographical information have been investigated in the area of Geographic Information Systems (GIS). GIS, like *ArcGIS*,[3] *GRASS*,[4] and *QGIS*[5] naturally evolved from the digitalisation of cartographic maps. GIS have efficient geographic visualisation and rendering capabilities, which can be combined with multi-layer spatial analysis functions like distance/buffer and topological queries, overlay operations, surface and terrain analysis. However, traditional GIS alone cannot deal with the increasing quantity of data, and do not suite the semantically-driven integration of heterogeneous information that involves both spatial and non-spatial attributes, or complex knowledge discovery tasks like supervised classification. All these tasks require an integrated representation of geographic and descriptive data features, and a greater flexibility in low-level data manipulation than the one offered by traditional GIS, which keep the management of geospatial data and other data types separated [43].

To address the needs of emerging spatial-driven applications, relational database systems, once geared towards providing efficient support for simple objects with discrete attributes, have been extended to represent and query geographic data in a natural way. For instance, *Oracle*[6] and *Postgres* (now *PostgreSQL*), were expanded with spatial modules, respectively *Oracle Spatial* [26] (now fully integrated in the main DBMS) and *PostGIS*. Nowadays, all the main DBMS, including *Microsoft SQL*,[7] *MySQL*,[8] *SQLite*[9] with *SpatiaLite*, provide support to cope with geographic data.

The data model of these DBMS has been extended with data types and structures for managing geometric data, relying on established international standards and recommendations.[10, 11] Spatial operations and types are fully integrated with the query language, and the DBMS engine is enhanced to map from the query language to the spatial features. Also, spatial indexing is provided for query optimisation. Most of the products include *B-Tree+* and *R-Tree* [17] indexes. *PostgreSQL* additionally implements the *Generalized Search Tree (GiST)* [18] to develop customised indexing, mixing, for instance, spatial and non-spatial optimisation.

---

[3] https://arcgis.com.

[4] https://grass.osgeo.org.

[5] https://qgis.org.

[6] https://www.oracle.com.

[7] https://www.microsoft.com/en-us/sql-server/sql-server-2017.

[8] https://www.mysql.com.

[9] https://sqlite.org.

[10] Open Geospatial Consortium (OGC) Simple Features Specification for SQL: Simple Feature Access—Part 1: Common Architecture https://www.opengeospatial.org/standards/sfa; Simple Feature Access—Part 2: SQL Option https://www.opengeospatial.org/standards/sfs.

[11] International Standard Organization's (ISO) ISO19107:2003 Geographic Information—Spatial Schema https://www.iso.org/standard/26012.html.

Most DBMS spatial extensions provide mature support for geometric data. Some products include also extensions for raster data (e.g. *Oracle Raster* and *pgraster*, which is integrated in *PostGIS*). Also, few products support 3D representations, including *PostGIS*.

This support can be further enhanced with novel abstract data types to represent spatial (and spatio-temporal) data, tailored to the application, analogously to *object-oriented DBMS* (OODBMS). These (object-) relational DBMS, taking advantage of both spatial and object-oriented functionalities, have been established as main players on the market for storing and querying spatial data. For this reason, nowadays most of GIS enable the use of a spatial DBMS as data storage layer. This hybrid solution combines the advantages of the efficient front-end of GIS and the robust and efficient storage and retrieval capabilities of spatial DBMS.

Few NoSQL, or *Not-only-SQL*, DBMS, i.e. all the DBMS that adopt alternative models to the (object-)relational one, offer spatial extensions. Among them, few research proposals exist to extend OODBMS to support spatial features, leveraging the model extensibility as described above. More recent NoSQL DBMS, like *MongoDB* and *DocumentDB*, which adopt a semi-structured document-based data model, enable the definition of spatial entities supporting *GeoJSON* objects and geo-spatial queries. *ArangoDB* is a multi-model (key value, graph and document) NoSQL DBMS that supports natively geo-spatial querying and indexing. *Rasdaman* (which stays for *raster data manager*) is a multi-dimensional DBMS for scientific data, which can be queried with an SQL-like language.

NoSQL DBMS relax some data modelling constraints to handle unstructured or semi-structured data, and improve *scalability* performance as required for big data and parallel processing. For instance, in order to process in parallel multiple vessel trajectories, and multiple areas, they enable to increase, seamlessly for the application, the number of nodes in the infrastructure that are used to host the database, and maintain replicas of the data. However, the gain in efficiency and flexibility comes at the price of relaxing some of the properties that, in traditional SQL databases, guarantee data consistency (namely, ACID properties: i.e. atomicity, consistency, isolation, and durability).

It should be noted that NoSQL solutions do not always outperform relational DBMS. For instance, the authors in [32] compare *PostgreSQL/PostGIS* and *MongoDB* performance on a maritime dataset of vessel positions, and show that *PostgreSQL/PostGIS* outperforms *MongoDB*. As shown in the *Knowledge base of relational and NoSQL DBMS*,[12] relational DBMS remain amongst the most popular DBMS.

---

[12]https://db-engines.com/en/ranking.

3.2.2   Spatial-Temporal and Movement Data in Databases

While widely studied, the representation of the temporal dimension of spatio-temporal and moving object data in databases remains rather limited or in the research phase [48]. In [19] the authors illustrate a spatio-temporal extension of the Object Data Management Group (*ODMG*) model, which is a *de facto* standard model for OODBMS, and of its query language, *OQL*. A renowned spatio-temporal model is the *SECONDO*'s one, which exploits abstract data types to model moving objects [57] and spatio-temporal operations among them.

PostgreSQL natively integrates a support for temporal data that nicely combines with the spatial extension of *PostGIS*. Additionally, the reader may consider two groups of alternatives. As a first option, a time-series DBMS can be used. For instance, *TimescaleDB*[13] is an open-source time-series DBMS developed as *PostgreSQL*'s extension, fully SQL and *PostgreSQL* compliant, which can integrate *PostGIS* to efficiently handle spatio-temporal time-series data like Internet-of-Things observations. Given an SQL table with temporal and location attributes, it transforms it and partitions the data according to the temporal and spatial dimensions and adds indexes for improving access performance. It also offers a small set of analytic functions (e.g. *first*, which provides the first value of time ordered series) that can be used jointly with *PostGIS*'s functions.

The second group of options implements a full spatio-temporal support. For instance, *PostGIS-T* extends *PostgreSQL* on the basis of a formal spatio-temporal algebra [50]. Analogously, *Pg-Trajectory* [28] which is developed by the Data Mining Lab of the Georgia State University, is a *PostgreSQL/PostGIS* extension designed for spatio-temporal data.

Other spatio-temporal proposals, still based on *PostgreSQL*, are *Hermes* and *MobilityDB*. *Hermes* [38] is an in-DBMS framework for data mining, more particularly for the processing of moving objects. It is a *Python* library, which is event-driven, failure-handling and *PostgreSQL*-talking, and enables an easy implementation of Python processes that require *PostgreSQL* communication. *MobilityDB* [55] is a *PostgreSQL* extension for mobility data management, which follows the OGC Moving Features Access specification[14] and defines the operations applicable to time-varying geometries.

3.3   Building a Maritime Information System

A Maritime Information System can be defined as a system for the collection, processing, storage and delivering of maritime information through a visualisation interface. The main purpose of a DBMS, when integrated to such a Maritime Information System, is to organise, store, access and process the maritime data. The database

---

[13]TimescaleDB https://www.timescale.com.

[14]OGC Moving Features Access http://www.opengis.net/doc/is/movingfeatures-access/1.0.

can be handled and queried through a management tool or an external programming environment (e.g. *Python*, *R*, *Java*). In order to visualise the data, dedicated applications and Application Programming Interface (API), e.g. Grafana,[15] Data Driven Documents (D3),[16] Kibana,[17] visual analytics tools (*cf.* the chapter of Andrienko et al. [2]) or a GIS software like QGIS can be used.

Action required

Download and install the following software to prepare the working environment (let us note that *PostGIS* and *pgAdmin* are in general part of the last *PostgreSQL* bundles):

- *PostgreSQL*: https://www.postgresql.org;
- *PostGIS*: https://postgis.net;
- *pgAdmin*: https://www.pgadmin.org;
- *QGIS*: https://www.qgis.org.

3.3.1 *PostgreSQL DBMS*

The DBMS employed in this chapter to illustrate the use of relational databases for maritime data is *PostgreSQL*. *PostgreSQL* is an open source, standard compliant and robust DBMS. It adopts and extends the *Structured Query Language (SQL)* [35] for data manipulation and querying. *PostgreSQL* runs on all major operating systems and can handle big datasets. *PostgreSQL* is highly extensible: the users can define their own data types, build customised new functions and interact with the DBMS through the query language *SQL*, using different programming languages like *C*, *Perl*, *Java*, *Python*, *R*, *JavaScript*, etc., and from shell scripts.

3.3.2 *PostGIS Spatial Extension*

The spatial component of the maritime dataset plays a fundamental role for maritime situational awareness. *PostGIS* is a powerful add-on to *PostgreSQL* that adds geospatial capabilities to the DBMS. *PostGIS* integrates both raster and vector types of data which are fully compliant with the *OGC Simple Features Specification for SQL*, and a large number of spatial operations. *PostGIS* also implements multiple spatial indexing methods, including *R-Tree* and *GiST*. Although *PostgreSQL* and *PostGIS* are usually used with two-dimension (2-D) geometries (coordinates X and Y), *PostGIS* supports the addition of a third dimension (Z), allowing to reason with three-dimensional geometries.

---

[15]Grafana https://grafana.com.

[16]Data Driven Documents (D3) https://d3js.org.

[17]Kibana https://www.elastic.co/products/kibana.

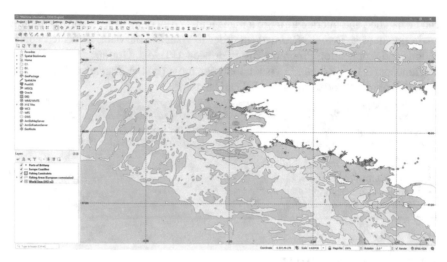

Fig. 3.1 The *QGIS* desktop visualising five geographical layers (all available in the dataset described in [42]): fishing areas (by [56]), ports of Brittany, Europe coastline, fishing constraints and world seas

3.3.3 PGAdmin Management Tool

Once the DBMS is installed, management tools are useful to setup and query the database. *pgAdmin* is a multi-platform software that is dedicated to manipulate *PostgreSQL* databases. Once connected to a *PostgreSQL* cluster (i.e. a collection of *PostgreSQL* servers), *pgAdmin* can create and query new databases and tables. Its powerful query tool supports colour syntax highlighting and graphical query plan displaying.

3.3.4 QGIS Visualization Tool

QGIS is the last component of our Maritime Information System. It is a user-friendly open-source GIS, which can retrieve, manage, display and analyse the geographic data stored in the *PostgreSQL/PostGIS* database. In this chapter, *QGIS* will be used to exemplify the visualisation and creation of maps to show the results of the spatial queries (see Fig. 3.1). Interested readers can refer to Anita Graser's blog, "Free and Open Source GIS Ramblings",[18] for practical illustrations on advanced functionalities for movement data. Moreover, the readers can refer to the *QGIS* online user guide[19] to get started with *QGIS*.

---

[18]"Free and Open Source GIS Ramblings" https://anitagraser.com/.

[19]https://docs.qgis.org/3.4/en/docs/user_manual/.

3.3.5   Getting the Maritime Dataset

The SQL examples of this chapter are based on a freely available open dataset that has been processed using *PostgreSQL* and its spatial extension *PostGIS*. The dataset is heterogeneous, and contains four categories of data:

- *Navigation data*, i.e. historical AIS positions of vessels navigating around a major shipping route;
- *Vessel-oriented data* (e.g. ship registers, i.e. lists of vessels with their characteristics);
- *Geographic data* (i.e. cartographic, topographic and regulatory context of vessel navigation);
- *Environmental data* (i.e. climatic and sea-state related information).

All data are temporally and spatially aligned to allow for efficient and advanced spatio-temporal analysis. The dataset covers a time period of six months (i.e. from October 1st 2015 to March 31st 2016) and provides around 18.6 million ship positions collected from 4,842 ships over the Celtic sea, the North Atlantic ocean, the English Channel, and the Bay of Biscay in France (Fig. 3.2). For additional details on the dataset, see [42].

> **Action required**
>
> Access the dataset at: https://zenodo.org/record/1167595

The AIS data included in the dataset aggregate different message types, and are stored in two separated files, which contain respectively the positioning messages of ships, and their nominative messages. Positioning messages are AIS messages of type ITU 1, ITU 2, ITU 3, ITU 18, and ITU 19, where ITU stands for International Telecommunication Union. Nominative messages are AIS messages of type ITU 5, ITU 19, and ITU 24, which contain static information about the emitting vessel, such as its name, dimensions or destination, identifiers like the International Maritime Organization—IMO—number and the Maritime Mobile Service Identity—MMSI—code. Two additional files, containing respectively search and rescue messages, i.e. ITU 9, and aids to navigation messages, i.e. ITU 21, have been also included.

> **Remark ! SQL queries**
>
> The SQL queries presented in this chapter focus on typical and useful features for maritime informatics. They complement the integration queries available in folder "[Q1] Integration Queries" of the maritime dataset. All the queries presented in this chapter have been prepared using local installations (on Windows x86-64) of *PostgreSQL* 11.5, *PostGIS* 2.5.3, and *pgAdmin*4 v4.18.

3.4   Creation of the Maritime Database

In this section, a new maritime database will be created and populated. Assuming that the software packages described in Sect. 3.3 have been installed, it is possible to connect to the *PostgreSQL* database server from *pgAdmin* interface, using the

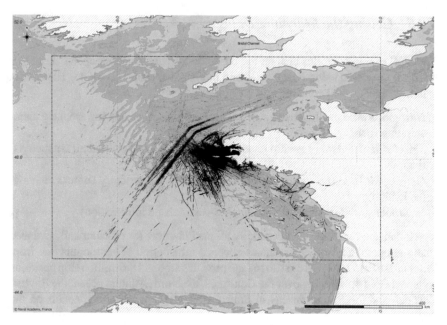

Fig. 3.2 AIS navigation data in the open dataset of [42]. The purple polygons represent fishing zones computed from navigation data by Vespe et al. [56]

credentials provided during the installation. Once a connection to the database is established, the following command can be executed (in menu *Tools → Query Tool*) to create a new empty database called `maritime_informatics`.

SQL	Q3.4.0.1

```
CREATE DATABASE "maritime_informatics";
```

Remark ! Important note about literals

PostgreSQL automatically converts to lower case all tables and fields names. If you want to force the use of capitals in tables or fields names, you must quote them (i.e. `"tableName"` or `"fieldName"`). To distinguish field names and strings in queries, *PostgreSQL* uses double quotes (" and ") for field names and single quotes (' and ') for string values.

3.4.1 The Database Schema and the Tables

A database is organised and structured using one or more *database schemas* and several *tables*. Each schema defines a workspace where new data types, functions and operators can be defined. Organising a database using different schemas is useful for various reasons:

- To organise database objects within logical groups and to ease their management;
- To isolate third-party applications and allow for duplicate objects or function names in different schemas;
- To grant users (and groups of users) appropriate access to data and functions that belong to different schemas.

Schema and Tables for AIS Data

AIS transceivers on board of vessels broadcast vessel positions and interesting information about static ship characteristics such as the vessel name, *callsign* (i.e. unique vessel identifier for radio broadcasts), IMO number and ship type. Each AIS transceiver has its own Maritime Mobile Service Identity (MMSI), which is registered and attributed by country of flag. This number identifies the source of the AIS message (`sourcemmsi`). Usually, AIS transceivers broadcast this information to all the ships in the vicinity. However, some AIS messages can be addressed to one specific AIS transceiver, identified using its MMSI (`destinationmmsi`). Ship's tenders can also have their own AIS transceiver. The tender's transceiver reports in the AIS messages the MMSI of the ship it belongs to (`mothership`).

For the examples presented in this chapter, AIS information will be organised within a schema named `ais_data`. The SQL instructions below show how to create the schema `ais_data` into the database and create, inside this schema, a new table `static_ships` to store voyage and nominative pieces of information. Note that each attribute has a specific type (`integer`, `text`, `double`, etc.).

```
SQL                                                                    Q3.4.1.1

CREATE SCHEMA "ais_data";

CREATE TABLE ais_data.static_ships(
    id bigserial, -- unique identifier of the row in the table
    sourcemmsi integer, -- MMSI identifier of the source ship's transceiver
    imo integer, -- IMO number of the ship, linked to the vessel structure
    callsign text, -- Callsign of the ship
    shipname text, -- Ship name
    shiptype integer, -- Type of the vessel, according to AIS
        specifications
    to_bow integer, -- Distance of the AIS transceiver from bow (front) of
        vessel, rounded to the nearest meter
    to_stern integer, -- Distance of AIS antenna form stern (back)
    to_starboard integer, -- Distance of AIS antenna form starboard (
        right)
    to_port integer, -- Distance of AIS antenna form port (left)
    eta text, -- Estimated Time of Arrival to destination
    draught double precision, -- Ship draught
    destination text, -- Declared ship destination
    mothershipmmsi integer, -- MMSI of the mothership
    ts bigint, -- Timestamp of the AIS frame
    CONSTRAINT static_ships_pkey PRIMARY KEY (id) -- Primary Key
);
```

3.4.2   Loading the Data into the Database

The new table can now be populated. Data can be manually inserted into the existing tables using the SQL statement INSERT INTO or, alternatively, directly from files using the *PostgreSQL* command COPY FROM. The following example shows how to load into the table static_ship the static ship information stored in the Comma Separated Values (CSV) file nari_static, which is included in the maritime dataset (in folder "[P1] AIS Data"). In the command below, <path_to_dataset> must be replaced with the local path to the file.

PostgreSQL	Q3.4.2.1

```
COPY ais_data.static_ships
    (sourcemmsi,imo,callsign,shipname,shiptype,to_bow,to_stern,
    to_starboard,to_port,eta,draught,destination,mothershipmmsi,ts)
FROM '<path_to_dataset>/[P1] AIS Data/nari_static.csv',
DELIMITER ',' CSV HEADER;
```

As soon as the database table is filled with *static data*, it can be queried using SQL commands. The query in the next example counts how many different ship names are listed in the table static_ships ([Answer: 4,824]).

SQL	Q3.4.2.2

```
SELECT DISTINCT shipname FROM ais_data.static_ships;
```

The table static_ships contains duplicated information about ships, transmitted at different timestamps, because nominative messages (ITU 5, ITU 19, and ITU 24) are automatically repeated at regular time intervals. The duplicated rows in this table can be grouped together using a SQL view, as shown in the following example.

SQL	Q3.4.2.3

```
CREATE OR REPLACE VIEW ais_data.ships AS
SELECT
    sourcemmsi, -- MMSI identifier of the ship, attributed by country of
        flag
    imo, -- IMO number of the ship, linked to the vessel structure
    callsign, -- Callsign of the ship
    shipname -- Ship name
FROM ais_data.static_ships
GROUP BY sourcemmsi, imo, callsign, shipname;
```

Using the view ais_data.ships, let assess some quality aspects of static AIS information, as proposed in the following exercise.

Do it yourself !	A3.4.2.1
How many ships have null, empty or space filled shipname ?	[Answer: 14]
How many MMSI numbers are used more than once?	[Answer: 327]

The view ais_data.ships and the queries proposed above show that the static ship information is not always accurate, because both fields shipname and sourcemmsi have duplicates and null or empty values. Indeed, this information can be manually modified. Being unreliable and inconsistent, none of these fields can used as a primary key in our tables.

In order to analyse the movement of ships, it is necessary to load their positions in the database. In the maritime dataset, vessel positions are stored in the file nari_dynamic.csv.

Do it yourself !	A3.4.2.2
Within the schema ais_data, create a new table named dynamic_ships. Fill it with the data of the file nari_dynamic.csv.	

The following query counts the number of the ships positions contained in the table dynamic_ships ([Answer: 19,035,630]).

SQL	Q3.4.2.4
SELECT COUNT(*) FROM ais_data.dynamic_ships;	

3.4.3   The Temporal Dimension of Data

PostgreSQL defines types, functions and operators to represent, handle and query dates and time based on Allen's interval algebra [1]. They enable, for instance, to test temporal relations and manipulate temporal events and intervals (e.g. the function overlaps checks if two temporal intervals overlap; "−" and "+" enable to add or subtract days or hours from a time or a timestamp value).

In the table dynamic_ships, the time of each vessel position is represented as an integer (in attribute ts), which expresses the corresponding epoch UNIX timestamp, i.e. the time, in seconds, elapsed since 1970-01-01 00:00:00 UTC (Coordinated Universal Time). In the following example, this numerical value is converted into timestamp, in order to facilitate its interpretation and to express time-oriented queries. In the following example, a new column is created to store the new representation.

SQL	Q3.4.3.1
ALTER TABLE ais_data.dynamic_ships ADD COLUMN t timestamp without time zone;	
UPDATE ais_data.dynamic_ships SET t=to_timestamp(ts);	

This timestamp column facilitates visualising the temporal component of the table dynamic_ships and expressing temporal queries. The query in the example below asks for the temporal range of the table (i.e. minimum and maximum timestamp) ([Answer: "2015-10-01 00:00:01"–"2016-03-31 23:59:59"]).

SQL	Q3.4.3.2
SELECT min(t), max(t) FROM ais_data.dynamic_ships;	

3.4.4 Make It Faster !

The queries of the previous section may take a while to complete. This is due to the size of the table `dynamic_ships` and because the table needs to be fully scanned for reading and sorting every timestamp in order to execute the queries. Querying very large tables can take a long time to execute. In order to optimise the queries, especially if they are executed frequently, the *PostgreSQL* commands EXPLAIN and ANALYZE can be used to understand how the DBMS plans to execute them. For example, the following command asks the *PostgreSQL* planner what is the execution plan for the last query in the previous section.

PostgreSQL	Q3.4.4.1

```
EXPLAIN ANALYZE SELECT min(t), max(t) FROM ais_data.dynamic_ships;
```

In the answer, the *PostgreSQL* planner indicates that it would perform a sequential scan (Seq Scan) of the table. The sequential scan can be avoided if the column t, which appears in the SELECT clause of the query, is previously sorted and indexed. Indexes can require time to build up, and use storage space on the *PostgreSQL* server, but they can drastically decrease the time required to execute a query.

In the SQL example below, a Binary Tree (BTREE) index is created on column t. This index is particularly efficient for reading attributes whose values have a linear ordering like temporal, numerical or string attributes.

SQL	Q3.4.4.2

```
CREATE INDEX idx_dynamic_ships_t  -- name of the index
ON ais_data.dynamic_ships  -- name of the table to index
USING btree (t) ;  -- indexing technique (columns to index)
```

After creating the index, a new investigation of the query plan (as for the example above) shows that the Index Scan of the query would take a few milliseconds.

Remark !	

In order to optimise the queries on very large tables, do not forget to define indexes on the columns that are often queried.

Do it yourself !	A3.4.4.1

Find how many different ships have broadcast their positions on AIS on January 1st 2016 and display their ship names. Which tables are required to answer this question? Do you require indexes on these tables?

[Answer: 79 rows, 78 different ship names]

3.4.5 Make It Geographic !

The geo-spatial component is an important feature of the maritime dataset. Ships move in a specific spatial context made of coastlines, straights, ports, mooring areas, restricted areas and bathymetry. This information should be considered when analysing the maritime situation. In order to support this variety of geographic objects, *PostgreSQL* can be extended with *PostGIS* that adds geographic types, functions and indexes to the DBMS. In order to activate the *PostGIS* extension on a *PostgreSQL* database, it is sufficient to execute the command below.

PostgreSQL	Q3.4.5.1

```
CREATE EXTENSION postgis;
```

Geographic Objects

A way to represent spatial objects, such as moving vessels, is to use simple geometric features such as points, lines and polygons along with extra alpha-numerical attributes, e.g. the name of the spatial features. This way of modelling spatial objects is called *vector* representation. The coordinates of spatial objects are expressed in a given Coordinate Reference System (CRS). Thanks to *PostGIS*, *PostgreSQL* can handle the following geometric data types:

- POINT, defined using spatial coordinates in a CRS;
- LINESTRING, ordered set of POINTs;
- POLYGON, defined based on a collection of rings, at least one outer and possibly inner rings (representing, e.g. enclaves in countries, reservoirs), given as LINESTRINGs;
- MULTIPOINT, MULTILINESTRING, MULTIPOLYGON, i.e. sets of geometric objects of the same base type;
- GEOMETRYCOLLECTION, a set of geometric objects of various nature.

In table dynamic_ships, vessel positions are given by their coordinates, i.e. longitude and latitude, specified in two separated columns, lon and lat, respectively. The *PostGIS* function ST_MakePoint creates a POINT from two coordinates, given their CRS. In the example below, the World Geodesic System 1984 (WGS84) is used. It is identified in *PostGIS* through its Spatial Reference Identifier (SRID) as defined by the European Petroleum Survey Group (EPSG:4326[20]). The World Geodetic System 1984 is extensively used to define worldwide locations by longitude and latitude. This CRS is used by the GPS satellite navigation system integrated into AIS transceivers. The *PostGIS* function ST_SetSRID is used to specify the CRS' SRID of the newly created column. As the table dynamic_ships has 19 million positions, the update process may take a while (about 10 min).

SQL	Q3.4.5.2

```
ALTER TABLE ais_data.dynamic_ships ADD COLUMN geom geometry(Point,4326);

UPDATE ais_data.dynamic_ships
SET geom=ST_SetSRID(ST_MakePoint(lon,lat),4326);
```

---

[20]https://epsg.io/4326.

The database must be aware of the data's CRS before executing the commands in the example above. Data's CRS must be listed in the system table `spatial_ref_sys`, which is included in the *PostgreSQL* database public schema. The public schema is automatically created in every *PostgreSQL* database and contains system tables and functions. It is also used to store all the user-defined tables that are created without referring to any specific schema.

> **Remark !**
>
> Geographic positions on Earth can also be projected on a flat plane, introducing some distortions. Each map projection preserves important properties (direction, angle, shape, area, distance, *etc.*) while distorting others. Once projected, distances between objects can be computed using the reference metric of the map projection (e.g. meters). Each CRS is usually customised for a specific area. The maritime dataset covers Europe and provides data expressed using CRS WGS84, and ETRS89/LAEA Europe (EPSG:3035[a]). The CRS ETRS89/LAEA Europe typically suits statistical mapping at all scales and covers the European Union (EU) countries onshore and offshore.
>
> ---
> [a]https://epsg.io/3035.

PostGIS can project geometric objects from one CRS to another. In the example below, the function `ST_Transform` is used to project vessel positions expressed in WGS84 into the projection ETRS89/LAEA Europe CRS (EPSG:3035).

SQL   *Q3.4.5.3*

```
ALTER TABLE ais_data.dynamic_ships ADD COLUMN geom3035 geometry(Point
    ,3035);

UPDATE ais_data.dynamic_ships
SET geom3035=ST_Transform(geom,3035);
```

3.4.6 Integrating Contextual Data

The maritime dataset contains also complementary data like ports' positions, restricted and protected areas, coastlines, weather and ocean conditions. These data have a different nature than AIS data, which are continuously streamed, and provide information useful to contextualise and understand the vessel movements. Conveniently, a dedicated database schema, with name `context_data`, can be created for these data.

> **Action required**
>
> Create a new schema `context_data` in the database `maritime_informatics`.

Loading Spatial *Shapefile* Data

All contextual (geographic) data in the dataset are in the same format, i.e. *Shapefile*, which is a well-known format for vector (i.e. geometric) data. *Shapefile* data are stored in multiple files, which have the same name and different extensions and represent different geometric aspects, as follows:

- SHP: is the shape of the geometric objects;
- DBF: contains alpha-numerical attributes;
- PRJ: is the coordinate reference system of the geometric objects;
- SHX: is a shape geometric index.

Do it yourself !	A3.4.6.1

Download the file "[C2] European Coastline.zip" from the dataset. Unzip the folder in your working directory and look at the different files. Have a look at the PRJ file. What is the coordinate reference system of the "Europe Coastline (Polygon)" shapefile (SHP)?

[Answer: GRS 1980]

Shapefile data can be imported in a spatially enabled *PostgreSQL* database using the *PostGIS* tool shp2pgsql. The example below uses shp2pgsql to load the *Shapefile* data stored in Europe Coastline (Polygone).shp. Data are loaded into the database maritime_informatics. A coordinate system (ETRS89/ETRS-LAEA, 3035) must be specified to correctly import them. Figure 3.3 shows the interface of the shp2pgsql tool.

Action required

In order to load the European coastline in the maritime_informatics database, execute the following steps:

1. start the shp2pgsql interface (shp2pgsql-gui);
2. load the file Europe Coastline (Polygone).shp;
3. rename the target table into coastlines;
4. select the schema context_data;
5. set the coordinate system to ETRS89/ETRS-LAEA (3035);
6. once the import parameters are set up, import the data.

Upon a successful upload, *pgAdmin* can be used to preview the first 100 rows of the newly populated table.

Action required

Using *pgAdmin*, to preview the first 100 rows of table coastlines:

1. right click on the table coastlines;
2. select *View/Edit data*, then *First 100 rows*.

Remark !

In *pgAdmin*, the "eye" button on a geometric column header enables to see the geometric shape of the first 100 rows in a spatial table.

Action required

In *pgAdmin*, click on the "eye" button on the column geom3035 of the table coastlines to visualise their geometric shape. Now, start *QGIS*, connect to *PostgreSQL* database using the *Data Source Manager* then, add the coastlines geographic layer in a new map.

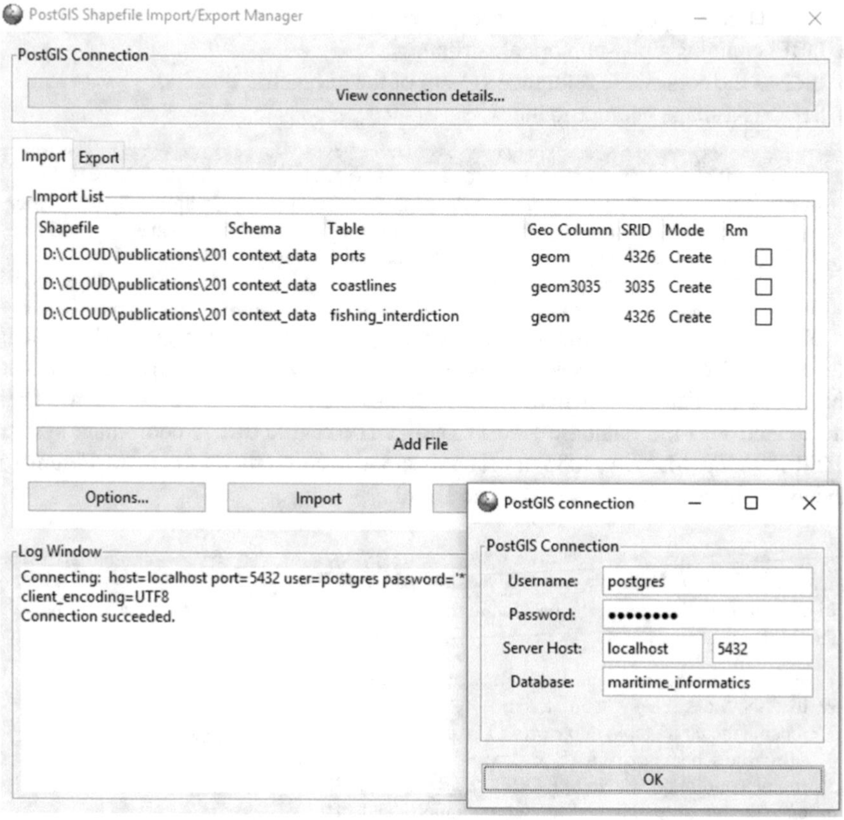

Fig. 3.3 The `shp2pgsql` interface is used to load *Shapefile* data into a spatially enabled *PostgreSQL* database

Readers having trouble to connect *QGIS* to *PostgreSQL* can refer to QGIS user manual.[21]

Do it yourself !	A3.4.6.2
Using *pgAdmin*, find the type of the column geom3035 in the table coastlines. What is the primary key of the table? [Answer: MULTIPOLYGON - column gid]	

---

[21] https://docs.qgis.org/3.4/en/docs/user_manual/managing_data_source/opening_data.html#database-related-tools.

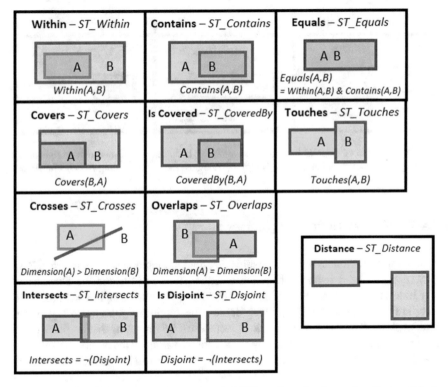

Fig. 3.4 *PostGIS* functions that output a numerical (distance) or a boolean value. On top of each
box, the *PostGIS* function and the corresponding spatial relation are reported (**relation**—*PostGIS*
function)

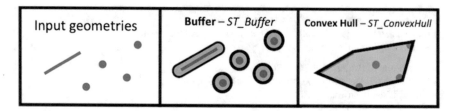

Fig. 3.5 Examples of *PostGIS* functions that output a spatial geometry (buffer and convex hull). Output geometries are shown in grey. On top of each box, the *PostGIS* function and the corresponding spatial function are reported

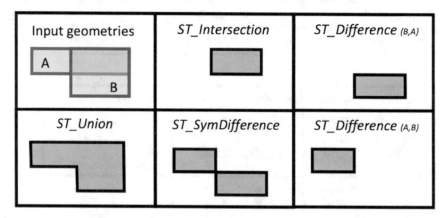

Fig. 3.6 Example of *PostGIS* functions that apply set-based operations and output a geometry. Output geometries are shown in grey. On top of each box, the corresponding *PostGIS* function is reported

3.4.7 Executing Spatial Queries

Spatial queries evaluate the relationships that hold among objects' geometries. *PostGIS* supports the 8 region connection calculus (RCC8) [40] and the Dimensionally Extended nine-Intersection Model (DE-9IM) [11], and complies with the OGC Simple Feature Access ([20]), which defines the supported routines to test spatial relationships between two geometric objects. *PostGIS* functions that output a value, either a distance (i.e. ST_Distance) or a Boolean value (all the others) are shown in Fig. 3.4. *PostGIS* functions that output a geometry are shown in Fig. 3.5. All functions can be applied between geometries of various types (i.e. POINTS, LINESTRINGS or POLYGONS), as far as the coordinates of these objects are defined in the same CRS. These spatial relationship functions can be combined with temporal operators, enabling to express complex spatio-temporal queries.

The functions ST_Equals, ST_Disjoint, ST_Intersects, ST_Touches, ST_Crosses, ST_Within, ST_Contains, and ST_Overlaps evaluate whether the corresponding spatial relationship holds between two input geome-

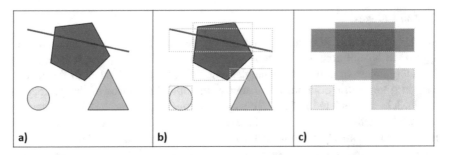

Fig. 3.7 Use of the GiST index for computing spatial queries. When geometries (**a**) are indexed (**c**), spatial queries use the geometries' bounding boxes to approximate the query

tries, and return a boolean value. The function `ST_Distance` returns a real number representing the minimum distance between the two geometries. Other functions return the geometries resulting from the application of geometric operations (`ST_Buffer` and `ST_ConvexHull`) (cf. Fig. 3.5) or set-based operations (e.g. `ST_Intersection`, `ST_Difference`, `ST_Union`, `ST_SymDifference`) (cf. Fig. 3.6). An extensive presentation of all these geometry processing functions is available on *PostGIS* website.[22]

The query in the following SQL example illustrates the use of such spatial functions. It searches for ships (name and unique identifier, i.e. the ship MMSI) that stopped within 500 m from a port in Brittany on January 1st, 2016. For each ship, the name of the port and the time spent by the vessel in the port is also shown.

Remark !

The distance between ship positions and port locations is calculated using the *PostGIS* function `ST_dWithin`. The distance threshold (500 m) can be visualised by creating a buffer around each port location using the function `ST_Buffer`.

SQL   *Q3.4.7.1*

```
SELECT port_name , mmsi, shipname , min_t , max_t , (max_t−min_t) as dur
FROM (
     SELECT libelle_po as port_name , mmsi, min(t) as min_t , max(t) as
     max_t
     FROM context_data.ports as q1 — ports location
     INNER JOIN ais_data.dynamic_ships as q2 — ship AIS positions
     ON (
          speed=0  — not moving
          AND t>='2016−01−01 .00:00:00'
          AND t<'2016−01−02 00:00:00'  — during Jan 1, 2016
          AND ST_dWithin(q1.geom3035 ,q2.geom3035 ,500)  — ships by 500m of
          port
     )
     GROUP BY libelle_po , mmsi ) as q3
LEFT JOIN ais_data.static_ships as q4 — ship names
ON (q3.mmsi=q4.sourcemmsi);
```

---

[22]https://postgis.net/docs/reference.html#Geometry_Processing.

Spatial Indexes

Similar to *classical* SQL queries, spatial queries can be optimised using indexes applied to spatial columns. *PostGIS* supports the *GiST* index, which indexes the *bounding boxes*, or minimum rectangles, enclosing the spatial extent of the geometries in a geometric column. For instance, the bounding boxes of the geometries in Fig. 3.7a are depicted in Fig. 3.7b as gray, dotted rectangles, and highlighted by coloured rectangles in Fig. 3.7c. The example below shows how to create a *GiST* index on the geometry column geom3035 of table dynamic_ships.

SQL	Q3.4.7.2

```
CREATE INDEX idx_dynamic_ships_geom3035
ON ais_data.dynamic_ships  -- table name
USING GiST (geom3035);  -- GiST Index on geometric column
```

Bounding boxes allow for a compact, while simplified, representation of geometries, that can be used for efficiently comparing the geometries in spatial queries. When a SQL query refers to an indexed geometry, a spatial relationship is firstly evaluated using the spatial extent of the geometry (Fig. 3.7c). Afterwards, only for the subset of the potentially matching geometries, the exact spatial relationship is processed and refined. In the example of Fig. 3.7, a query on (a) searching for intersecting geometric objects would exclude the yellow circle, because its bounding box does not intersect with any other geometric object's bounding box. The precise computation of the intersection is performed only between the blue pentagon and both the red line and green triangle (as they are the only boxes to overlap). With the index, the computation of the complex polygon intersection requires only two comparisons (first between the blue pentagon and the red line, then between the blue pentagon and the green triangle). Checking every geometric object with each other would have required $\binom{4}{2} = 6$ comparisons.

Do it yourself !	A3.4.7.1

Execute again the SQL query of the previous example (looking for the ships that stopped at less than 500 m from a port in Brittany on January 1st, 2016). After creating a spatial index on column geom3035, you should get the answer faster than without the spatial index. Then, search for the vessels (names and identifiers) that fished more than 15 min on January the 22nd 2016 within any fishing area. Usually, a fishing vessel fishes at a speed between 2.5 and 3.5 knots.

[Answer: 7 vessels].

3.4.8 Extending *PostgreSQL* Functions

As described in the previous section, *PostGIS* defines functions to manipulate and query geographic data. These functions extend the set of data management functions provided by *PostgreSQL*. Novel functions may be created using procedural programming languages (PL) like *PL/pgSQL, PL/Python, PL/Tcl, PL/Perl, PL/Java, PL/R, PL/sh.*[23]

> **Remark !**
>
> You can view the *PostgreSQL* and *PostGIS* functions in the folder "Functions" which is included in every database schema.

In the following example, a *PL/pgSQL* function is defined to search for the name and the identifier, i.e. the IMO number, of a vessel, given its MMSI. The function returns the result in formatted text.

SQL Q3.4.8.1

```
CREATE OR REPLACE FUNCTION get_vessel_info( -- name of the fuction
    mmsi integer -- list of input argument with types
)
RETURNS text AS $$ -- type of the returned value
DECLARE
    vessel_imo integer; -- integer local variable
    vessel_name text; -- text local variable
BEGIN
    SELECT shipname, imo INTO vessel_name, vessel_imo
    FROM ais_data.static_ships
    WHERE sourcemmsi=mmsi LIMIT 1;
    RETURN '[' || vessel_imo || '] ' || vessel_name; -- return value
END;
$$ LANGUAGE plpgsql; -- programming language used
```

Once defined, the function can be called in a query, as shown below.

SQL Q3.4.8.2

```
SELECT get_vessel_info(227705102);
```
[Answer: "[262144] BINDY"]

3.5 Understanding Vessel Movements with Trajectory-Based Queries

A trajectory can be defined as "a record of the evolution of the position (perceived as a point) of an object moving in space during a given time period" [52]. Position-based queries (i.e. relying on geometric data of type POINT) are easy to formalise and can provide meaningful information and statistics about these movement data. However, they have several limitations. First, there is a computational limit, as such spatial queries are very expensive. Second, the information they provide is sometimes limited by the update rate or the coverage of the sensor that measures the object's position.[24] In the maritime domain, for instance, it is difficult to identify with certainty

---

[23] https://www.postgresql.org/docs/11/xplang.html.

[24] In the case of AIS, vessel positions are reported only in the areas covered by AIS receivers, and at sparse time intervals. The AIS data in the open dataset are collected from a terrestrial receiver.

a vessel that has crossed a narrow passage, in order to check whether it has entered a restricted area or to calculate exactly the minimum distance to the coast.

The notion of *trajectory*, for instance as discussed in [36], whilst sometime complex to implement, has been introduced to address these limitations. It underlies the use of filtering and clustering techniques that make it possible to clearly define the starting, intermediate and ending points of a trajectory.

In this section we will create vessel trajectories by connecting the points of the same ship between them, in the form of polylines. This is a simple implementation of a *stop-move* model of trajectories [52]. Ship trajectories are segmented using time intervals. During these time intervals, the ship positions can stay still (stop) or change (move).

In *PostGIS*, the *move* part of the trajectory can be represented as a LINESTRING, which connects sequences of positions. In order to derive stationary (*stop*) areas, multiple AIS positions of anchored ships may be spatially grouped together, which can be spatially represented by cluster centroids. The overall ship movements can be modeled with a graph, whose nodes depict the stop locations (e.g. ports, mooring areas), and edges link stops. Ship trajectories can be grouped along edges and aggregated using statistics. This *node-edges* model can itself be manipulated, queried, analysed, for instance to analyse ships' life cycle as shown in [21].

In order to store the trajectories and the results of the data analysis queries that will be presented in the following sections, the creation of a new schema named "data_analysis" is required.

> **Action required**
>
> Define a schema data_analysis to store all the results of the data analysis queries.

3.5.1 Characterising Port Areas Through Spatial Partitioning

Spatial partitioning enables to understand the essential characteristics of movement data. With this aim, Andrienko and Andrienko [3] partition movement data according to a *Voronoi tessellation* [4]. A Voronoi diagram is a partition of a plane into regions (Voronoi cells) around a set of seed points. The voronoi cell around a seed point encompass all points of the plane closer to that seed point.

In this section, a Voronoi tessellation will be used to get an approximation of the area of competence of each port. In most official databases, ports are represented by geometrical points. This is also the case of the data in table ports in schema context_data. This representation is not effective, in extracting, for instance, from the database the vessels that are stopped in a port at a certain instant in time. In order to answer this query, ports should be represented by geometrical regions, on which e.g. the *PostGIS* function ST_Within could be applied. In order to derive ports' regions, a space partitioning must be created. Note that, since ports are not equally distributed along the coast, a uniform partition of the space would not be accurate to answer the query above.

Fig. 3.8 Voronoi tessellation of the geographical space, partitioning sea and land to identify the nearest port of any maritime location in Gulf of Morbihan. The Voronoi cells have random colours. Land is depicted in white, with the black line representing the coastline. Yellow stars represent ports' positions

By constructing a Voronoi tessellation of the space based on port locations, vessel positions can be dynamically associated to ports. The Voronoi Polygons around a set of points can be computed using *PostGIS* ST_VoronoiPolygons function. For instance, given the tessellation in Fig. 3.8, stopped vessels (i.e. with null or negligible speed over ground) can be associated to the closest ports using the spatial operators described above. Afterwards, the distance between the stop location and the port can be computed.

In the example below, a new table (ports_voronoi) is created to contain the Voronoi tessellation based on port locations. A *GiST* index is created to optimise the access to the newly created geometry.

```
SQL                                                                        Q3.5.1.1
CREATE TABLE data_analysis.ports_voronoi AS
    SELECT por_id as port_id, libelle_po as port_name, geom3035,
    voronoi_zone3035
    FROM context_data.ports
    LEFT JOIN (
        SELECT (ST_Dump(ST_VoronoiPolygons(ST_Collect(geom3035)))).geom
            as voronoi_zone3035
        FROM context_data.ports) as vp
    ON (ST_Within(ports.geom3035, vp.voronoi_zone3035));

CREATE INDEX idx_ports_voronoi_zone ON data_analysis.ports_voronoi
USING gist (voronoi_zone3035);
```

The following SQL query creates a new table (non_moving_positions) dedicated to the storage of stopped vessel positions. Thanks to the Voronoi tessellation in table ports_voronoi, each stop position can be matched to a unique port. Once

matched, the distance between a vessel stop and the associated port can be computed
using the built-in function `ST_Distance`.

```
SQL                                                                         Q3.5.1.2
CREATE TABLE data_analysis.non_moving_positions AS
    SELECT id, mmsi, t, q1.geom3035, port_id, port_name,
    ST_Distance(q1.geom3035, ports_voronoi.geom3035) as port_dist
    FROM (
        SELECT *
        FROM ais_data.dynamic_ships
        WHERE speed=0 -- non moving ship positions only
        ) as q1
    LEFT JOIN data_analysis.ports_voronoi
    ON ST_Within(q1.geom3035, ports_voronoi.voronoi_zone3035); -- ships in
       voronoi area
```

With the execution of the following queries, additional indexes are created to optimise the access to this new table `non_moving_positions` (on position geometry, position identifiers and associated timestamp).

```
SQL                                                                         Q3.5.1.3
CREATE INDEX idx_non_moving_positions_geom3035
ON data_analysis.non_moving_positions
USING gist (geom3035);

CREATE INDEX idx_non_moving_positions_port_id
ON data_analysis.non_moving_positions
USING btree(port_id);

CREATE INDEX idx_non_moving_positions_t
ON data_analysis.non_moving_positions
USING btree (t);
```

```
Do it yourself !                                                            A3.5.1.1
Using the results of table non_moving_positions, compute the average distance of vessel stops
in the Voronoi area of the Brest port.
                                                             [Answer: 1434.75 m]
```

3.5.2   Detecting and Clustering Ship Stops

As it will be illustrated in the chapter of Andrienko et al. [2], clustering techniques
can be applied to group vessels stops outside ports and for detecting the different
docking areas within a port. In the example of the previous section, many vessel
stops are located outside the ports, in areas which are likely mooring areas.

3.5.2.1   Detecting Ship Stops

The analysis of vessel movement begins with the identification of stops. A stop can
be defined as a temporally ordered sequence of vessel positions with speed (i.e. speed
over ground, SOG) equal to zero or below a very low threshold. The queries presented

in the next examples create, for each vessel, the sequences of stops extracted from the history of the vessel's positions.

To build this sequence, first, an auxiliary table of successive position pairs, namely `segments` is created. For each ship, this table enables us to connect with a line segment every consecutive pair of AIS positions. The table is ordered according to the MMSI number and the position timestamp (ORDER BY `mmsi, t`), and the *PostgreSQL* function LEAD will be used to access the data in the next row of an ordered table.

From this table, for each vessel, stop events can be extracted and filtered according to the vessel speed that is associated to the preceding and the following positions of each segment. This table is also useful to detect transmissions gaps between vessel positions (*cf.* the chapter of Patroumpas [37]) and GPS malfunctions (e.g. detecting unfeasible speeds. A new index combining the vessels' MMSI and the timestamp of the vessels' positions is created to optimise the query. The SQL commands for creating the index and the table are given in the example below.

```
SQL                                                              Q3.5.2.1

CREATE INDEX idx_dynamic_ships_mmsi_t ON ais_data.dynamic_ships
USING btree (mmsi,t);

CREATE TABLE data_analysis.segments AS
    SELECT mmsi, -- ship identifier
        t1, t2, -- starting and ending timestamps
        speed1, speed2, -- starting and ending speeds
        p1, p2, -- starting and ending points
        st_makeline(p1,p2) as segment, -- line segment connecting points
        st_distance(p1,p2) as distance, -- distance between points
        extract(epoch FROM (t2-t1)) as duration_s, -- timestamps in
    seconds
        (st_distance(p1,p2)/extract(epoch FROM (t2-t1)))
            as speed_m_s -- speed in m/s
    FROM (
        SELECT mmsi, -- ship identifier
            LEAD(mmsi) OVER (ORDER BY mmsi, t) as mmsi2, -- next MMSI
            t as t1, -- starting time
            LEAD(t) OVER (ORDER BY mmsi, t) as t2, -- ending time
            speed as speed1, -- initial speed
            LEAD(speed) OVER (ORDER BY mmsi, t) as speed2, -- final speed
            geom3035 as p1, -- initial point
            LEAD(geom3035) OVER (ORDER BY mmsi, t) as p2 -- final point
        FROM ais_data.dynamic_ships ) as q1
    WHERE mmsi=mmsi2; -- filter out different mmsi

CREATE INDEX idx_segments_speed ON data_analysis.segments
USING btree (speed1,speed2);
```

Once the table `segments` is created, stop events can be detected. Aligned with the chapter of Andrienko et al. [2], each stop starts immediately after a move, and ends as soon as the vessel starts moving again. Starting from the data in table `segments`, segments whose initial speed is above the speed threshold, and whose final speed is below the speed threshold, represent the beginning of stops (i.e. the vessel decreases its speed and goes steady). The end of a stop event is detected conversely.

The following example creates the auxiliary tables `stop_begin` and `stop_end`, which contain the potential starting and ending positions of vessel stops, selected as

just described. In this example, a speed threshold of 0.1 knot is used. To optimise
the future access to these tables, indexes are created.

```
SQL                                                                    Q3.5.2.2

CREATE TABLE data_analysis.stop_begin AS -- first stop position
    SELECT mmsi, t2 as t_begin -- stop starts at first steady position
    FROM data_analysis.segments
    WHERE speed1 >0.1 AND speed2 <=0.1; -- speed threshold is 0.1 kn

CREATE INDEX idx_stop_begin_mmsi_t ON data_analysis.stop_begin
USING btree (mmsi,t_begin);

CREATE TABLE data_analysis.stop_end AS -- last stop position
    SELECT mmsi, t1 as t_end -- stop ends at last steady position
    FROM data_analysis.segments
    WHERE speed1 <=0.1 AND speed2 >0.1; -- speed threshold is 0.1 kn

CREATE INDEX idx_stop_end_mmsi_t ON data_analysis.stop_end
USING btree (mmsi,t_end);
```

Afterwards, the table `stops` is created by coupling, for each ship, `stop_begin`
and `stop_end`, as illustrated in the following example (note that stops are ordered
by time, that is, ORDER BY t_end).

```
SQL                                                                    Q3.5.2.3

CREATE TABLE data_analysis.stops AS
    SELECT
        mmsi, -- ship identifier
        t_begin, -- start of the stop event
        t_end, -- end of the stop event
        extract(epoch FROM (t_end-t_begin)) as duration_s -- stop in
seconds
    FROM data_analysis.stop_begin INNER JOIN LATERAL (
        -- keep only stops that have an end
        SELECT t_end
        FROM data_analysis.stop_end
        WHERE stop_begin.mmsi=stop_end.mmsi AND
            t_begin<=t_end -- stop follows the beginning
        ORDER BY t_end LIMIT 1 -- select only the first stop end
    ) AS q2 ON (true);
```

In the example above, LATERAL is a reserved *PostgreSQL* keyword that enables
cross-references between the main query and the subquery (at the right of LATERAL.
In this case, it is used to compare the time of `stop_begin.t_begin` with
`stop_end.t_end`.

Not all the detected stops are meaningful. For instance, stops that last for a short
time, or are spatially too spread, could be filtered out. Similarly, stops with few vessel
positions can be discarded. In the example below, the centroid of each vessel' stop
cluster and the number of associated positions are computed.

```
SQL                                                                    Q3.5.2.4

ALTER TABLE data_analysis.stops ADD COLUMN centr3035 geometry(Point,3035)
  ;
ALTER TABLE data_analysis.stops ADD COLUMN nb_pos integer;

-- compute the centroid and number of positions
UPDATE data_analysis.stops SET (centr3035, nb_pos) = (
    SELECT st_centroid(st_collect(geom3035)), -- centroid of a multipoint
        count(*) as nb -- number of points
    FROM ais_data.dynamic_ships
    WHERE mmsi=stops.mmsi AND
        t>=stops.t_begin AND t<=stops.t_end); -- timestamp range
```

In the following example, indicators on the dispersion of the vessel's positions around the centroid of the cluster they belong to are computed. Once aggregated, this information offers some intuition about the nature of the stop. Additional columns are created in the table stops to store the computed indicators (average and maximum distances from the centroid).

```
SQL                                                                    Q3.5.2.5

ALTER TABLE data_analysis.stops ADD COLUMN avg_dist_centroid numeric;
ALTER TABLE data_analysis.stops ADD COLUMN max_dist_centroid numeric;

-- compute the distance of the position cluster to the centroid
UPDATE data_analysis.stops
SET (avg_dist_centroid, max_dist_centroid) = (
    SELECT avg(d), max(d)
    FROM (
        SELECT st_distance(centr3035,geom3035) as d -- distance to
    centroid
        FROM ais_data.dynamic_ships
        WHERE mmsi=stops.mmsi AND
            t>=stops.t_begin AND t<=stops.t_end -- timestamp range
    ) as q1);
```

In order to illustrate to use of the structures built above, the next query calculates how many of the vessel' stops lasted more than 5 min, involved more than 5 consecutive positions and occurred in average within 10 m of the centroid of the vessel' stop cluster.

```
SQL                                                                    Q3.5.2.6

SELECT count(*)
FROM data_analysis.stops
WHERE duration_s >=(5*60) -- 5 minutes long
    AND nb_pos>5 -- minimum 5 consecutive positions
    AND avg_dist_centroid <=10; -- within 10m from centroid

                                                        [Answer: 22,987]
```

3.5.2.2 Clustering Ship Stops

The auxiliary structures created in the previous section calculated, for each vessel, its stop clusters, i.e. the areas where the vessel was steady. In this section a density-based algorithm will be used to detect stationary areas.

Density-based clustering algorithms [34, 56] can be used to extract from data stationary areas, turning points in trajectories, activities areas, and so on. *PostGIS* implements the well-known *Density-Based Clustering of Applications with Noise* algorithm (DBSCAN) [12]. This algorithm clusters geometric objects together provided there are more than a minimum number (n) of neighbouring objects located within a threshold distance (d). These parameters should be tailored to the area under consideration.

In the following example, DBSCAN is used to cluster stop centroids in order to detect frequent stop areas. In the example, n defines the minimum number of stops to form a cluster; d defines the minimum distance between stops belonging to two different clusters. The smaller the d and n parameters, the bigger the number of resulting clusters. If d is large, then multiple mooring areas on a same dock will be clustered together. A similar approach is also presented in the chapter of Andrienko et al. [2] to detect stopping areas.

In the example below, groups of stops lasting more than one minute and having at least 5 different occurrences within 50 m distance range are clustered and stored in table `cluster_stops`. DBSCAN generated 354 different clusters that can be visualised using *QGIS*. These clusters are precise enough to detect different mooring locations within big ports like the one of Brest as illustrated in Fig. 3.9.

```
SQL                                                                  Q3.5.2.7

CREATE INDEX idx_stops_centroid ON data_analysis.stops
USING btree (centr3035);

CREATE TABLE data_analysis.cluster_stops AS
    SELECT *, ST_ClusterDBSCAN(centr3035, eps := 50, minpoints := 5) OVER
    () AS cid
    FROM data_analysis.stops
    WHERE duration_s >=60; -- 1 minute long
```

In order to to define the location of frequently used docking and mooring areas, the *spatial hull* of these clusters can be computed.

The convex (respectively, concave) hull of a set of points represents the minimum convex (respectively, concave) geometry that encloses all the points within the set. The Minimum Bounding Circle represents the smallest circle that fully contains all the points of the set. All these polygons can be computed using various *PostGIS* functions.

The area of the spatial hull can also give hints about the clustering process quality. Clusters with high spatial dispersion will have a bigger polygon surface. This can be acceptable for mooring areas outside a port, but within a port smaller clusters are likely more adequate to match docking areas.

The following query shows how to compute different spatial hulls of vessel stops. For each cluster, statistics are also included. In Fig. 3.10, the convex hulls of vessel stops in the port of Brest are visualised with *QGIS*.

Fig. 3.9 DBSCAN clustering of vessel stops (random colours, one per cluster) in Brest

```
SQL                                                                    Q3.5.2.8
CREATE TABLE data_analysis.clusters_stops_hulls AS
    SELECT cid , -- cluster id
        ST_ConvexHull(st_collect(centr3035)) as convex_hull ,
        ST_ConcaveHull(st_collect(centr3035) ,0.75) as concave_hull ,
        ST_MinimumBoundingCircle(st_collect(centr3035)) as
                bounding_circle ,
        ST_Centroid(st_collect(centr3035)) as centroid ,
        count(*) as nb_stops , -- number of stops in this cluster (area)
        sum(nb_pos) as nb_pos , -- sum of stops in the cluster
        count(DISTINCT mmsi) as nb_ships , -- num of unique ships
        min(duration_s) as min_dur ,
        avg(duration_s) as avg_dur ,
        max(duration_s) as max_dur
    FROM data_analysis.cluster_stops
    WHERE cid IS NOT NULL -- exclude outliers
    GROUP BY cid ; -- group all the stops centroids within the same
        cluster
```

3.5.3 Extracting Trajectory Tracks and the Navigation Graph

Maritime traffic can be modelled as a graph whose nodes correspond to the stationary
areas (e.g. mooring or docking areas) and edges represent the vessel movements
between these areas. Relying on such a graph, vessel movement can be analysed [10,
21, 29]. For example, it is possible to count and identify the incoming and outgoing
destinations from each stationary area, to calculate the traffic density, and to identify
the most frequently used tracks.

In the examples that follow, and similarly to the computation of stops, vessel
tracks will be created considering the successive positions of the same ship between
a *track start* and a *track stop*. By definition:

Fig. 3.10 Convex hulls of vessel stop clusters (random colours) in the area of the Brest port

- the *start* of a vessel track is the end of the previous stop event; and
- the *end* of a track is the start of the next stop event.

In the following example, a new table tracks is created and populated with stop events selected from the table cluster_stops, which already contains all the filtered stops with associated cluster identifiers. These will be used as the nodes of the vessel traffic graph. An index is added to optimise the selection of temporally ordered points.

SQL　　　　　　　　　　　　　　　　　　　　　　　　　　　　　　　*Q3.5.3.1*

```sql
CREATE INDEX idx_cluster_stops_mmsi_tbegin ON data_analysis.cluster_stops
USING btree (mmsi,t_begin);

CREATE TABLE data_analysis.tracks AS
    SELECT q1.mmsi, -- ship id
        q1.cid as start_cid, -- start cluster node id of the track
        q3.cid as end_cid, -- end cluster node id of the track
        q1.t_end as t_start, -- track start is end of the previous stop
        q3.t_begin as t_end, -- track end is begin of the next stop
        extract(epoch FROM (q3.t_begin-q1.t_end))
            as duration_s -- track duration in seconds
    FROM (
        SELECT mmsi, cid, t_end
        FROM data_analysis.cluster_stops
        WHERE cid IS NOT NULL ) as q1
    INNER JOIN LATERAL (
        -- track that are in between two clustered stops area
        SELECT q2.cid, q2.t_begin
        FROM data_analysis.cluster_stops as q2
        WHERE q2.cid IS NOT NULL -- stop must be in a clustered area
        AND q1.mmsi=q2.mmsi -- same ship
        AND q2.t_begin>q1.t_end -- search for the next stop event
        ORDER BY q2.t_begin LIMIT 1 ) as q3 ON (true);
```

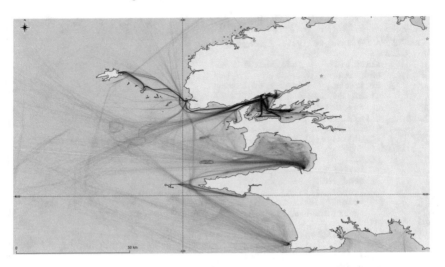

Fig. 3.11 Vessel trajectory tracks in the Brest area (yellow stars are ports of Brittany)

As before, the reserved word LATERAL enables to cross reference the elements of the first sub-query with alias q1 in the second sub-query q2 (q1.mmsi=q2.mmsi AND q2.t_begin>q1.t_end).

In the next query, the geometry of each *track move* is represented as a segment connecting consecutive track stops. Ship positions between the t_start and t_end timestamps of the trajectory are selected, ordered with respect to time and connected to form a LINESTRING using the ST_makeline function of *PostGIS*. Figure 3.11 shows the computed trajectory tracks.

```
SQL                                                                          Q3.5.3.2
ALTER TABLE data_analysis.tracks
ADD COLUMN track geometry(LineString,3035);

UPDATE data_analysis.tracks SET track = (
    SELECT st_makeline(geom3035)
    FROM (
        SELECT geom3035
        FROM ais_data.dynamic_ships
        WHERE mmsi=tracks.mmsi AND t>=tracks.t_start AND t<=tracks.t_end
        ORDER BY t) as q1); -- order points by time
```

The SQL example that follows creates a new table (graph_edges) to represent the traffic graph of the Brest area. For each graph edge, statistics on the underlying traffic, like the number of associated vessel tracks, the number of ships navigating it, the average trajectory duration and length, are extracted. These indicators are stored in the table and can be visualised for analysis. The traffic along the edges of the graph is oriented.

```
SQL                                                                    Q3.5.3.3
CREATE TABLE data_analysis.graph_edges AS
    SELECT
        start_cid , -- from node
        end_cid , -- to node
        nb_tracks , -- number of tracks
        q1.nb_ships , -- number of different ships
        q1.avg_duration , -- average transit time
        q1.avg_length , -- average transit length
        q1.min_length , -- minimum transit length
        st_makeline(c1.centroid,c2.centroid) as straight_edge -- edge
    FROM (
    SELECT
        start_cid , -- from node
        end_cid , -- to node
        count(*) as nb_traj , -- number of vessel trajectories
        count(distinct mmsi) as nb_ships , -- number of unique ships
        avg(duration_s) as avg_duration , -- average trajectory
duration
        avg(st_length(track)) as avg_length , -- average length
        min(st_length(track)) as min_length -- shortest length
    FROM data_analysis.tracks
    GROUP BY start_cid , end_cid ) as q1
    LEFT JOIN data_analysis.clusters_stops_hulls as c1
        ON (c1.cid=q1.start_cid)
    LEFT JOIN data_analysis.clusters_stops_hulls as c2
        ON (c2.cid=q1.end_cid);
```

In order to improve the visualisation of the traffic statistics on a map, the lines of the graph edges can be bent using a custom *PL/pgSQL* function. Figure 3.12 illustrates the maritime traffic graph obtained using this function. The edges thickness is proportional to the number of trajectories associated to the track (i.e. thickest edges represent the most frequent routes between stationary areas). The navigation graph shown in Fig. 3.12 is a simplified, summarised version of the density map shown in Fig. 3.11.

3.5.4 Managing Data Quality Using Constraints

The previous sections highlight typical *PostgreSQL* and *PostGIS* features to process moving object data. The readers should be aware that the results of these queries may be affected by the quality of data. Apart from errors and irrelevant messages, AIS data in the given maritime dataset have been provided as received, including duplicates and other veracity issues. The coverage of the data is also not uniform, with 70.5% of vessel positions located in a range of 10 km from the AIS receiver [41]. Erroneous or missing AIS positions influence the quality of results but raise interesting algorithmic and processing challenges, therefore these veracity issues have been maintained in the dataset, which realistically represents the situation in the area and the period it refers to.

The accuracy of maritime clustering and trajectory tracks depend on data quality. For instance, the readers may compare the movement of passenger ships within the Brest roadstead versus the movement of passenger ships travelling to the Ushant

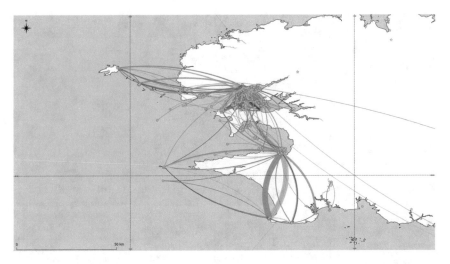

Fig. 3.12 Traffic graph. Graph nodes (green hexagons) represent stationary areas (e.g. ports, mooring areas), while edges (curved lines, one random colour per line) approximate tracks of vessels moving between stops

and Molène islands where the data quality is lower. Pre-filtering, error-checking algorithms can affect the results as much as the variation of the analysis parameters (e.g. the speed threshold). Reasoning on intermediate structures can provide meaningful data quality information. For instance, the table `segments` can help detecting abnormal situations that can be ignored, like movements with extremely long duration, excessive distances between positions, or unfeasible speeds. Trajectories with unknown or erroneous MMSI identifiers could also be flagged as abnormal and discarded by the analysis.

In the database, *integrity constraints* may be used to manage data quality issues. Figure 3.13, generated using pgModeler[25] presents two of the database schemas we built along the chapter, i.e. `ais_data` and `context_data`. The structure of the tables in these schemas is reported, with column names and types. Adding relations and constraints which are another essential part of database model design allows to manage data properties and to encompass the aforementioned quality issues in accordance with the database application. For example, integrity constraints may prevent that a ship appears travelling in two locations at the same time; may ensure that the vessel speed is always a positive value; and may guarantee that the vessel heading has either a default value (i.e. correctly setting it to 511), or is between 0 and 360 degrees.

Column constraints express integrity rules on the data in the specified columns. For instance, a NOT NULL constraint prevents the data in a column to assume the NULL value. *Table-level constraints* define additional rules that apply to all the data in a table. For instance, a PRIMARY KEY constraint combines a NOT NULL and a

---

[25]https://pgmodeler.io.

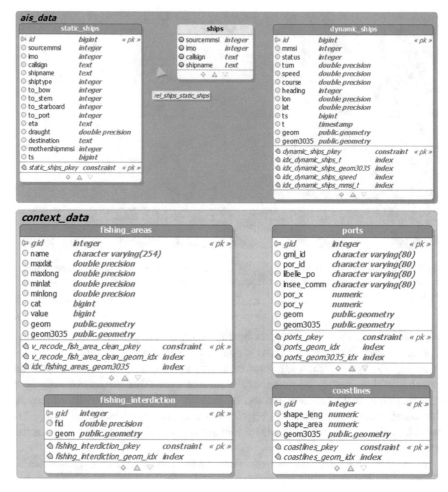

Fig. 3.13 Entities' structure (tables and views) of the database schemas `ais_data` and `context_data`. The data model is extracted from the *PostgreSQL* maritime database using pgModeler

UNIQUE constraints, ensuring that all the data in a column (or in multiple columns, in combination) have unique values over the table. Figure 3.13 includes PRIMARY KEY constraints in both `ais_data` and `context_data` schema.

Remark ! Modelling database constraints

Incorporating constraints in the database requires altering the tables' structure. Preferably, this modelling step should be accomplished altogether with the generation of the database, preceding the data insertion. Constraints can also be integrated along an iterative process, throughout the creation and the subsequent modification of the database.

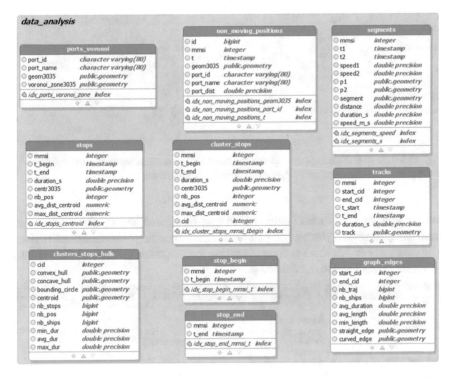

Fig. 3.14 Entities' structure (tables and views) of the database schema `data_analysis`. The data model is extracted from the *PostgreSQL* maritime database using pgModeler

Figure 3.14 presents the entities created to analyse the data along the chapter, and stored in the schema `data_analysis`. In the following, we are going to extend this part of the maritime database data model with constraints that preserve the data integrity when new operations (e.g. `INSERT`, `UPDATE`, `DELETE`) on tables and columns are executed.

Cardinality constraints define how many entities may participate in a relationship. A *many-to-many* relationship is realised by a table that connects the entities in two other tables. *One-to-many* relationships are achieved with *foreign key constraints*, which maintain the referential integrity of data between two related tables. A `FOREIGN KEY` specifies that the values of the data in one or more columns must exist in a related table.

In the following example, a foreign key constraint is defined to ensure that all the ports for which a Voroni cell has been calculated, i.e. the ports in the table `data_analysis.ports_voronoi`, exist in table `context_data.ports`. In the example, after defining the tables' primary keys, a `FOREIGN KEY` constraint is specified altering the table `data_analysis.ports_voronoi`. This constraint realises a *one-to-many* relation between the two tables, as illustrated in Fig. 3.15. The condition `ON DELETE CASCADE` in the SQL example avoids the creation of

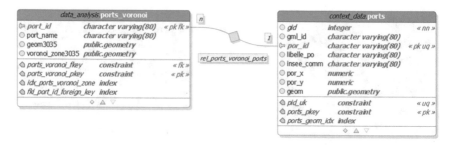

Fig. 3.15 Use of foreign keys to implement a one-to-many relation between ports and Voronoi tessellation (extracted from the *PostgreSQL* maritime database using pgModeler)

inconsistent, orphan Voronoi cells, which may be created when ports are removed from table `context_data.ports` but the corresponding Voronoi cell still exists. The constraint avoids this situation triggering the automatic deletion of the Voronoi cell that is associated to the deleted port.

```
SQL                                                                                Q3.5.4.1
--- Table ports
ALTER TABLE context_data.ports
    DROP CONSTRAINT ports_pkey; --- Drop the old primary key which was
        automatically generated when importing the ports shapefile

ALTER TABLE context_data.ports
    ADD CONSTRAINT ports_pkey PRIMARY KEY (por_id); --- Column por_id is
        now the primary key of the table

--- Table ports_voronoi
ALTER TABLE data_analysis.ports_voronoi
    ADD CONSTRAINT ports_voronoi_pkey PRIMARY KEY (port_id); --- Column
        port_id is now the primary key of the table

ALTER TABLE data_analysis.ports_voronoi
    ADD CONSTRAINT ports_voronoi_fkey FOREIGN KEY (port_id)
    REFERENCES context_data.ports (por_id) ON DELETE CASCADE; --- Column
        port_id references the ports table using a foreign key.
```

As a result of the activation of the foreign key constraint in the example, when creating a Voronoi cell for a port, this port must exist, i.e. all the ports in the table `data_analysis.ports_voronoi` must match some port in the referenced table `context_data.ports` (cf. keyword REFERENCES). However, the contrary may not hold, i.e. adding a new port without recomputing the Voronoi tessellation is allowed. In order to implement the dual constraint, a *one-to-one* relation must be defined. In this case, adding a new port would require generating the associated Voronoi cell.

> **Remark ! A note on *one-to-one* relations**
>
> *One-to-one* relations are not well represented in standard SQL, because they lead to a logical union of two tables. As a solution, reciprocal foreign keys may be used, i.e. each table in the relation has a foreign key reference to the primary key of the other table. This solution creates a circular dependency that may block the insertion of new data in the tables. Luckily, constraints may be deferred, i.e. lazily validated, allowing to temporarily violation of the integrity of the relation to enable the data insertion. The option `DEFERRABLE` can be used with this aim when defining the constraint.

As an alternative to the definition of reciprocal foreign keys, a *one-to-one* relation may be obtained by defining a *database trigger* to enforce it. This solution is illustrated in the following example. The trigger uses the database function `context_data.rebuild_table_ports_voronoi()`, which automatically generates a voronoi tessellation on the basis of the ports existing in table `context_data.ports`. The trigger `trigger_ports _voronoi` is defined for table `context_data.ports` to execute the function, rebuilding the Voronoi tessellation, whenever the table `context_data.ports` is modified (`AFTER INSERT OR UPDATE OR DELETE OR TRUNCATE on context_data.-ports`).

> **Remark ! A note on triggers**
>
> Database triggers are SQL procedures that are executed upon the occurrence of a monitored database event, such as the insertion of a row in a table or view. In *PostgreSQL*, triggers are functions, which are automatically invoked by the DBMS whenever an insert, update, delete or truncate event occurs on a specified table.

```
SQL                                                                          Q3.5.4.2
— Create a function to rebuild the Voronoi tessellation
CREATE OR REPLACE FUNCTION context_data.rebuild_table_ports_voronoi()
    RETURNS TRIGGER AS $$ — this function return a trigger
BEGIN
— Same as Q5.1.1
DROP TABLE IF EXISTS data_analysis.ports_voronoi;
CREATE TABLE data_analysis.ports_voronoi AS
    SELECT por_id as port_id, libelle_po as port_name, geom3035,
    voronoi_zone3035
    FROM context_data.ports
    LEFT JOIN (
    SELECT (ST_Dump(ST_VoronoiPolygons(ST_Collect(geom3035)))).geom as
    voronoi_zone3035 FROM context_data.ports) as vp
    ON (ST_Within(ports.geom3035, vp.voronoi_zone3035));

— Add primary and foreign key constraints
ALTER TABLE data_analysis.ports_voronoi ADD CONSTRAINT ports_voronoi_pkey
    PRIMARY KEY (port_id); — primary key

ALTER TABLE data_analysis.ports_voronoi ADD CONSTRAINT ports_voronoi_fkey
    FOREIGN KEY (port_id) REFERENCES context_data.ports (por_id)
— Column port_id references ports.por_id using a foreign key
    ON DELETE CASCADE; — Deletes on ports are propagated

CREATE INDEX idx_ports_voronoi_zone ON data_analysis.ports_voronoi
USING gist (voronoi_zone3035);
    RETURN NEW; — return the updated rows
END;
$$ LANGUAGE plpgsql; — programming language used

— Create a trigger to execute the function whenever ports is updated
DROP TRIGGER IF EXISTS trigger_ports_voronoi ON context_data.ports;
CREATE TRIGGER trigger_ports_voronoi
    AFTER — the trigger is executed after a table update
    INSERT OR UPDATE OR DELETE OR TRUNCATE — any type of update
    ON context_data.ports — the monitored table
    FOR EACH STATEMENT — after each statement
    EXECUTE PROCEDURE context_data.rebuild_table_ports_voronoi(); —
    execute the function
```

The trigger can be tested by deleting a port from the database, as in the following example. As a result of the following SQL, the Voronoi table should be automatically updated.

```
SQL                                                                          Q3.5.4.3
DELETE FROM context_data.ports WHERE libelle_po='Sein'; — delete the
    port of the Sein Island
```

Action required

Study the database tables for identifying potential relations among them. Study the table columns, their types and expected values (cf. the files README in the maritime dataset) in order to define additional constraints using the aforementioned NOT NULL, UNIQUE, PRIMARY KEY, FOREIGN KEY keywords. Consider also the CHECK constraint, which allows to evaluate and check a condition when inserting new data or updating existing ones, and DEFAULT, which permits to assign a default value when a new row is inserted in a table.

3.6 Summary and Conclusion

More than two decades ago, in an essay entitled *Marine Informatics: a new discipline emerges*, Roger Bradbury postulated that the scientific community was on the way to creating a new holistic discipline addressing the challenges of marine data integration and analysis [9]. One of the major issues he identified at that time was the lack of regularity and sparsity in the data collected, which were also temporally scattered. Nowadays, the large variety of maritime sensors together with scientific background and techniques for the monitoring, analysis, and visualisation of sea data has revolutionized the domain (*cf.* the chapter of Bereta et al. [7]). Positioning data correlated with contextual heterogeneous data as provided by the maritime dataset used in this chapter makes Bradbury's vision concrete, at least with respect to ships' movements.

This chapter, in line with Bradbury's vision, takes advantage of the navigation data brought by the maritime dataset to illustrate the benefits of relational database for *maritime informatics*. Specifically, the chapter addresses the design, storage and querying of maritime data through a spatio-temporal DBMS. In particular, the functionalities of a relational DBMS have been illustrated, due to the ability thereof to find, define, sort, modify, link and transform data in complex databases, while guaranteeing the user a robust layer for spatial analysis.

The open-source relational DBMS *PostgreSQL*, enhanced with the extension *PostGIS* for manipulating spatial features, was used to exemplify the concepts presented in the chapter. It is worth noticing that the proposed technical environment is open source, freely available and standard based, and is used in many academic and professional applications. As such, the reader may easily find additional material to extend the examples proposed herein. Furthermore, this technical choice is not to be considered as a limitation, because the examples of commands proposed in the chapter can find an easy correspondence in other DBMS that offer a spatial support.

The spatial representation, analysis and visualisation techniques presented in this chapter are a useful basis to understand the analytics approaches that are presented in the rest of the book, in particular in the chapters of Tampakis et al. [54] and of Andrienko et al. [2], which discuss data analytics and visual analytics, respectively.

3.7 Bibliographical Notes

The interested reader can refer to some additional material to get practical illustrations on the functionalities offered by spatio-temporal DBMS. The authors in [43] provide an introduction on the theoretical aspects of spatial databases, useful to better understand the different spatial object models and the available formats. An overview of the most important aspects of spatio-temporal databases is given in [27], which presents the main research results of the CHOROCHRONOS project.[26] For the last

---

[26]CHOROCHRONOS project http://chorochronos.datastories.org.

research updates, the reader may refer to the series of the *International Symposium on Spatial and Temporal Databases (SSTD)*.

For additional material on OODBMS, the works extending the SECONDO DBMS offer many examples on how to extend an OODBM model to support complex spatio-temporal queries like group spatio-temporal patterns [47] and range-queries [58]. Reference [16] shows also an extension of the SECONDO model to support symbolic trajectories. The reader interested in distributed computation can have a look to Parallel SECONDO,[27] developed to improve the performance of mobility data analysis that supports a specialised version of the *R-Tree* index called *TM-RTree*.

Practical examples of the combined used of *PostgreSQL/PostGIS* and *QGIS* for movement data analysis are available in Anita Graser's blog.[28] We also refer to the official *PostgreSQLPostGIS* documentation for a detailed description of spatial and temporal database functionalities. Note that, although *PostgreSQL* and *PostGIS* are mainly used for two dimensions geometries (X and Y coordinates), *PostGIS* also supports the handling of three-dimensional (3-D) geometries. This extra dimension, namely Z, is added to each vertex in the geometry, and the geometry type itself is enhanced accordingly, to enable the correct interpretation of the additional dimension. For instance, the 2-D geometries `Point`, `Linestring` and `Polygon` become 3-D `PointZ`, `LinestringZ` and `PolygonZ`, respectively. Also, the use of the 3-D only `Polyhedralsurface` makes possible the generation of volumetric objects in the database. Special *PostGIS* functions for spatial relationships have been adapted to 3-D geometries. The reader can refer to the official *PostGIS* documentation.[29]

The NoSQL Database management website[30] provides a useful overview of the capabilities of existing NoSQL systems, including spatial extensions. Recently, [6] described a proposal to extend column-based *Cassandra* stores to the spatial dimension. Another solution to spatio-temporal objects persistence for large-scale data and geo-spatial analytics is *GeoMesa*,[31] an open-source suite of tools that interfaces with NoSQL databases like *Google Bigtable* and *Cassandra*, among others. *GeoMesa* supports near-real time analytics for streaming data and distributed data processing, and relies on *GeoServer*,[32] a well known data server for geographic data, and OGC application programming interfaces for map server integration.

Port and stationary areas detection, which is discussed in the chapter, is a hot topic in maritime related research. Most of the proposed approaches use unsupervised learning. For instance, Millefiori et al. use a data-driven approach (i.e. kernel density estimation (KDE) on AIS data) to define the extended areas of operation of seaports [34]. Similarly, Vespe et al. [56] also use density estimates on AIS data to map fishing activities at European scale. This topic is also addressed in the chapter of

---

[27]Parallel SECONDO http://dna.fernuni-hagen.de/secondo/ParallelSecondo/.

[28]Anita Graser's blog https://anitagraser.com/.

[29]*PostGIS* 3-D https://postgis.net/workshops/postgis-intro/3d.html.

[30]NoSQL Database management website http://nosql-database.org/.

[31]GeoMesa https://www.geomesa.org/.

[32]Geoserver http://geoserver.org/.

Andrienko et al. [2] in this book. Taking advantage of the identified stationary areas, vessel trajectory analysis and prediction [13, 59], identification of human activities at sea [49], anomaly detection [30, 31, 45, 51] are also widely developed in the literature.

The reader interested in uncertainty representation and reasoning in maritime data and information fusion can also refer to [5, 15, 46] and the chapter of Jousselme et al. [23].

References

1. Allen, J.F.: Maintaining knowledge about temporal intervals. Commun. ACM **26**(11), 832–843 (1983). https://doi.org/10.1145/182.358434
2. Andrienko, N., Andrienko, G.: Visual analytics of vessel movement. In: Artikis, A., Zissis, D. (eds.) Guide to Maritime Informatics, chap. 5. Springer, Berlin (2021)
3. Andrienko, N.V., Andrienko, G.L.: Spatial generalization and aggregation of massive movement data. IEEE Trans. Vis. Comput. Graph. **17**, 205–219 (2011)
4. Aurenhammer, F.: Voronoi diagrams: a survey of a fundamental geometric data structure. ACM Comput. Surv. **23**(3), 345–405 (1991). https://doi.org/10.1145/116873.116880
5. Battistello, G., Koch, W.: Knowledge-aided multi-sensor data processing for maritime surveillance. GI Jahrestag. **2**, 796–799 (2010)
6. Ben Brahim, M., Drira, W., Filali, F., Hamdi, N.: Spatial data extension for cassandra nosql database. J. Big Data **3**(1), 11 (2016). https://doi.org/10.1186/s40537-016-0045-4
7. Bereta, K., Chatzikokolakis, K., Zissis, D.: Maritime reporting systems. In: Artikis, A., Zissis, D. (eds.) Guide to Maritime Informatics, chap. 1. Springer, Berlin (2021)
8. Borkowski, P.: The ship movement trajectory prediction algorithm using navigational data fusion. Sensors **17**(6), 1432 (2017)
9. Bradbury, R.: Marine informatics: a new discipline emerges. Marit. Stud. **1995**(80), 15–22 (1995). https://doi.org/10.1080/07266472.1995.10878407
10. Carlini, E., Monteiro, V., Soares, A., Etemad, M., Machado, B., Matwin, S.: Uncovering vessel movement patterns from ais data with graph evolution analysis. In: Proceedings of the MASTER Workshop, 23rd International Conference on Extending Database Technology (EDBT), pp. 1–7 (2020)
11. Egenhofer, M.: A mathematical framework for the definition of topological relations. In: Proceedings the 4th International Symposium on Spatial Data Handing, pp. 803–813 (1990)
12. Ester, M., Kriegel, H.P., Sander, J., Xu, X.: A density-based algorithm for discovering clusters in large spatial databases with noise. In: Proceedings of 2nd International Conference on Knowledge Discovery and Data Mining, pp. 226–231 (1996)
13. Gan, S., Liang, S., Li, K., Deng, J., Cheng, T.: Ship trajectory prediction for intelligent traffic management using clustering and ann. In: 2016 UKACC 11th International Conference on Control (CONTROL), pp. 1–6. IEEE (2016)
14. Grasso, R.: Ship classification from multi-spectral satellite imaging by convolutional neural networks. In: Proceedings of the 27th European Signal Processing Conference, A Curuna, Spain, 2–6 Sep 2019
15. Guerriero, M., Willett, P., Coraluppi, S., Carthel, C.: Radar/ais data fusion and sar tasking for maritime surveillance. In: 2008 11th International Conference on Information Fusion, pp. 1–5. IEEE (2008)
16. Güting, R.H., Valdés, F., Damiani, M.L.: Symbolic trajectories. ACM Trans. Spatial Algorithms Syst. **1**(2), 7:1–7:51 (2015). https://doi.org/10.1145/2786756
17. Guttman, A.: R-trees: A dynamic index structure for spatial searching. In: Proceedings of the 1984 ACM SIGMOD International Conference on Management of Data, SIGMOD '84, pp. 47–57. ACM, New York, NY, USA (1984). https://doi.org/10.1145/602259.602266

18. Hellerstein, J.M., Naughton, J.F., Pfeffer, A.: Generalized search trees for database systems. In: Proceedings of the 21th International Conference on Very Large Data Bases, VLDB '95, pp. 562–573. Morgan Kaufmann Publishers Inc., San Francisco, CA, USA (1995). http://dl.acm.org/citation.cfm?id=645921.673145

19. Huang, B., Claramunt, C.: STOQL: An ODMG-based spatio-temporal object model and query language. In: Richardson, D.E., van Oosterom, P. (eds.) Advances in Spatial Data Handling, pp. 225–237. Springer, Berlin (2002)

20. ISO Central Secretary: Geographic information. simple feature access. sql option. Standard ISO 19125-2:2006, International Organization for Standardization, Geneva, CH (2006)

21. Itani, A., Ray, C., Falou, A.E., Issa, J.: Mining ship motions and patterns of life for the EU common information sharing environment (CISE). In: OCEANS 2019, Marseille, France, pp. 1–6 (2019)

22. Johansen, T.A., Cristofaro, A., Perez, T.: Ship collision avoidance using scenario-based model predictive control. IFAC-PapersOnLine **49**(23), 14–21 (2016)

23. Jousselme, A.L., Iphar, C., Pallotta, G.: Uncertainty handling for maritime route deviation. In: Artikis, A., Zissis, D. (eds.) Guide to Maritime Informatics, chap. 9. Springer, Berlin (2021)

24. Jugović, A., Hess, S., Poletan Jugović, T.: Traffic demand forecasting for port services. Promet-Traffic Transp. **23**(1), 59–69 (2011)

25. Kim, S., Lee, J.: Small infrared target detection by region-adaptive clutter rejection for sea-based infrared search and track. Sensors (Basel) **14**, 13210–13242 (2014). https://doi.org/10.3390/s140713210

26. Kothuri, R., Ravada, S.: Oracle Spatial, Geometries, pp. 821–826. Springer US, Boston (2008). https://doi.org/10.1007/978-0-387-35973-1_936

27. Koubarakis, M., Sellis, T., Frank, A., Grumbach, S., Güting, R., Jensen, C., Lorentzos, N., Manolopoulos, Y., Nardelli, E., Pernici, B., Schek, H.J., Scholl, M., Theodoulidis, B., Tryfona, N. (eds.): Spatio-Temporal Databases - The CHOROCHRONOS Approach. Lecture Notes in Computer Science, vol. 2520. Springer, Berlin (2003)

28. Kucuk, A., Hamdi, S.M., Aydin, B., Schuh, M.A., Angryk, R.A.: Pg-trajectory: A postgresql/postgis based data model for spatiotemporal trajectories. In: Proceedings of the 2016 IEEE International Conferences on Big Data and Cloud Computing (BDCloud), Social Computing and Networking (SocialCom), Sustainable Computing and Communications (SustainCom), 8-10 Oct 2016, Atlanta, GA, USA, pp. 81–88 (2016). https://doi.org/10.1109/BDCloud-SocialCom-SustainCom.2016.23

29. Laddada, W., Ray, C.: Graph-based analysis of maritime patterns of life. In: Proceedings of the GAST Workshop, 20th Journées Francophones Extraction et Gestion des Connaissances (EGC), pp. 1–14 (2020)

30. Lane, R.O., Nevell, D.A., Hayward, S.D., Beaney, T.W.: Maritime anomaly detection and threat assessment. In: 2010 13th Conference on Information Fusion (FUSION), pp. 1–8. IEEE (2010)

31. Liu, B., de Souza, E.N., Hilliard, C., Matwin, S.: Ship movement anomaly detection using specialized distance measures. In: 2015 18th International Conference on Information Fusion (Fusion), pp. 1113–1120. IEEE (2015)

32. Makris, A., Tserpes, K., Spiliopoulos, G., Anagnostopoulos, D.: Performance Evaluation of MongoDB and PostgreSQL for Spatio-temporal Data. In: 2nd International Workshop on Big Mobility Data Analytics (BMDA2019), Lisbon, Portugal (2019). https://doi.org/10.5281/zenodo.2649876

33. Mazzarella, F., Alessandrini, A., Greidanus, H., Alvarez, M., Argentieri, P., Nappo, D., Ziemba, L.: Data fusion for wide-area maritime surveillance. In: Proceedings of the COST MOVE Workshop on Moving Objects at Sea, Brest, France, pp. 27–28 (2013)

34. Millefiori, L.M., Zissis, D., Cazzanti, L., Arcieri, G.: Scalable and distributed sea port operational areas estimation from ais data. In: 2016 IEEE 16th International Conference on Data Mining Workshops (ICDMW), pp. 374–381. IEEE (2016)

35. Organisation, I.S.: Iso/iec 9075:2016 information technology database languages sql

36. Parent, C., Spaccapietra, S., Renso, C., Andrienko, G., Andrienko, N., Bogorny, V., Damiani, M.L., Gkoulalas-Divanis, A., Macedo, J., Pelekis, N., et al.: Semantic trajectories modeling and analysis. ACM Comput. Surv. (CSUR) **45**(4), 42 (2013)

37. Patroumpas, K.: Online mobility tracking against evolving maritime trajectories. In: Artikis, A., Zissis, D. (eds.) Guide to Maritime Informatics, chap. 6. Springer, Berlin (2021)

38. Pelekis, N., Theodoridis, Y., Vosinakis, S., Panayiotopoulos, T.: Hermes - A framework for location-based data management. In: Advances in Database Technology - EDBT 2006, 10th International Conference on Extending Database Technology, Munich, Germany, 26–31 Mar 2006, Proceedings, pp. 1130–1134 (2006). https://doi.org/10.1007/11687238_75

39. Petrossian, G.A.: Preventing illegal, unreported and unregulated (iuu) fishing: a situational approach. Biol. Conserv. **189**, 39–48 (2015)

40. Randell, D.A., Cui, Z., Cohn, A.G.: A spatial logic based on regions and connection. In: Proceedings of the 3rd International Conference on Principles of Knowledge Representation and Reasoning, pp. 165–176. Morgan Kaufmann Publishers Inc. (1992)

41. Ray, C., Dréo, R., Camossi, E., Jousselme, A.L.: Heterogeneous integrated dataset for maritime intelligence, surveillance, and reconnaissance (version 0.1) [data set] (2018). https://doi.org/10.5281/zenodo.1167595

42. Ray, C., Dréo, R., Camossi, E., Jousselme, A.L., Iphar, C.: Heterogeneous integrated dataset for maritime intelligence, surveillance, and reconnaissance. Data Brief **25**, 104141 (2019). https://doi.org/10.1016/j.dib.2019.104141. http://www.sciencedirect.com/science/article/pii/S2352340919304950

43. Rigaux, P., Scholl, M., Voisard, A.: Spatial Databases with Application to GIS. Morgan Kaufmann Publishers Inc., San Francisco (2002)

44. Riveiro, M., Pallotta, G., Vespe, M.: Maritime anomaly detection: a review. Wiley Interdisciplinary Reviews: Data Mining and Knowledge Discovery e1266 (2018). https://doi.org/10.1002/widm.1266

45. Roberts, S.: Anomaly detection in vessel track data. Ph.D. thesis, Oxford University, UK (2014)

46. Roy, J., Bosse, E.: Sensor integration, management and data fusion concepts in a naval command and control perspective. Technical report, Defence Research Establishment Valcartier (Québec) (1998)

47. Sakr, M.A., Güting, R.H.: Group spatiotemporal pattern queries. Geoinformatica **18**(4), 699–746 (2014). https://doi.org/10.1007/s10707-013-0198-7

48. Siabato, W., Manso-Callejo, M., Camossi, E.: An annotated bibliography on spatiotemporal modelling trends. Int. J. Earth Environ. Sci. **2**(135), 26 (2017)

49. Silveira, P., Teixeira, A., Soares, C.G.: Use of ais data to characterise marine traffic patterns and ship collision risk off the coast of Portugal. J. Navig. **66**(6), 879–898 (2013)

50. Simoes, R., Queiroz, G., Ferreira, K., Vinhas, L., Câmara, G.: Postgis-t: towards a spatiotemporal postgresql database extension. In: XVII Brazilian Symposium on Geoinformatics (GeoInfo 2016) (2016)

51. Soleimani, B.H., De Souza, E.N., Hilliard, C., Matwin, S.: Anomaly detection in maritime data based on geometrical analysis of trajectories. In: 2015 18th International Conference on Information Fusion (Fusion), pp. 1100–1105. IEEE (2015)

52. Spaccapietra, S., Parent, C., Damiani, M.L., de Macedo, J.A., Porto, F., Vangenot, C.: A conceptual view on trajectories. Data Knowl. Eng. **65**(1), 126 – 146 (2008). https://doi.org/10.1016/j.datak.2007.10.008. http://www.sciencedirect.com/science/article/pii/S0169023X07002078. Including Special Section: Privacy Aspects of Data Mining Workshop (2006) - Five invited and extended papers

53. Strobl, C.: PostGIS, pp. 891–898. Springer US, Boston, MA (2008). https://doi.org/10.1007/978-0-387-35973-1_1012

54. Tampakis, P., Sideridis, S., Nikitopoulos, P., Pelekis, N., Theodoridis, Y.: Maritime data analytics. In: Artikis, A., Zissis, D. (eds.) Guide to Maritime Informatics, chap. 4. Springer, Berlin (2021)

55. Vaisman, A., Zimányi, E.: Mobility data warehouses. ISPRS Int. J. Geo-Inf. **8**(4) (2019). https://doi.org/10.3390/ijgi8040170

56. Vespe, M., Gibin, M., Alessandrini, A., Natale, F., Mazzarella, F., Osio, G.C.: Mapping eu fishing activities using ship tracking data. J. Maps **12**(sup1), 520–525 (2016)

57. Xu, J., Güting, R.H.: A generic data model for moving objects. Geoinformatica **17**(1), 125–172 (2013). https://doi.org/10.1007/s10707-012-0158-7
58. Xu, J., Lu, H., Guting, R.H.: Range queries on multi-attribute trajectories. IEEE Trans. Knowl. Data Eng. **30**(6), 1206–1211 (2018). https://doi.org/10.1109/TKDE.2017.2787711
59. Xu, T., Liu, X., Yang, X.: Ship trajectory online prediction based on bp neural network algorithm. In: 2011 International Conference on Information Technology, Computer Engineering and Management Sciences (ICM), vol. 1, pp. 103–106. IEEE (2011)
60. Yu, H., Fang, Z., Lu, F., Murray, A.T., Zhao, Z., Xu, Y., Yang, X.: Massive automatic identification system sensor trajectory data-based multi-layer linkage network dynamics of maritime transport along 21st-century maritime silk road. Sensors **19**(19) (2019). https://doi.org/10.3390/s19194197. https://www.mdpi.com/1424-8220/19/19/4197

Chapter 4
Maritime Data Analytics

Panagiotis Tampakis, Stylianos Sideridis, Panagiotis Nikitopoulos,
Nikos Pelekis, and Yannis Theodoridis

Abstract The goal of mobility data analytics is to extract valuable knowledge out of a plethora of data sources that produce immense volumes of data. Focusing on the maritime domain, this relates to several challenging use-case scenarios, such as discovering valuable behavioural patterns of moving objects, identifying different types of activities in a region of interest, estimating fishing pressure or environmental fingerprint, etc. In this chapter, we focus on the exploration, preparation of data and application of several offline maritime data analytics techniques. Initially, we present several methods that assist an analyst to explore and gain insight of the data under analysis. Subsequently, we study several preprocessing techniques that aim to clean, transform, compress and partition long GPS traces into meaningful portions of movement. Finally, we overview some representative maritime knowledge discovery techniques, such as trajectory clustering, group behaviour identification, hot-spot analysis, frequent route or network discovery and data-driven predictive analytics methods.

4.1 Introduction

In the recent years, there has been observed an "explosion" of trajectory data production due to the proliferation of GPS-enabled devices, such as mobile phones and tablets. In parallel, location broadcast has been made mandatory in several trans-

P. Tampakis (✉) · S. Sideridis · P. Nikitopoulos · N. Pelekis · Y. Theodoridis
University of Piraeus, Karaoli & Dimitriou St. 80, Piraeus, Greece
e-mail: ptampak@unipi.gr

S. Sideridis
e-mail: ssider@unipi.gr

P. Nikitopoulos
e-mail: nikp@unipi.gr

N. Pelekis
e-mail: npelekis@unipi.gr

Y. Theodoridis
e-mail: ytheod@unipi.gr

© Springer Nature Switzerland AG 2021
A. Artikis and D. Zissis (eds.), *Guide to Maritime Informatics*,
https://doi.org/10.1007/978-3-030-61852-0_4

portation means, such as vessels and aircrafts, for safe navigation purposes. This massive-scale data generation has posed new challenges in the data management community in terms of storing, querying, analyzing and extracting knowledge out of such data. Knowledge discovery from mobility data is essentially the goal of every mobility data analytics task. Especially in the maritime domain, this relates to challenging use-case scenarios, such as discovering valuable behavioural patterns of moving objects, identifying different types of activities in a region of interest, estimating fishing pressure or environmental fingerprint, etc. However, transforming this kind of data into an appropriate form for the application of data analytic tasks is not a straightforward task. During this transformation procedure, many aspects arise, such as noise elimination, compression and trajectory identification. After getting the data to an appropriate form, the analyst can apply several knowledge discovery techniques, such as trajectory clustering, group behaviour identification, hot-spot analysis, frequent route or network discovery and data-driven predictive analytics methods.

4.1.1 Mobility Data Analytics

Mobility data analytics aim to describe the mobility of objects, to extract valuable knowledge by revealing motion behaviours or patterns, to predict future mobility behaviours or trends and in general, to generate various perspectives out of data, useful for many scientific fields and applications, like communication science, public health, transportation, statistics and insurance, environment, etc.

At this point, it is important to provide some basic definitions, namely trajectory (point-based), trajectory (segment-based), and subtrajectory.

Trajectory (point-based). A trajectory r is a sequence of timestamped locations $\{r_1, \ldots, r_N\}$. Each $r_i = (p_i, t_i)$, where $p_i = (x_i, y_i)$, represents the ith sampled point of trajectory r, where N denotes the length of r (i.e. the number of points it consists of). The pair (x_i, y_i) and t_i denote the 2D location in the xy-plane and the time coordinate of point r_i respectively.

In order to simulate the continuous movement of objects, a different representation of a trajectory can be adopted, where a trajectory is represented by a 3-dimensional polyline.

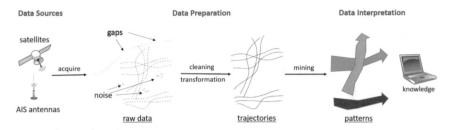

Fig. 4.1 Steps for mobility data analytics

Trajectory (segment-based). A trajectory r is a sequence of 3D line segments $\{r_1, \ldots, r_{N-1}\}$, where the 3rd dimension is time. Each $r_i = ((p_i, t_i), (p_{i+1}, t_{i+1}))$ represents the movement between (p_i, t_i) and (p_{i+1}, t_{i+1}), assuming linear interpolation.

Subrajectory. A subtrajectory $r_{i,j}$ is a subsequence of timestamped locations or 3D line segments $\{r_i, \ldots, r_j\}$ of trajectory r, which represents the movement of the object between t_i and t_j where $i < j$.

To serve its purpose, mobility data analytics follows a series of steps (Fig. 4.1). Having assured the collection and efficient storage of mobility data from various sources (satellites, AIS antennas, etc.), the analyst then should be familiarized with the mobility data in hand. For this purpose, the analyst may use a number of techniques, from statistical analysis to data visualization, to form a compact and complete picture of the available mobility (raw) data.

In the next step, the analyst, depending on the application requirements, proceeds to the appropriate preprocessing steps. The goal is to bring mobility data in a form that aids the usage of various processes and algorithms in order to answer a given question (e.g. which are the most frequent paths in a specific area of interest?). Data preparation is essential for successful mobility data analytics, since low-quality data typically result in incorrect and unreliable conclusions.

Mobility data are now ready for data mining methods that will satisfy the given application requirements. There are already several analytical methods and algorithms available from the scientific community (e.g. clustering, frequent pattern mining, prediction) and an analyst has the capability to employ some of the existing techniques or implement some ad-hoc solutions that better serve the problem needs.

Finally, the analyst, with the help of domain experts or with evaluation metrics, confirms and interprets the produced results. This step of evaluation and interpretation is necessary because each method or algorithm may have some "weak" points that may lead to erroneous results.

4.1.2 What Is Special About Maritime Data

Maritime data describe the movement of vessels in the sea. A vessel, equipped with AIS message transponders, periodically broadcasts messages that include *kinematic* information, such as its location, speed, course, heading, rate of turn, destination, estimated arrival, etc., as well as *static* information about the vessel characteristics, cargo, destination, ship type, etc.

A first observation about maritime data is the **reporting frequency** of messages. As far as analytics is concerned, low reporting frequency may result in difficulty to identify areas of interest (e.g. areas where vessel stays or stops for long time, or specific events take place) and thus limit the knowledge that can be gained. At the same time, a high reporting frequency creates a scalability problem. The frequency can vary a lot, from quite low (e.g. when they sailing at open seas) to high frequencies.

Regarding **timestamping** of messages, which is used as time reference for each incoming message, it sometimes occurs that the messages do not arrive chronologically ordered. It is presumed that this imperfection is an inherent problem caused by collecting data from various sources (terrestrial or satellite) or by combining input streams that are out of sync. Since several data processing algorithms absolutely rely on proper time ordering amongst incoming messages, this issue has to be tackled so as to provide coherent trajectory representations.

Vessel Identification. The attribute *MMSI* is present in every AIS message and serves as the identifier for each vessel and its respective trajectory. A 9-digit number is expected that will be given according to specifications. Certain errors exists in the *MMSI* code, such as when an operator inputs a simple, meaningless sequence like 123456789, or when the *MMSI* contains just the first three digits (representing the country of origin) followed by zeros. Having duplicate or invalid *MMSI* identifiers (which may result in a vessel appearing in two different locations at the same time) is a real scenario and problematic cases or inconsistencies, such as the above, should be resolved.

The **geolocation** of vessels in messages should be given in a consistent way using a common geo-reference system (e.g. WGS 1984) by all message sources. When this is not the case then locations may not be interpreted correctly and vessels may appear moving in the mainland (Fig. 4.2). Maritime data analytics may fail to produce valuable and reliable results if the issue of data reliability is not dealt adequately in a preprocessing step.

Deduplication of messages is another concern. It often appears that a position is reported twice from a vessel, i.e., identical coordinates at the same timestamp. Another problematic case is when the same message is received by two base stations with deferred clocks, so it may get different timestamps. Such problems necessitate additional filters against the incoming messages in order to discard duplicates.

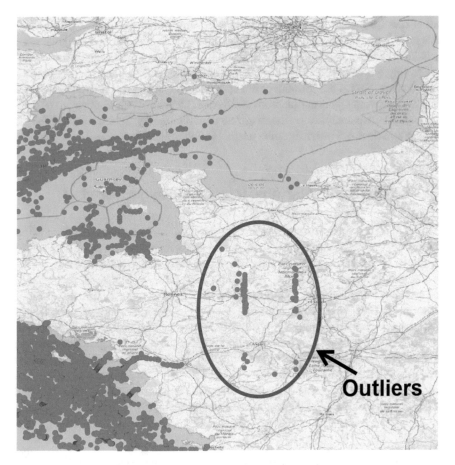

Fig. 4.2 Vessel position signals in the area of Brest, France [35]

Finally, another point to mark is that there is no fixed underlying network that vessels strictly follow (as in urban road networks). This, in combination with the aforementioned issues, may drive the analyst to various misleading assumptions.

4.1.3 Chapter Structure

The rest of the chapter is structured as follows. In Sect. 4.2, we discuss various methods for preparing the data for analytical purposes. In more detail, Sect. 4.2.1 presents some essential data exploration methods whose goal is to help the analyst to gain insight over the data of interest, while Sect. 4.2.2 focuses on cleaning the data and transforming them in an appropriate form for analytical purposes. Subsequently, Sect. 4.2.3, familiarizes the reader with ways to identify "meaningful" trajectories out of raw GPS traces. Section 4.3, presents several techniques for extracting valuable knowledge out of mobility data. More specifically, Sect. 4.3.1, overviews some popular trajectory clustering algorithms and Sect. 4.3.2 presents a group of methods aiming to discover several types of collective behaviour among moving objects. Furthermore, Sect. 4.3.3 introduces hot-spot analysis, Sect. 4.3.4 frequent route and maritime network discovery methods and Sect. 4.3.5 presents the application of data-driven predictive analytics on the maritime domain. Finally, Sect. 4.4 concludes the chapter and Sect. 4.5 provides some useful bibliographical notes.

4.2 Preparing the Maritime Data for Analytics Purposes

Getting maritime mobility data into an appropriate form, in order to perform analytical tasks, is not straightforward. To elaborate more, many aspects arise, such as the treatment of inaccurate or noisy data which can compromise the correctness of the results and the conclusions drawn from analytical tasks. Moreover, the reduction of the size of the data, without significant loss in quality and representational power, is important in order to deal with the storage and computational complexity challenges that may arise when applying many advanced data analytics tasks. Finally, the identification of trajectories, i.e. meaningful portions of movement (e.g. in the maritime domain from port to port), from raw sequences of sampled positions of vessels is crucial when performing several analytics tasks, such as frequent route discovery. The issues of noise in maritime data and stop detection, among others, are discussed at a more practical level in the chapter of Etienne et al. [13].

4.2.1 Getting Familiar with the Data

Before proceeding in any kind of preprocessing, the analyst has to "get a glimpse" and gain insight of the data under analysis. The first idea that comes naturally in mind when one is asked to explore a dataset that involves movement of objects is to visualize the data on a map layer. For this purpose, we may utilize the V-Analytics tool [4]. By doing this, we can quickly identify obvious spatiotemporal patterns and erroneous situations, such as errors in the reported location or timestamp and get a general idea of the dataset under consideration. In Figs. 4.3 and 4.4 we investigate two vessels that seem to have some abnormality, since they appear to travel over land.

In more detail, we can see in Fig. 4.3a that the highlighted point falls very far away from the route of the vessel. This can be depicted even better in Fig. 4.3b, where we also have the temporal dimension in the z-axis. In fact we can see that the highlighted point is very "close" temporally from its previous and next point but very "far" spatially, which indicates that we have an error in the reported location.

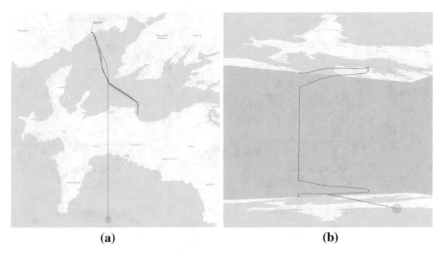

Fig. 4.3 An example trajectory with error in the reported location illustrated **a** in 2D and **b** in 3D (time as a third dimension in the z-axis)

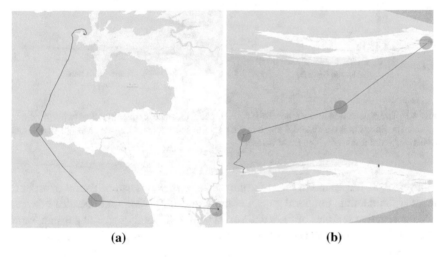

Fig. 4.4 An example trajectory with large temporal and spatial gap illustrated **a** in 2D and **b** in 3D (time as a third dimension in the z-axis)

Furthermore, Fig. 4.4a depicts that the consecutive highlighted points falls very far away spatially from each other. By examining also the temporal dimension in Fig. 4.4b, we can see that the highlighted points have very large temporal gaps in between them, which justifies the large spatial gaps.

However, simply visualizing a dataset is not enough. In order to gain real insight in a dataset, several statistics need to be calculated, such as the distribution of the temporal (dt) and/or spatial (ds) distance between two consecutive sampled positions

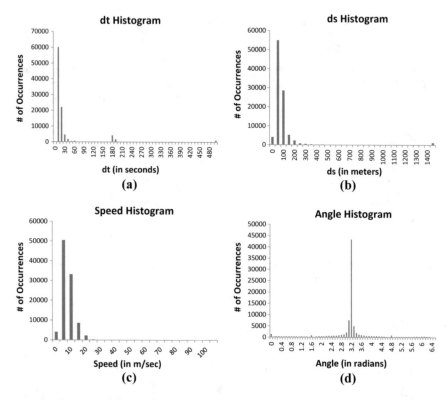

Fig. 4.5 The distribution of **a** temporal distance, **b** spatial distance, **c** speed between two consecutive sampled positions of a vessel and **d** angle made by three consecutive sampled positions of a vessel, as derived from a subset of the Brest dataset

of a vessel, as illustrated in Fig. 4.5a and b. These statistics can be useful in order to identify "unnatural" temporal gaps, i.e. more than 480 s temporal distance between two consecutive samples of the same vessel (Fig. 4.4a), and/or spatial gaps i.e. more than 1400 m spatial distance between two consecutive samples of the same vessel (Fig. 4.4b).

An additional interesting statistical measure is the distribution of the speed between two consecutive samples of a vessel which can assist the analyst to identify speeds that might be caused by erroneous situations (GPS/timestamping errors). Figure 4.5c illustrates this measure and it is obvious that speed higher than 30 m/s can be considered erroneous for this dataset. Another useful statistical measure is the distribution of the angle made by three consecutive samples, which can help the analyst to identify sharp turns that are not natural for maritime data and might be caused by errors. Figure 4.5d depicts that the majority of angles is between 2.4 and 5 rads (137°–229°), which is something anticipated since it shows that sharp turns (e.g., lower that 90° or higher than 270°) is a rare phenomenon.

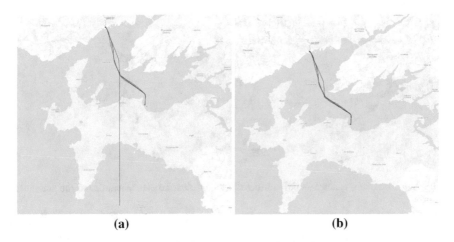

(a)   (b)

Fig. 4.6 An example trajectory with **a** a systematic error in the reported location and **b** with the error removed

4.2.2   Data Cleansing and Transformation

As already mentioned, having noisy and erroneous data is not a rare phenomenon when dealing with maritime data. However, when the goal is to perform analytic operations over this kind of data, having a cleansed and noise-free dataset is a pre-requisite. Errors, on the spatial position can be categorized into two types: **systematic errors** and **random errors**. **Systematic errors** are invalid GPS positioning readings, compared to the actual location of the vessel. This kind of errors occur, mainly, due to the insufficient number of satellites in view. A widely acceptable method to deal with systematic errors is the design and implementation of automated "filters" which identify and remove such "outliers". In this context, a parametric approach can be adopted, that filters out noisy positions by taking into account the maximum allowed speed of a vessel, V_{max}, in order to avoid being considered as noise. This can be achieved by utilizing a threshold in order to decide whether a reported timestamped position should be considered as noise or not. Such a threshold can be set by employing statistics, similar to the distribution of speed between two consecutive position signals, that was presented in Sect. 4.2.1. For example, recalling Fig. 4.5c we observe that a "normal" speed between two consecutive positions reaches up to 30 m/s (hence $V_{max} = 30$ m/s). Now, let us consider again the example displayed in Fig. 4.3, which is a systematic error. If we apply the aforementioned filter, then, as depicted in Fig. 4.6, the error is fixed due to the removal of the outlier point (actually the speed of the erroneous point was approximately 5000 m/s).

On the contrary, **random errors** are small distortions from the true values, up to a few meters that are caused by the satellite orbit, clock or receiver deficiency. Hence, the removal of such erroneous data is not necessary since their influence can be decreased by smoothing. These smoothing techniques may be classified based on

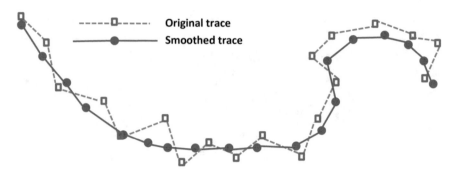

Fig. 4.7 Smoothing example—in blue (dotted) the original trajectory and in red the smoothed version

their statistical background. A first such approach targets to smooth out the trajectory by relying on a kernel-based smoothing method which adjusts the probability of occurrences in the data stream to modify outliers [41], as illustrated in Fig. 4.7.

Another approach smoothes data points by recursively modifying error values. To this end, a Kalman filter is employed, which uses measurements observed over time, that is the positions coming in the GPS receiver, and predicts positions that tend to be closer to the true values of the measurements. Eventually, the Kalman filter smoothes a position by computing the weighted average of the predicted position and the measured position. Figure 4.8 illustrates a smoothing example using Kalman filter, where the red trajectory is the original one and the blue (dotted) is the smoothed version of it.

The fact that mobility data analytics are, in general, computationally "heavy" operations can lead into a "bottleneck" in terms of storing, processing and extracting knowledge out of them. For this reason, the need for mechanisms that compress trajectories without losing their representational power is imperative. Generally, trajectory compression algorithms can be classified into four groups: top-down, bottom-up, sliding window, and open window. The top-down methods are based on recursively splitting the sequence of positions and only keeping the representative positions in each sub-sequence. The bottom-up algorithms start from the finest possible representation (i.e. the raw positions), and merge the successive points until some ending conditions are met. In the sliding window approach, data are compressed in a fixed window size; whilst in open window methods a dynamic and flexible data segment size is used. Both of the aforementioned window approaches can be directly applicable to streaming data.

A representative example of a top-down method is the Douglas–Peucker (DP) algorithm [11], with many subsequent extensions. Its main functionality is the reduction of the number of points in a curve that is approximated by a significantly smaller series of points, by maximizing the similarity between the original and the simplified version of a trajectory. Here, similarity is defined based on the maximum distance between the original curve and the simplified curve. Figure 4.9, illustrates a simpli-

Fig. 4.8 Smoothing example—in red the original trajectory and in blue (dotted) the smoothed one

fication example using the Douglas–Peucker method, where the original trajectory is depicted with blue and the simplified one with red. In this particular example, the original trajectory consists of 393 points, whereas the simplified one consists of 130 points only, thus we achieve a compression of 33% in the size of the trajectory.

An example of a bottom-up approach can be found in [19], where the goal is to identify and keep only the points of a trajectory where the behavior changes rapidly, called *characteristic* points. In order for this to be achieved, the minimum description length (MDL) principle is employed. In more detail, each trajectory is scanned point by point and at each point two measures are calculated, the MDL_{par}, that corresponds to making the hypothesis that this point is indeed a *characteristic* point and MDL_{nopar} that corresponds to making the hypothesis that this point is not a *characteristic* one. If $MDL_{par} \geq MDL_{nopar}$, then the specific point is considered *characteristic*. For more details on trajectory simplification methods, please refer to the chapter of Patroumpas [29].

Fig. 4.9 Simplification example using the Douglas–Peucker method- in blue the original trajectory and in red the simplified version

4.2.3 Getting from Locations to Trajectories

The input data of each vessel are long sequences of sampled positions (their duration might even be equal to the duration of the whole dataset i.e. several months or years). However, in order to extract valuable knowledge from mobility data it is crucial to partition these long sequences into meaningful portions of movement (e.g. from port to port in the maritime domain). The main problem to be tackled here is to define realistic identification rules in order to detect splitting points (p_i, t_i) that can divide the GPS feed into successive trajectories. Trajectory identification can be classified as follows.

Identification via raw gap: A sequence of spatiotemporal points can be divided into trajectories with respect to the raw GPS gaps, i.e. when the spatial distance or temporal duration between two consecutive timestamped positions exceeds a distance threshold G_{sp} or temporal threshold G_t, respectively. The specification of these parameters directly affects the trajectory identification procedure and may indirectly affect the results and the conclusions drawn from the data analytics methods that will be applied to the specific dataset. In order to determine these thresholds, we can utilize the statistics that where presented in Sect. 4.2.1. In this example, we observe with a naked eye that G_{sp} is approximately 500 m and G_t is 200 s. By applying the raw gap approach with the specific thresholds to the trajectory illustrated in Fig. 4.6b, we end up with the two trajectories depicted in Fig. 4.10.

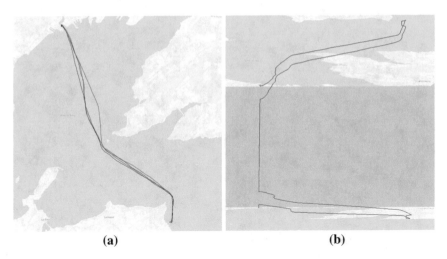

Fig. 4.10 Two identified trajectories (in red and blue) via the raw gap method **a** in 2D and **b** in 3D visualization

Identification by prior knowledge: Another approach is to utilize some existing knowledge that might be useful in the trajectory identification procedure. Such knowledge can be predefined time intervals or predefined spatial regions. Examples include vessels that perform scheduled routes during predefined intervals (e.g. week days from 8 a.m to 4 p.m) or trajectories from port to port.

Correlation-based identification: Correlation-based techniques can also be applied to discover spatiotemporal trajectories. Here, the partitioning of a cleansed trajectory solely depends on the correlations between the data points rather than prior knowledge. Hence, conventional time series segmentation methods could be applied for the identification of division points between two successive trajectories that correspond to the same vessel. To exemplify, assume a trajectory in a transformed space, where instead of a time series of spatial positions we have a time series of relationships between spatial positions, like dt, ds, speed or angle presented in Sect. 4.2.1. We can then apply traditional time series segmentation techniques (e.g. [27, 28]) in order to segment trajectories based on the specific feature.

Identification via stop detection: A different approach that tries to identify meaningful portions of movement out of sequences of spatiotemporal points is inspired by the notion that a trajectory is defined as the movement of an object between two "stops". Hence, the problem of trajectory identification is transformed to a stop detection problem. A stop is considered to be a long lasting event rather than an instantaneous one, since a stop with insignificant duration, less than threshold δt, might be coincidental. Having this in mind, given a sequence of spatiotemporal points of a vessel, one may identify portions of movement where speed v is significantly low for a duration of at least δt. Figure 4.11, illustrates the application of the

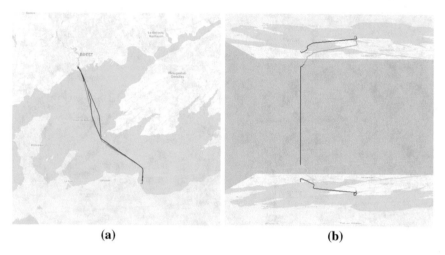

(a)   (b)

Fig. 4.11 Five identified trajectories via the stop detection method **a** in 2D and **b** in 3D visualization

trajectory identification via stop detection over the data of Fig. 4.6b. There are five trajectories identified, in red, yellow, blue, green and purple.

Another line of research utilizes density-based spatial clustering algorithms, such as DBSCAN [12] and OPTICS [5], in order to identify areas of large concentration of points within a vessel's movement. These areas may be considered as stops, since the spatial projection of objects moving with minimal speed for significantly long periods of time resembles a spatial cluster of points. Figure 4.12 illustrates such an example, where we apply the OPTICS algorithm over the trajectory of Fig. 4.12a and result in discovering two clusters/stops which define some regions (in this example these regions are located near ports). By taking advantage of this knowledge we can segment the trajectory, as in Fig. 4.11b, whenever a vessel arrives or leaves such a region. The advantage of this line of techniques is that they do not rely on user defined parameters and do not require knowledge of the statistics of the dataset. However, since they do not take into account the duration of a stop, they might falsely identify as stops periods where a vessel moves with low speed and reports its position very frequently.

4.3 Discovering Knowledge from Maritime Data

Knowledge discovery from mobility data is essentially the goal of every mobility data analytics task. Discovering valuable behavioural patterns of vessels that may be exploited in many fields, including the maritime domain, is of great value. In order to unveil this "hidden" knowledge, mobility data mining may be performed. The kind of mining procedure depends on the application scenario and the type of knowledge the analyst targets to identify. A line of research aims at identifying

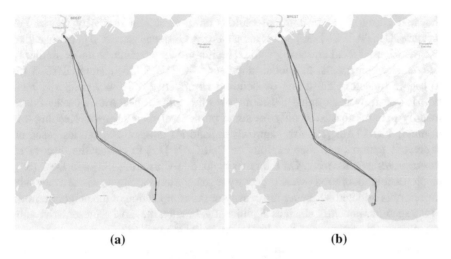

(a)   **(b)**

Fig. 4.12 Identification of stops through the application of OPTICS

various types of clusters of vessels. Some methods identify clusters that are valid for the entire lifespan of the vessels, while others try to discover patterns that are valid only for a portion of their lifespan. Another branch of related work aims to discover several types of collective behaviour among vessels, forming groups of objects that move together for a certain time period. In a different line of research, hot-spot analysis is the process of partitioning the spatiotemporal space in uniform 3D cells and identifying statistically significant clusters. Especially in the maritime domain, this relates to significant challenging use-case scenarios, such as identifying different types of activities in a region of interest, estimating fishing pressure or environmental fingerprint, etc. Although clustering-oriented approaches prevail in the literature, there are many other interesting techniques that aim to discover valuable knowledge from mobility data; among them, in this chapter we discuss frequent route discovery, network discovery and predictive analytics.

4.3.1 Cluster Analysis

Cluster analysis aims at identifying clusters of vessels and may assist the analyst to detect vessels that demonstrate abnormal behaviour. Discovering clusters of vessels is an important operation since by doing so, the underlying hidden patterns of collective behaviour can be unveiled.

A typical approach to address the trajectory clustering problem is to either transform trajectories to vector data and directly apply some off-the-shelf clustering algorithm (e.g. k-means) or define an appropriate trajectory similarity function, which is the basic building block of every clustering approach, and then utilize some off-the-shelf clustering algorithm that exploits on the resulting similarity matrix. For instance, CenTR-IFCM [30] builds upon a Fuzzy C-Means variant to perform a kind of time-focused local clustering using a region-growing technique under similarity and density constraints. In another approach, T-OPTICS [23] adapts OPTICS [5] to enable trajectory clustering by defining an effective trajectory similarity metric. However, by performing clustering over the entire trajectories we may fail to identify patterns that exist only for some portions of their lifespan. Keeping this in mind, TRACLUS [19] first segments trajectories into sub-trajectories based on geometric features and then exploits on DBSCAN [12] to cluster the discovered sub-trajectories clustering. On the other hand, S^2T-Clustering [32] and DSC [39], employ a neighbourhood-aware trajectory segmentation technique, where a new sub-trajectory is initiated whenever the density or the composition of the neighbourhood of the specific trajectory changes significantly. Finally, the most representative sub-trajectories are selected and the clusters are built "around" them.

T-OPTICS: One way to perform trajectory clustering is to define a similarity function, as a variant of the DISSIM [15] distance function, and use it to group trajectories by employing some well-known generic clustering algorithm. A typical case of such an approach is T-OPTICS which defines a similarity measure between trajectories and utilizes OPTICS, a well-known density-based clustering algorithm. To exemplify, let us apply the T-OPTICS algorithm over the dataset of Fig. 4.13a, which is prepared for analytic purposes according to the techniques presented in Sect. 4.2. The resulting clusters of trajectories discovered by T-OPTICS are illustrated in Fig. 4.13b.

TRACLUS: Initially, the algorithm partitions trajectories to directed segments (i.e., subtrajectories) whenever the behaviour of a trajectory changes rapidly, called *characteristic* points, by employing the minimum description length (MDL) principle, which was discussed in Sect. 4.2.2. Subsequently, the resulting sub-trajectories are grouped by employing a modified version of the DBSCAN algorithm, which is applicable to directed segments. Finally, for each identified cluster the algorithm calculates a fictional representative trajectory, which is a (more or less) smooth linear trajectory that best describes the corresponding cluster.

S^2**T-Clustering:** The objective of the S^2T-Clustering method is to partition trajectories into sub-trajectories and then form groups of similar ones, while at the same time separate the ones that fit into no group, called outliers. It consists of two phases:

- a Neighborhood-aware Trajectory Segmentation (NaTS) phase, where the trajectories are split to sub-trajectories by applying a voting and segmentation process that detects homogeneous sub-trajectories w.r.t. the density of their "neighborhood".
- a Sampling, Clustering and Outlier (SaCO) detection phase, where the most representative sub-trajectories are selected to serve as the seeds of the clusters, around which the clusters are formed (also, the outliers are isolated).

To exemplify, let us apply the S^2T-Clustering algorithm over the dataset of Fig. 4.13a. In Fig. 4.13c and d, we present the resulting clusters after the application of the S^2T-Clustering algorithm over the entire trajectories and the subtrajectories, respectively.

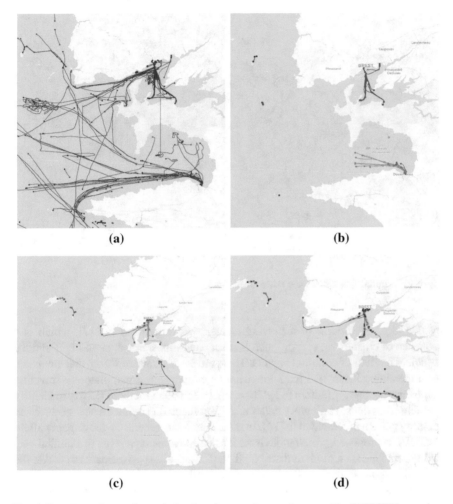

Fig. 4.13 **a** Input dataset for analytics, **b** trajectory clusters discovered by T-OPTICS, **c** entire-trajectory and **d** sub-trajectory clusters discovered by S^2T-Clustering

4.3.2 Group Behaviour

A similar line of work, aims at discovering several types of collective behaviour among vessels, forming groups of objects that move together for a certain time period. One of the first approaches for identifying such collective mobility behaviour is the so-called flock pattern [6], which takes into account the spatial proximity and the direction of the vessels. For such a pattern to be captured, a minimal number of objects that satisfy such constraints are required. Among the most related to this work, in [18], the authors define various mobility behaviours around the idea of the flock pattern, such as the meeting, convergence and encounter patterns. Inspired

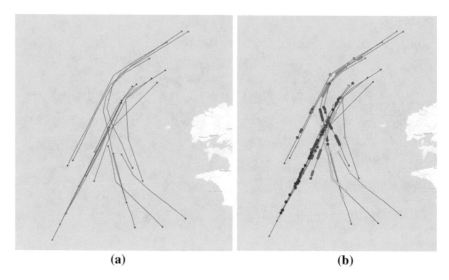

(a) (b)

Fig. 4.14 **a** Initial Trajectories and **b** Identified Flocks

by this idea, the notion of a moving cluster was introduced in [17], which is a sequence of clusters $c_1, \ldots, c_k$, such that for each timestamp i, c_i and c_{i+1} share a sufficient number of common objects. There are several related works that emanated from the above ideas, like the approaches of convoys, swarms, platoons, traveling companion, gathering pattern [42]. There are several other methods that try to identify frequent (thus dense) trajectory patterns. An approach that defines a new generalized mobility pattern is presented in [14]. In more detail, the general co-movement pattern (GCMP), is proposed, which models various co-movement patterns in a unified way and is deployed on a modern distributed platform (i.e., Apache Spark) to tackle the scalability issue. Recently, GCMP was adapted [9] to detect the general co-movement pattern in an online fashion considering a real-time and streaming environment, by utilizing Apache Flink.

Flocks: A flock consists of at least m objects that moved together within a disk of radius r for a time interval I, which spans for at least k successive timepoints. Figure 4.14a depicts a subset of the dataset upon which we apply the flock discovery algorithm and Fig. 4.14b depicts the identified flocks in different color each.

Moving Clusters: Another interesting pattern is the so-called "moving cluster", which is actually a generalization of the flock pattern. As mentioned, the flock pattern requires that the same m entities stay together during the entire interval I. On the other hand, if the entities change during the given interval, a kind of varying-flock is formed, which is called a moving cluster. Let us consider a set of snapshots of the positions of vessels. A snapshot S_i corresponds to the vessels' positions at time t_i, which gets clustered by employing a typical density-based spatial clustering algorithm, like DBSCAN, in order to discover dense groups of vessels in S_i. By doing this also for

S_{i+1}, we are able to identify two spatial clusters at two consecutive snapshots that have a large percentage of common vessels, which is considered as a moving cluster between these t_i and t_{i+1}.

Convoys: Having a disk of radius r in the definition of flocks poses some restrictions that concern the shape and the extend of the discovered patterns. In order to overcome these restrictions a number of follow-up techniques have been proposed, such as convoys. A convoy is a group of objects that has at least m objects, which are density-connected with respect to a distance threshold e, during k consecutive timepoints. The difference between convoys and flocks is that flocks are not restricted to a disk of radius r, rather they are density-connected in a DBSCAN fashion. Furthermore, convoys are different from moving clusters since they require that the same m objects should be found in consecutive clusters.

4.3.3   Hot-Spot Analysis

Hot-spot analysis is the process of identifying statistically significant clusters, i.e. clusters which have unusually high or low concentration of a specific attribute's values. In geospatial hot-spot analysis, a statistically significant geographic region is defined as a region which has a very low probability of containing a specific observation. For example, a region is identified as a hot-spot if it contains a high concentration of vessels which are moving in unusually high speeds. Hot-spot analysis may be of interest in the maritime domain, since it can discover regions with unusually high intensity of fishing activities, or measure the environmental fingerprint on regions with unusually high density of moving vessels.

Hot-spot analysis should not be confused with clustering when examining the density of groups of data. The computation of density provides information where clusters in the data exist, but not whether the clusters are statistically significant; that is the outcome of hot-spot analysis. Studies in geospatial hot-spot analysis, use spatial statistics like Moran's I [22] or Getis-Ord Gi* [26] to discover statistically significant regions. These statistics aim to calculate z-scores for a set of predetermined regions of interest, by measuring their spatial autocorrelation based on their attribute values. These z-scores can then be converted to p-values which quantify the probability of regions to be statistically significant; a small p-value (typically less than 0.05) indicates strong evidence against the null hypothesis, so that the region can be characterized as statistically significant.

The hot-spot analysis problem has recently been studied for spatio-temporal data [16, 20], where a hot-spot is defined as a spatial region having low probability value of occurrence for a certain amount of time. Efficient hot-spot analysis algorithms for spatio-temporal point data have been proposed in [21, 25]. Both studies apply the Getis-Ord formula over point spatio-temporal data, to efficiently discover hot-spots in parallel. These algorithms calculate an attribute value for each spatio-temporal region. Then they apply the Getis-Ord statistic, which calculates

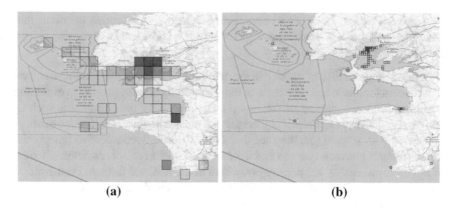

(a)   (b)

Fig. 4.15 Hot-spots on **a** large regions and **b** small regions

a z-score for each cell, by aggregating the neighboring regions' attribute values. Higher z-scores express higher probability of a spatio-temporal region to be a hot-spot. Hence, based on the Getis-Ord statistic, the statistical significance of a region, is not only dependent on the region's attribute value, but also on the attribute values of its neighborhood.

Trajectory Hot-Spot: More recently, Hot-Spot analysis has been used to discover statistically significant regions based on trajectory data [24], instead of spatio-temporal data. This approach identifies hot-spots, by evaluating trajectory congestion in spatio-temporal regions, based on a user-defined partitioning of the spatio-temporal data space in 3D cells; the attribute value of a cell, equals to the sum of the lifespans of moving vessels inside that cell. Hence, a cell has higher attribute value if more moving objects are present inside that particular 3D cell. Moreover, the effect of a cell's attribute value to the z-score of another cell, may decay based on their distance; cells that are located far away from each other, have small effect to each others' z-scores, when compared to closer cells. In [24] two parallel solutions are proposed: THS which exhaustively aggregates the attribute values of all cells to calculate their z-scores, thus providing exact results with increased computation and communication cost; and aTHS, which is an approximate but efficient solution with error guarantees.

Figure 4.15 demonstrates the top-50 hot-spots discovered by the THS algorithm for a data set covering the Brest area, based on a user-defined grid. Each hot-spot is a region defined by a rectangular cell, which is part of the grid provided for the entire data set. The size of the cells in Fig. 4.15a is 0.05 degrees in both longitude and latitude dimensions, while in Fig. 4.15b the cell size is configured to be 0.01 degrees. The color of the cells corresponds to the probability of the cell to be a statistically significant region, based on the total lifespans of the containing moving vessels; a red cell has strong evidence, while a more opaque cell has weaker evidence (but still stronger than most other cells of the rest of the grid).

4.3.4   Frequent Route—Network Discovery

The availability of vessel's trajectories, derived from maritime mobility data as described in previous sections, can be used for extracting the frequent vessels routes. A domain expert can exploit these patterns of motion as a tool to evaluate nautical maps or take further actions to protect particularly sensitive sea areas (like the Canary Islands, Spain or the Strait of Bonifacio, France and Italy) or even to predict how, why and where future routes will take place and set, proactively, necessary rules such as prevention of pollution or traffic rules.

A naive approach would be to apply a clustering algorithm on trajectories in order to identify frequent patterns of movement. However such an algorithm should hold some properties, such as the ability to group trajectories that differ in the temporal dimension (e.g. different days), which excludes spatio-temporal trajectory clustering algorithms. On the other hand, purely spatial trajectory clustering algorithms are not sufficient either, due to the fact that they do not take into account the speed of the objects. Hence, it is obvious that in order to be able to identify frequent route patterns by employing trajectory clustering is not a trivial task, since a tailor-made trajectory clustering technique should be devised and utilized in order to achieve this.

Another approach [1], cartographic generalization methods can be used, like aggregation, where several trajectories or sub-trajectories are put together and represented as a single unit. Aggregation reduces the number of trajectories or sub-trajectories from further processing, which is very helpful in case of numerous trajectories. At the same time, it does not just omit some items, but transforms the original items into a smaller number of constructs that summarize the properties of the original items. This method is relatively fast and provides a good overview of the routes most vessels are using.

Consider the vessel trajectories in a specific square area in Brest, France, illustrated in Fig. 4.16a. The number of distinct displayed trajectories in the specific area of interest is 950. To apply the above method of aggregation, first characteristic points[1] are extracted (more details about this methodology in [1]) from trajectories. Then characteristic points are grouped by spatial proximity. This grouping of characteristic points is realized by applying a density-based clustering algorithm, a modified version of DBSCAN [12] or OPTICS [5]. From these groups of points, centroids (average points) are extracted and used as generating points for a Voronoi tessellation.

A Voronoi tessellation is a partitioning of a plane into regions based on distance to points in a specific subset of the plane. That set of points (called seeds, sites, or generators) is specified beforehand, and for each seed there is a corresponding region consisting of all points closer to that seed than to any other. These regions are called Voronoi cells.

The resulting Voronoi cells (Fig. 4.16b) are used for partitioning the territory into appropriate disjoint areas (the cells found) and with that partition, trajectories

---

[1]Characteristic points of trajectories include their start and end points, the points of significant turns, and the points of significant stops (pauses in the movement). If a trajectory has long straight segments, it is also necessary to take representative points from these segments too.

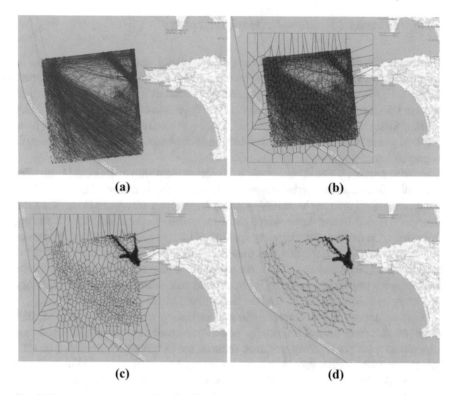

Fig. 4.16 Frequent (aggregated) routes discovery

are divided into segments. Finally, for every pair of such areas, trajectory segments are aggregated (Fig. 4.16c). The result of the aggregated trajectory segments (after setting the number of moves, that is the number of transitions between pair of areas just found, between 15 and 350 as a filter) is visualized in Fig. 4.16d.

A more robust approach, providing a solid foundation on the sequential nature of the route patterns to be found and adding more insights, is to apply a mining technique that can reveal frequent sequential patterns. Trajectories of vessels are essentially composed of sequences of multidimensional points. These points may additionally hold spatial, temporal and semantic information. So the challenge here is; having all these trajectories, how to discover or reveal frequent motion patterns which then constitute frequent routes.

Sequential pattern mining is an important data mining problem with broad applications. Its goal is, given a set of sequences where each sequence consists of a list of items and given a user-specified *min_support* threshold, find all frequent sub-sequences. Specifically, the sub-sequences whose occurrence frequency in the original set of sequences is no less than *min_support*. Additionally, further constraints may be applied to the sequential pattern mining process to allow finding more inter-

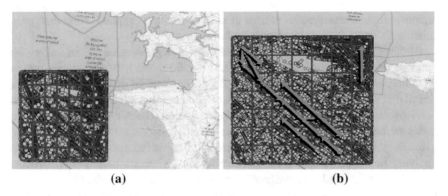

Fig. 4.17 Patterns found, with *min_support* over 80

esting patterns or to indicate more precisely the types of pattern to be found (e.g. time constraints in between two consecutive items in a pattern).

One such method follows the pattern-growth approach, in which the search space is explored using a depth-first search, like in [8]. The approach starts from sequential patterns of length one and proceeds by recursively appending items to patterns to create longer pattern sequences (pattern-growth).

The aforementioned method of finding frequent sequential patterns can be applied to the maritime domain also, as trajectories are by nature sequences of points, but not directly. The reason is that points of different trajectories are distinct from each other depending on the accuracy of their coordinates (i.e. it is extremely rare that points of different trajectories have exactly the same coordinates), so there are no common starting and destination "regions" (group of points) but rather trajectory points might be found scattered in an area and thus no common motion patterns between pairs of starting and destination regions can be directly found. Additionally, vessel position signals are usually numerous and thus make the application of the process quite slow.

To deal with this issue one may use "abstraction". In general, with abstraction we can see a very detailed level from a higher perspective with less details and without losing necessary information for our purposes. For example a very detailed and complicated map where all vessel position points are shown, can be seen as a map with less details as grid cells where many points are assigned into one cell. Then the size of these cells determine the information loss we can afford.

Assigning points to regions (when it comes to spatial or spatio-temporal coordinates) or to more general semantic categories (for example "near-port", "open-sea", "near-land", etc.) when it comes to semantic attributes, hides complicated details and reveals common "areas" for points from different trajectories and thus the method of sequential pattern mining becomes applicable.

Consider again, vessel trajectories in an area of Brest (as defined in Sect. 4.1.1). In Fig. 4.17a all points of the area are illustrated on top of a regular grid of equally sized cells, that assigns each point to a larger region (its cells). With that abstraction, trajectories of points are transformed from sequences of points into sequences of regions, where the sequential pattern mining algorithm is applied. Note that, more suitable abstractions may be chosen for different cases.

Another benefit of frequent route discovery is the revelation of the underlying network that vessels follow. In Sect. 4.2.3, we discussed how trajectories can be identified via stop detection. These stops could then constitute the nodes of a network. Recall that stops are clusters of points that can potentially replace the grid cells of the previous method. Moreover, if mainland information (coastline maps) is available, nodes can be further categorized as ports-stops and sea-stops, to provide more insights to the domain expert. It then comes natural that frequent routes of vessels, found by applying the frequent sequential mining methodology, become the edges of a network as now instead of grid cells we have stop clusters as "cells" and the trajectories are transformed from sequences of vessel position points into sequences of stop clusters.

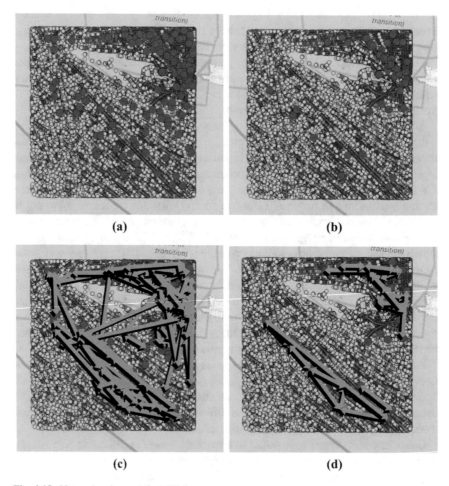

(a)   (b)

(c)   (d)

Fig. 4.18 Network using weight 1500 for stops

Again, in the same area of Brest, applying a density-based clustering algorithm such as DBSCAN or OPTICS, as described in Sect. 4.2.3, stops are detected. In Fig. 4.18a, stops found are illustrated (the red vertices) by using weight of 500 (the weight here is the cluster membership, that is, clusters that have less members are discarded and not taken as valid stops) and in Fig. 4.18b by using weight of 1500. After the network nodes are found, edges are discovered by applying a frequent sequential pattern method and a network is formed. Similar to the weight constraint imposed on stops clusters, an analogous constraint in the discovery of edges can be imposed (which reflects the *min_support* of the pattern). Note, that in practice a hierarchy of networks is formed depending on the weight parameters.

4.3.5 Data-Driven Predictive Analytics

Predictive analytics is a scientific domain that aims at the extraction of valuable knowledge from data and the utilization of it in order predict future behavioural patterns and trends. When dealing with data that represent the movement of objects, predictive analytics can be of great importance since it can assist an analyst to predict events, such as collision, traffic, etc. Especially, in the maritime domain, vessels do not always move over the network, thus the need for having tools that can predict the future location of a vessel is very important.

Towards this direction there have been several efforts, such as [33, 34, 40], that try to predict the future location of an object by utilizing extracted mobility patterns from historical data. The general approach that these future location prediction techniques follow, is first to identify popular mobility patterns, either global (for the whole dataset) or local (for each vessel separately), by employing some clustering technique

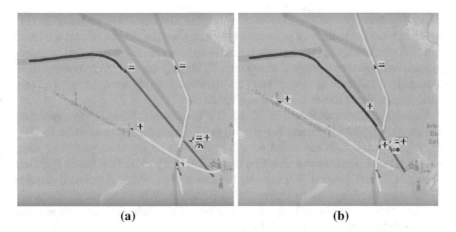

(a)   (b)

Fig. 4.19 Future location prediction of a vessel at different temporal snapshots. In red the actual movement and in grey the predicted

that also provides the cluster representatives, such as [32, 39] or the techniques presented in Sect. 4.3.4. Then, when some new positions of a vessel are reported, the goal is to "match" the new portion of movement with the most similar historical patterns and employ this pattern in order to predict its future location. Figure 4.19 illustrates an example of future location prediction of a trajectory at different temporal snapshots. The actual (predicted) movement of the object is depicted in red (grey, respectively).

4.4 Conclusions

In this chapter, we focused on the preparation of maritime data and the presentation and actual application of several maritime data analytics techniques. Initially, the particularities of maritime data, in terms of their usage for data analytic purposes, were presented. Then, we overviewed several ways that an analyst can familiarize with and gain insight in the data under analysis. Then, we showed ways on how to eliminate noise, compress, and reconstruct trajectories. Thereafter, we focused on ways to extract knowledge out of such data. In more detail, we presented several trajectory clustering, collective behaviour identification techniques, hot-spot analysis, frequent route/network discovery and data-driven predictive analytics methods. The overall message is that maritime data is rich in underlying knowledge and several techniques can be found out there in order for this knowledge to be discovered and be exploited in real-world applications.

4.5 Bibliographical Notes

Regarding data preparation, the Douglas–Peucker algorithm was proposed in [11] which reduces the number of points and thus simplifies the amount of data needed to describe a trajectory. In [19] an approach to keep only characteristic points of trajectories is developed as a pre-processing step. To identify areas of large concentration, that can be considered as "stops" and thus identify where trajectories begin and end, clustering methods like in [5, 12] are utilized.

Regarding cluster analysis, in [5, 12, 23] density based clustering algorithms are formed to provide interesting groupings of whole-trajectories, while in [19, 31, 32, 38] groupings are based on sub-trajectories or segments providing more refined groups. Towards the direction of sub-trajectory clustering but in Big Data setting, [39] builds upon [37] in order to provide a a scalable distributed solution by utilizing the MapReduce programming paradigm. When uncertainty occurs in the data, techniques like in [15] are more appropriate. Clustering algorithms using various distance functions as described in [15].

Behavioral explanation of movement gains more insights when groups of moving objects are taken into account together. Various such mobility behaviors are defined in [18] and explored in [42]. The concept of moving clusters was introduced in [17] and that of collocation patterns in [7]. Algorithms for discovering most popular routes from trajectories proposed in [10] and online approaches also proposed, like in [36].

Hot-spot analysis discovers statistically significant clusters. Spatial statistics were used in [22] and [26] while the spatio-temporal case has been studied in [16] and [20]. Furthermore, efficient parallel hot-spot analysis algorithms on point spatio-temporal hot-spots are proposed in [21, 25]. Recently, in [24] statistics were used to discover hot-spots on trajectory data.

In frequent route discovery, visual methods are described in [3], while more advanced mining techniques for discovering frequent sequential patterns, like in [2, 8], utilized to get a deep understanding of frequent motion patterns.

Predictive analytics to extract valuable knowledge for future location prediction of moving vessels are truly needed by domain experts to take pro-active actions in various scenarios. Efforts like [40] try to predict future location based on mobility patterns or profiles.

References

1. Adrienko, N., Adrienko, G.: Spatial generalization and aggregation of massive movement data. IEEE Trans. Visual. Comput. Graph. **17**(2), 205–219 (2011)
2. Agrawal, R., Srikant, R.: Mining sequential patterns. In: Proceedings of ICDE, pp. 3–14 (1995)
3. Andrienko, G., Andrienko, N., Bak, P., Keim, D., Wrobel, S.: Visual Analytics of Movement. Springer Publishing Company, Incorporated (2013)
4. Andrienko, G.L., Andrienko, N.V., Bak, P., Keim, D.A., Wrobel, S.: Visual Analytics of Movement. Springer, Berlin (2013)
5. Ankerst, M., Breunig, M.M., Kriegel, H., Sander, J.: OPTICS: ordering points to identify the clustering structure. In: Proceedings of the ACM SIGMOD, pp. 49–60 (1999)
6. Benkert, M., Gudmundsson, J., Hübner, F., Wolle, T.: Reporting flock patterns. Comput. Geom. **41**(3), 111–125 (2008)
7. Cao, H., Mamoulis, N., Cheung, D.W.: Discovery of collocation episodes in spatiotemporal data. In: Proceedings of ICDM, pp. 823–827 (2006)
8. Chen, H.P., Dayal, U., Hsu, M.: Prefixspan: mining sequential patterns efficiently by prefix-projected pattern growth. In: Proceedings of ICDE (2001)
9. Chen, L., Gao, Y., Fang, Z., Miao, X., Jensen, C.S., Guo, C.: Real-time distributed co-movement pattern detection on streaming trajectories. Proc. PVLDB **12**(10), 1208–1220 (2019)
10. Chen, Z., Shen, H.T., Zhou, X.: Discovering popular routes from trajectories. In: Proceedings of ICDE, pp. 900–911 (2011)
11. Douglas, D.H., Peucker, T.K.: Algorithms for the reduction of the number of points required to represent a digitized line or its caricature. Cartographica **10**(2), 112–122 (1973)
12. Ester, M., Kriegel, H., Sander, J., Xu, X.: A density-based algorithm for discovering clusters in large spatial databases with noise. In: Proceedings of KDD, pp. 226–231 (1996)
13. Etienne, L., Ray, C., Camossi, E., Iphar, C.: Maritime data processing in relational databases. In: Artikis, A., Zissis, D. (eds.) Guide to Maritime Informatics (chap. 3). Springer, Berrlin (2021)
14. Fan, Q., Zhang, D., Wu, H., Tan, K.: A general and parallel platform for mining co-movement patterns over large-scale trajectories. Proc. PVLDB **10**(4), 313–324 (2016)
15. Frentzos, E., Gratsias, K., Theodoridis, Y.: Index-based most similar trajectory search. In: Proceedings of ICDE, pp. 816–825 (2007)

16. Hong, L., Zheng, Y., Yung, D., Shang, J., Zou, L.: Detecting urban black holes based on human mobility data. In: Proceedings of SIGSPATIAL (2015)
17. Kalnis, P., Mamoulis, N., Bakiras, S.: On discovering moving clusters in spatio-temporal data. In: Proceedings of SSTD, pp. 364–381 (2005)
18. Laube, P., Imfeld, S., Weibel, R.: Discovering relative motion patterns in groups of moving point objects. Int. J. Geograph. Inf. Sci. 19(6), 639–668 (2005)
19. Lee, J., Han, J., Whang, K.: Trajectory clustering: a partition-and-group framework. In: Proceedings of the ACM SIGMOD, pp. 593–604 (2007)
20. Lukasczyk, J., Maciejewski, R., Garth, C., Hagen, H.: Understanding hotspots: a topological visual analytics approach. In: Proceedings of SIGSPATIAL (2015)
21. Makrai, G.: Efficient method for large-scale spatio-temporal hotspot analysis. In: Proceedings of SIGSPATIAL (2016)
22. Moran, P.: Notes on continuous stochastic phenomena. Biometrika 37(1), 17–23 (1950)
23. Nanni, M., Pedreschi, D.: Time-focused clustering of trajectories of moving objects. J. Intell. Inf. Syst. 27(3), 267–289 (2006)
24. Nikitopoulos, P., Paraskevopoulos, A., Doulkeridis, C., Pelekis, N., Theodoridis, Y.: Hot spot analysis over big trajectory data. In: Proceedings of IEEE BigData, pp. 761–770 (2018)
25. Nikitopoulos, P., Paraskevopoulos, A.I., Doulkeridis, C., Pelekis, N., Theodoridis, Y.: BigCAB: distributed hot spot analysis over big spatio-temporal data using Apache Spark. In: Proceedings of SIGSPATIAL (2016)
26. Ord, J.K., Getis, A.: Local spatial autocorrelation statistics: distributional issues and an application. Geograph. Anal. 27(4), 286–306 (1995)
27. Panagiotakis, C., Kokinou, E., Vallianatos, F.: Automatic p-phase picking based on local-maxima distribution. IEEE Trans. Geosci. Remote Sens. 46(8), 2280–2287 (2008)
28. Panagiotakis, C., Tziritas, G.: A speech/music discriminator based on RMS and zero-crossings. IEEE Trans. Multimedia 7(1), 155–166 (2005)
29. Patroumpas, K.: Online mobility tracking against evolving maritime trajectories. In: Artikis, A., Zissis, D. (eds.) Guide to Maritime Informatics (Chap. 6). Springer, Berlin (2021)
30. Pelekis, N., Kopanakis, I., Kotsifakos, E.E., Frentzos, E., Theodoridis, Y.: Clustering uncertain trajectories. Knowl. Inf. Syst. 28(1), 117–147 (2011)
31. Pelekis, N., Tampakis, P., Vodas, M., Doulkeridis, C., Theodoridis, Y.: On temporal-constrained sub-trajectory cluster analysis. Data Min. Knowl. Discov. 31(5), 1294–1330 (2017)
32. Pelekis, N., Tampakis, P., Vodas, M., Panagiotakis, C., Theodoridis, Y.: In-dbms sampling-based sub-trajectory clustering. In: Proceedings of EDBT, pp. 632–643 (2017)
33. Petrou, P., Nikitopoulos, P., Tampakis, P., Glenis, A., Koutroumanis, N., Santipantakis, G.M., Patroumpas, K., Vlachou, A., Georgiou, H.V., Chondrodima, E., Doulkeridis, C., Pelekis, N., Andrienko, G.L., Patterson, F., Fuchs, G., Theodoridis, Y., Vouros, G.A.: ARGO: a big data framework for online trajectory prediction. In: Proceedings of SSTD, pp. 194–197 (2019)
34. Petrou, P., Tampakis, P., Georgiou, H., Pelekis, N., Theodoridis, Y.: Online long-term trajectory prediction based on mined route patterns. In: MASTER workshop in conjuction with ECML/PKDD (2019)
35. Ray, C., Dreo, R., Camossi, E., Jousselme, A.L., Iphar, C.: Heterogeneous integrated dataset for maritime intelligence, surveillance, and reconnaissance. Data in Brief p. 104141 (2019)
36. Sacharidis, D., Patroumpas, K., Terrovitis, M., Kantere, V., Potamias, M., Mouratidis, K., Sellis, T.K.: On-line discovery of hot motion paths. In: Proceedings of EDBT, pp. 392–403 (2008)
37. Tampakis, P., Doulkeridis, C., Pelekis, N., Theodoridis, Y.: Distributed subtrajectory join on massive datasets. ACM Trans. Spatial Algorithms Syst. 6(2), 8:1–8:29 (2020)
38. Tampakis, P., Pelekis, N., Andrienko, N., Andrienko, G., Fuchs, G., Theodoridis, Y.: Time-aware sub-trajectory clustering in hermes@postgresql. In: 34th IEEE International Conference on Data Engineering, ICDE 2018, Paris, France, April 16–19, 2018, pp. 1581–1584 (2018)
39. Tampakis, P., Pelekis, N., Doulkeridis, C., Theodoridis, Y.: Scalable distributed subtrajectory clustering. In: 2019 IEEE International Conference on Big Data (Big Data), pp. 950–959 (2019)
40. Trasarti, R., Guidotti, R., Monreale, A., Giannotti, F.: Myway: Location prediction via mobility profiling. Inf. Syst. 64, 350–367 (2017)

41. Yan, Z., Parent, C., Spaccapietra, S., Chakraborty, D.: A hybrid model and computing platform for spatio-semantic trajectories. In: Proceedings of ESWC, pp. 60–75 (2010)
42. Zheng, Y.: Trajectory data mining: An overview. ACM TIST **6**(3), 29:1–29:41 (2015)

Chapter 5
Visual Analytics of Vessel Movement

Natalia Andrienko and Gennady Andrienko

Abstract Visual analytics techniques support the process of data analysis, reasoning, and knowledge building performed by a human analyst. The techniques combine interactive, human-controllable visual displays with interactive operations for data querying and filtering, data transformations, calculation of derived data, and application of computational techniques for analysis and modelling. We demonstrate the use of visual analytics techniques and procedures for analyzing Automatic Identification System (AIS) data. We begin with showing how visual analytics approaches can help in exploring properties of the data, detecting problems, and finding ways to clean and improve the data. Then we describe two analysis scenarios focusing on the events of vessel stopping and on the vessel traffic through the strait between the bay of Brest, France, and the outer sea. Thereby we show how different techniques are applied and combined.

5.1 Introduction

Human reasoning plays a crucial role in data analysis and problem solving. By means of reasoning, humans build and/or update knowledge in their mind. The knowledge includes understanding of data and understanding of the phenomena reflected in the data. Reasoning requires conveying information to the human's mind, and visual representations are best suited for this. Visual analytics, which is defined as "the science of analytical reasoning facilitated by interactive visual interfaces" [18, p.4], develops approaches combining visualizations, interactive operations, and computational processing to support human analytical reasoning and knowledge building.

N. Andrienko · G. Andrienko (✉)
Fraunhofer Institute IAIS, Sankt Augustin, Germany
e-mail: gennady.andrienko@iais.fraunhofer.de

N. Andrienko
e-mail: natalia.andrienko@iais.fraunhofer.de

University of London, London, UK

© Springer Nature Switzerland AG 2021
A. Artikis and D. Zissis (eds.), *Guide to Maritime Informatics*,
https://doi.org/10.1007/978-3-030-61852-0_5

In this chapter, we demonstrate examples of using visual analytics approaches for exploration and analysis of Automatic Identification System (AIS) data [8]. Our example dataset consists of trajectories of vessels that moved between the bay of Brest, France, and the outer sea [14]. We first investigate the properties of the data and then focus on revealing and understanding patterns of vessel movements. The example analysis task is to study when, where, and for how long the vessels were stopping and to understand whether the events of stopping may indicate waiting for an opportunity to enter or exit the bay (through a narrow strait) or the port of Brest.

5.2   Exploration of the Data Properties

Knowing properties of the data that need to be analyzed is essential for performing valid analysis and drawing valid conclusions. Hence, before focusing on the primary analysis goal, it is necessary to explore the data for gaining understanding of their properties, identifying quality problems, and finding ways to solve or mitigate them. Possible quality problems in movement data [3, 17] include, apart from errors in spatial positions of objects and missing records for long time intervals, gaps in spatio-temporal coverage, low temporal and/or spatial resolution, use of the same identifiers for multiple objects, and others.

To explore the properties of the data we are going to analyze, we use aggregated representations of original and derived attributes. Thus, a frequency histogram of the lengths of the temporal spaces between consecutive records [17] shows us that the most frequent spacing is around 10 s, and smaller intervals also occurred quite frequently. Smaller peaks around 20, 30, 40, 60, 180, and 360 s corresponds to the required frequencies of position reporting depending on the vessel status (moving or stationary), movement speed, type of the vessel, and positioning equipment. Longer gaps between recorded positions may correspond to equipment malfunctioning or being off, or to periods when the vessels were out of the area covered by the data.

When trajectories are represented by lines on a map (Fig. 5.1, top), long straight line segments can be noticed. Most of these segments correspond to *spatio-temporal gaps* in the trajectories, i.e., absence of recorded vessel positions during long time intervals. Hence, such segments must be excluded when it is necessary to analyze the paths of the vessels or to aggregate the trajectories into overall traffic flows; otherwise, the results will be wrong and misleading. A suitable way to exclude spatio-temporal gaps is to *divide the trajectories* by these gaps: the point preceding a gap is treated as the end of the previous trajectory, and the following point is treated as the beginning of the next trajectory. A gap is defined by choosing appropriate thresholds for the spatial and temporal distances between consecutive trajectory points. Suitable thresholds are chosen based on the statistics of the distances in the data. Thus, for the data presented in Fig. 5.1, 78.6% of the spatial distances are below 10 m, 12.1% are from 10 to 50 m, and 6.4% from 50 to 100 m. Only 0.5% of the distances exceed 250 m, 0.2% exceed 500 m, 0.12% are over 1 km, and 0.06% are over 2 km. Hence, a suitable spatial threshold may be from 0.25 to 2 km, depending on the intended spatial

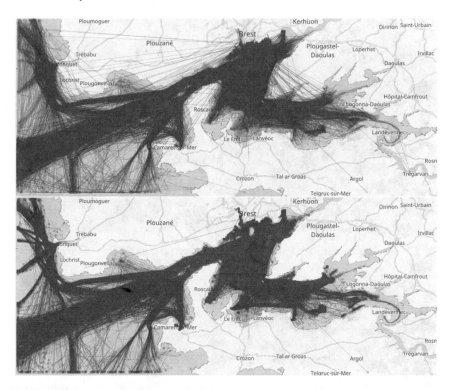

Fig. 5.1 Top: Long straight line segments in trajectories correspond to *spatio-temporal gaps*, i.e., long time intervals in which position records for the vessels are missing. Bottom: The result of dividing the trajectories by the spatio-temporal gaps in which the spatial distance exceeded 2 km and the time interval length exceeded 30 min

scale of the analysis. Only 0.4% of the temporal differences exceed 10 min, 0.22% exceed 20 min, and 0.19% exceed 30 min. Taking the spatial threshold of 2 km and temporal threshold of 30 min defines 1,852 spatio-temporal gaps in the data, which is 0.018% of the total number 10,446,156 of the available position records for the territory shown in Fig. 5.1. The image at the bottom of Fig. 5.1 shows the result of dividing the trajectories by these gaps. From the original 392 trajectories, the division produced 8,227 trajectories.

As any real data, the data we need to analyze have quite many errors, particularly, wrong positions. Some of these errors, such as positions on the land far from the sea, are very easy to spot visually (Fig. 5.2). In the trajectories presented in Fig. 5.1, such obvious errors have been already cleaned. Other positioning errors may be more difficult to detect, especially in a large dataset. A good indication of a recorded position being out of the actual path of a vessel is an unrealistically high value of the *computed speed* in the previous position. The speed is computed as the ratio between the distance to the next position and the length of the time interval between the positions. The computed speed may differ from the measured speed, which is

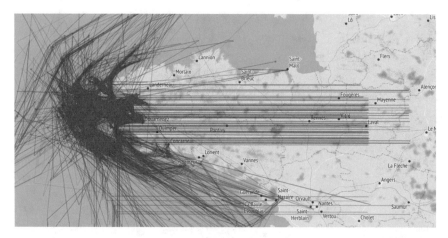

Fig. 5.2 Some trajectories include positions located on land far from the sea

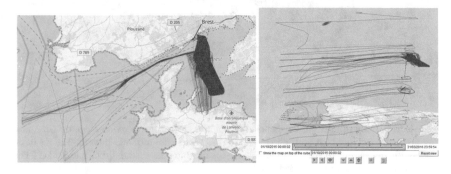

Fig. 5.3 A trajectory with an extremely high number of outlying points

recorded in the data. In the data shown in Fig. 5.1, the maximal recorded value of the measured speed is 102.2 knots (189.2744 km/h) and most of the values (88.5%) are below 36 knots (66.67 km/h), whereas the computed speed values reach as high as 31626 km/h. There are about 4000 points (0.1% of all) with the computed speed values exceeding 200 km/h (about 108 knots). Among these 4000 points, the median being 5161 and the lower and upper quartiles being 2432 and 9863, respectively. These values undoubtedly indicate positioning errors in the data records. Occasionally occurring singular outlying positions are easy to identify and exclude from the trajectories; however, there may be more difficult cases. It is useful to have a close look at trajectories containing many points with extremely high speed values.

An example of a trajectory containing extremely many points with very high values of computed speed is shown in Fig. 5.3. The trajectory has a very long duration—6 months. By looking at the map only, it is hard to understand what is shown. It is useful to look at this trajectory in a space-time cube (STC), which is a three-dimensional displays with two dimensions in the cube base representing space and the third, ver-

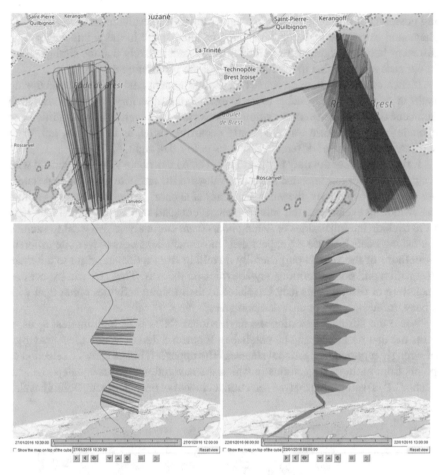

Fig. 5.4 Detailed investigation of the periods with high numbers of positioning errors

tical dimension representing time. The STC on the right of Fig. 5.3 shows that there
were a few short periods during this time in which the trajectory looks strange. In
Fig. 5.4, the appearance of the trajectory in two such periods is shown in more detail
on maps (upper images) and in STC views (lower images). The images indicate that
correctly recorded positions alternate with erroneous recording. Moreover, the wrong
positions seem to be displaced with respect to the correct positions in a systematic
manner rather than randomly. In the part of the trajectory shown on the upper and
lower left, the displacements occurred with varying temporal frequency, while the
directions and distances of the displacements were similar. The part of the trajectory shown on the upper and lower right has a much more complicated shape. The
displacement distances increase and decrease in a periodic manner, while the angles
change gradually. Such a pattern may mean that two or more simultaneously moving
vessels might refer to the same vessel identifier in reporting their positions. Arrang-

ing positions of different vessels in a single trajectory results in an unrealistic path shape and extremely high values of the computed speed in the points due to high distances between consecutive points. Still, it is not unlikely that the strange shape in our case emerged due to malfunctioning of the positioning device.

If such errors may have an impact on the subsequent analysis (e.g., in analyzing paths or flows), it is reasonable to try to exclude the trajectory fragments with high numbers of errors, or even the whole trajectories. However, it is quite difficult to exclude in an automatic way frequently occurring shifts, as in Fig. 5.4, left, or to separate movements of different vessels, as in Fig. 5.4, right.

Errors may occur not only in positions but also in attribute values associated with the positions. For our intended study, the values of the attribute 'navigational status' are relevant. Particularly, the value 1 means "at anchor", which may help us to find the anchoring events, and the value 7 means "engaged in fishing", which may help us to exclude the trajectories of fishing boats from our analysis. We want to exclude the fishing boats because we expect their movement behaviors to be quite different from those of the vessels purposefully travelling from a certain origin to a certain destination and not performing any activities on the way. Thus, we can expect that anchoring of fishing boats may be related to their fishing activities rather than with a busy traffic situation or crowded port area.

While the attribute reporting the navigational status is of high interest to us, it turns out that its values may be unreliable. Figure 5.5 demonstrates a few examples of wrongly reported navigational statuses. The upper left image shows a selection of points from multiple trajectories in which the navigational status equals 1, i.e., "at anchor". It is well visible that the selection includes not only stationary points but also

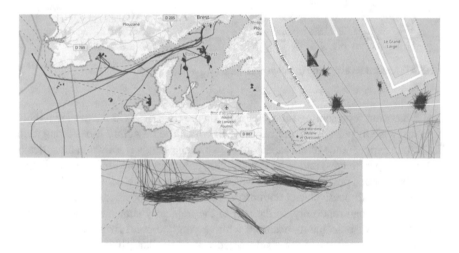

Fig. 5.5 Fragments of trajectories with wrongly reported navigation status. The reported status in the upper left and bottom images is "at anchor", whereas the vessels were actually moving. On the upper right, the reported status is "under way using engine", while the vessels remained at the same places and should have reported "at anchor"

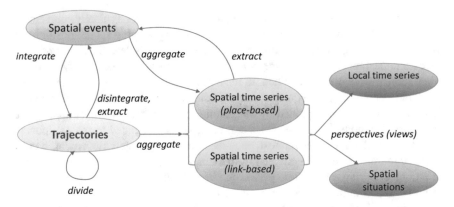

Fig. 5.6 Different representations of movement data and transformations between them (*Source* [1])

point sequences arranged in long traces, which means that the vessels were moving rather than anchoring. On the opposite, the upper right image demonstrates several trajectory fragments that look like hairballs. Such shapes are typical for stops, when the position of a moving object does not change but the tracking device reports each time a slightly differing position due to unavoidable errors in the measurements. The locations of the hairballs also signify that the vessels were at anchor or moored at the shore. However, the recorded navigational status is 0, which means "under way using engine". The lower image shows trajectory fragments of several vessels. The character of their movements (back and forth repeated multiple times) indicates that they were fishing, i.e., the navigational status should be 7, but the value attached to the positions is 0, i.e., "under way using engine".

Hence, in our study, we should not fully rely on the attribute values, and should also find ways to mitigate the possible impacts of the other errors we have detected.

5.3 Transformations of Movement Data

In the following analysis, we apply multiple transformations of the data. Movement data can be considered from several complementary perspectives [1]: trajectories, spatial events, dynamically changing situations over a territory, variation of presence of moving objects in selected places, or aggregated movements (flows) between places. For example, a sequence of positions of a vessel can be treated as its trajectory. An anchorage with its spatial position and temporal interval is an event. A situation can represent a spatial distribution of vessels at a given time or movement flows of vessels over a time interval. These diverse perspectives are supported by techniques for transformations between different possible representations of movement data, as illustrated Fig. 5.6.

The transformation scheme can be explained as follows. Typically, movement data are originally available as collections of records specifying spatial positions of moving objects (e.g. vessels) at different times. Such records describe *events* of presence (or appearance) of the moving objects at certain locations and specify the times when these events occurred. When all records referring to the same moving object are put in a chronological sequence, they together describe a trajectory of this object. Hence, trajectories are obtained by integrating spatial events of object appearance at specific locations. The trajectories can be again disintegrated to the component events. Particular events of interest, such as stops or zigzagged movement, can be detected in trajectories and extracted from them. A trajectory describing movements of an object during a long time period can be divided into shorter trajectories, for example, representing different trips of the object.

Having divided the space into *compartments* (shortly, places) and time into *intervals*, it is possible to aggregate either spatial events or trajectories by places and time intervals. Place-based aggregation involves counting for each pair of place and time interval (1) the events that occurred in this place during this interval, or (2) the number of visits of this place by moving objects and the number of distinct objects that visited this place or stayed in it during the interval. Additionally, various summary statistics of the events or visits can be calculated, for example, the average or total duration of the events or visits. The result of this operation is time series of the aggregated counts (e.g. counts of stops or counts of distinct visitors) and statistical summaries associated with the places.

Link-based aggregation summarizes movements (transitions) between places and, thus, can be applied to trajectories. For each combination of two places and a time interval, the number of times when any object moved from the first to the second place during this interval and the number of the objects that moved are counted. Additionally, summary statistics of the transitions can be computed, such as the average speed or the duration of the transitions. The result of this operation is time series of the counts and statistical summaries associated with the pairs of places. The time series characterize links between the places; therefore, they can be called link-based. The term "link between place A and place B" refers to the existence of at least one transition from A to B.

Both place-based and link-based time series can be viewed in two complementary ways: as spatially distributed local time series (i.e., each time series refers to one place) and as temporal sequence of spatial situations, where each situation is a particular distribution of the counts and summaries over the set of places or the set of links. These perspectives require different methods of visualisation and analysis. Thus, the first perspective focuses on the places or links, and the analyst compares the respective temporal variations of the attribute values such as counts of distinct vessels in ports over days. The second perspective focuses on the time intervals, and the analyst compares the respective spatial distributions of the values associated with the places or links.

5.4   Detection and Analysis of the Anchoring Events

The goal of our analysis is to investigate when, where, and for how long the vessels were stopping and to understand whether the stops may indicate waiting for an opportunity to enter or exit the bay of Brest through a narrow strait or the port of Brest. If many stops might have happened due to waiting, it may mean that the management of the traffic through the straight or the port services is sub-optimal and requires improvement.

5.4.1   Data Selection and Preparation

For our study, we selected the trajectories of the vessels that passed the strait connecting the bay of Brest to the outer sea at least once. We excluded the vessels that had, at some point, the navigational status 7, i.e., "engaged in fishing", for the reasons explained at the end of Sect. 5.2. We also excluded the vessels that never (during the time period covered by the data) had the navigational status 0, i.e., "under way using engine". The resulting selection consists of trajectories of 346 vessels.

From these trajectories, we selected only the points located inside the bay of Brest, in the strait, and in the area extending to about 20 km west of the strait. The shapes of the resulting trajectories are shown in Fig. 5.1. As we explained previously (Sect. 5.2), the long straight line segments correspond to periods of position absence. To exclude these segments, we divided the trajectories into sub-trajectories by the spatio-temporal gaps with distance thresholds 2 km in space and 30 min in time. This means the following: if there is a trajectory point such that its distance to the next point exceeds 2 km and the time difference exceeds 30 min, the trajectory is divided into two smaller trajectories. The first point is taken as the end of the first trajectory, and the next point is taken as the beginning of the second trajectory. The division gave us 6,346 smaller trajectories. To clean these trajectories from outliers, we removed the points whose distances from the previous and next points were more than 2 km. We further divided the trajectories by stops (segments with low speed) within the Brest port area and selected from the resulting trajectories only those that passed through the strait and had duration at least 15 min. As a result of these selections and transformations, we obtained 1,718 trajectories for further analysis. Of these trajectories, 945 came into the bay from the outer area, 914 moved from the bay out, and 141 trajectories include incoming and outgoing parts.

5.4.2   Detection and Extraction of Anchoring Events

Our study aims at the detection of anchoring events in the trajectories. As explained in Sect. 5.2, we cannot rely on the values of the attribute 'navigational status', because

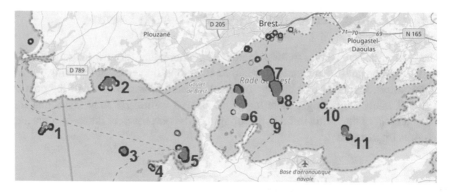

Fig. 5.7 Delineation of anchoring zones

they may be wrong. Assuming that, despite the errors we detected (Fig. 5.5), many vessels reported their anchoring correctly, we take the following approach. We extract from the database all trajectory points where the navigation status equals 1 (i.e., "at anchor") and the speed over ground is less than 2 knots. The extracted points are represented in Fig. 5.7 by purple circles drawn with a high level of transparency (90%). The circle symbols appear dark and thick where the points are clustered in space. We define 11 anchorage areas relevant to our study by creating buffer zones around the point clusters, excluding the clusters on the main traffic lanes (these clusters consist of points of singular vessels and may thus result from errors in data or indicate abnormal situations), within the port or attached to the shore. The zones, labelled by numbers from 1 to 11, are shown in Fig. 5.7.

Having defined the anchorage zones, we marked the points of the trajectories as *probable anchoring* if, first, they belong to one of the zones and, second, have the speed over ground below 2 knots. There are 158 trajectories that contain points satisfying these conditions. Pitsikalis et al. [13] describe another approach to detecting anchoring events, which can be used for online detection of such events, while the analysis described in this chapter is performed offline.

Figure 5.8, top, shows the previously selected set of 1,718 selected trajectories. The segments corresponding to probable anchoring are colored in red. The lower left image shows the shapes of the trajectory segments corresponding to probable anchoring in more detail. In the lower right, only those 158 trajectories that contain probable anchoring events are shown in an STC view. The trajectories have been aligned in time by putting together their starting times. Such transformation of the absolute time references to relative makes the segments corresponding to probable anchoring better visible (they appear as vertical lines in the cube) and their duration (which is represented by the lengths of the vertical lines) easier to compare.

We extract the probable anchoring events (i.e., the segments identified as probable anchoring) from the trajectories to a separate dataset. We obtained 327 events in total with the duration ranging from 1 min to 130.5 h. We deem it unlikely that a vessel would actually stay at anchor for a very short time, since the anchoring and unanchor-

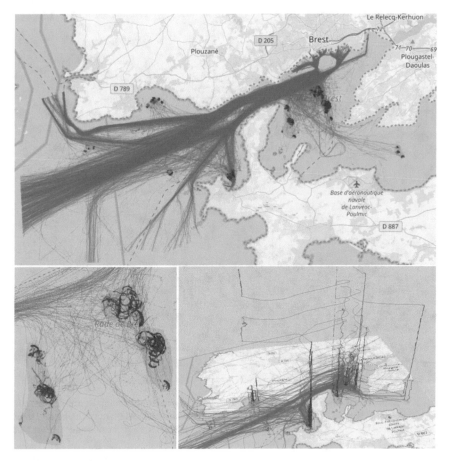

Fig. 5.8 The trajectories selected for analysis with the segments corresponding to probable anchoring marked in red. The lower left image shows an enlarged map fragment in more detail, and the STC on the right shows only the trajectories that contain probable anchoring; the trajectories have been temporally aligned

ing procedures are not instantaneous. So, we want to exclude unrealistically short events. To determine the minimal realistic duration, we find the shortest event that included points with the reported navigational status 1, i.e., "at anchor". Its duration was slightly less than 6 min. This gives us a ground to assume that the events shorter than 5 min may not correspond to real staying at anchor. After filtering these events out, we obtain 212 events with the median duration 134 min and the lower and upper quartiles being 18 min and 896 min (14.9 h), respectively. These events happened in 126 trajectories (7.33% of the 1,718 trajectories under study).

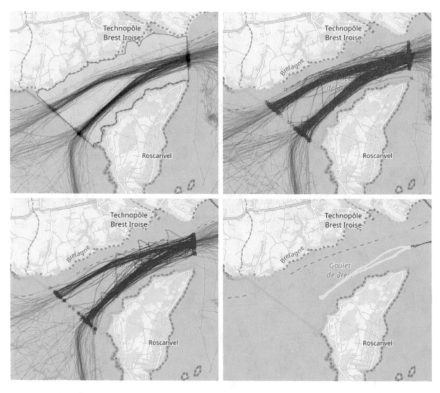

Fig. 5.9 Top left: Finding the trajectory parts corresponding to strait passing. Remaining images: Identifying the directions of the strait passing. Top right: inward, bottom left: outward, bottom right: a vessel entered the strait from the bay and returned back before reaching the outer side of the strait. This trajectory is labelled "in2in"

5.4.3 Exploration of the Anchoring Events in Relation to Strait Passing

Since we want to know how the anchoring events are related to passing the strait between the bay and the outer sea, we find the part corresponding to strait passing in each trajectory. Figure 5.9, left, demonstrates how they have been identified. We outlined interactively the area of the strait. We applied a spatial computation operation that determined for each trajectory point whether it lies inside the outlined area. The parts of the trajectories lying within the strait area are marked in a darker grey shade in Fig. 5.9, top left. These parts of the trajectories will be further referred to as *strait passing events*. The duration of these events ranges between 7.8 and 88 min, the median being 16.7 min.

For each event of strait passing, we identified the direction of the vessel movement by means of spatial queries involving the pair of interactively defined areas corresponding to the outer and inner sides of the strait; see Fig. 5.9, right (the areas are painted in light green). The trajectory fragments that passed first the outer side and then the inner side received the label 'inward' (colored in red in Fig. 5.9, top right), and the fragments that passed the areas in the opposite order were labelled as 'outward' (colored in blue on the bottom left). After performing this operation, we detected a fragment that was not labelled. We inspected it separately and found that it belongs to a vessel that entered the strait at the inner side but then returned back into the bay. This fragment is colored in yellow in Fig. 5.9, bottom right. We labelled this fragment 'in2in', which means "from inner area back to inner area".

Next, we determined the temporal relationships of the stops to the strait passing events of the same vessels. We used temporal queries to find for each stop the nearest straight passing event of the same vessel that happened in the past and in the future with respect to the time of the stop. Then we categorized the stops according to the directions of the past and future straight passing; see the legend on the right of Fig. 5.10. The most common category (105 stops) is 'inward;none', which means that the anchoring took place after passing the strait in the inward direction and there was no other strait passing after the anchoring, i.e., the vessels finally came in the port of Brest. There were 36 'outward;inward' stops, i.e., the vessels exited the bay through the strait, anchored in the outer area, and then moved back into the bay. 34 stops took place before entering the bay ('none;inward'), 18 happened after exiting the bay ('outward;none') and 11 before exiting the bay ('none;outward'). In 7 cases, vessels entered the bay from the outside, anchored, and then returned back without visiting the port ('inward;outward'), and there was one stop that happened after entering the strait at the inner side and returning back ('in2in;none').

The sizes of the pie charts in Fig. 5.10 are proportional to the total counts of the stops in the respective zones; the largest chart corresponding to 70 events is located inside the bay. The pie segments represent the proportions of the stops according to

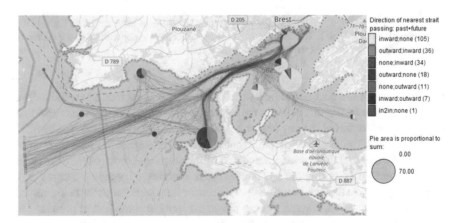

Fig. 5.10 The pie charts represent the counts of the stops in the anchoring zones categorized according to the temporal relationships to the strait passing by the vessels. The largest pie area represents 70 stops (see the legend in the lower right corner)

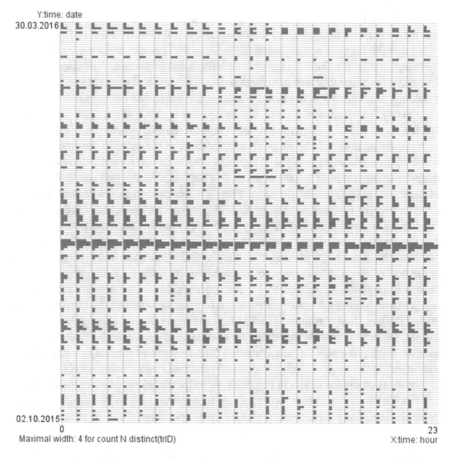

Fig. 5.11 The temporal distribution of the stops by the dates (rows of the matrix) and hours of the day (columns of the matrix). The bar lengths in the matrix cells are proportional to the counts of the anchored vessels; the maximal length represents 4 vessels

the directions of the past and future strait passing. The stops that happened before entering the bay and/or after exiting are located in the outer area, and the events that happened after entering the bay and/or before exiting are located inside the bay.

We see that the majority of the events (yellow pie segments) happened after entering the bay and, moreover, a large part of the stops that took place in the outer area happened after exiting the bay and before re-entering it (orange pie segments). It appears probable that the vessels stopped because they had to wait for being served in the port. Most of them were waiting inside the bay but some had or preferred to wait outside. Hence, the majority of the anchoring events can be related to waiting for port services rather than to a difficult traffic situation in the strait.

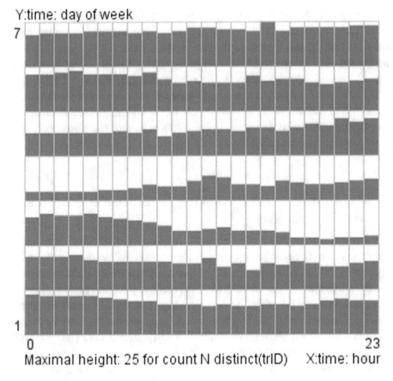

Fig. 5.12 The temporal distribution of the stops by the days of the week (rows of the matrix; 1 corresponds to Monday and 7 to Sunday) and hours of the day (columns of the matrix). The bar heights in the matrix cells are proportional to the counts of the anchored vessels; the maximal height represents 25 vessels

5.4.4   Temporal Distribution of the Anchoring Events

Apart from the spatial distribution, we look at the distribution of the stops over time using two-dimensional histograms (Figs. 5.11 and 5.12) the dimensions of which correspond to time components: hours of the day represented by the horizontal dimension versus dates (Fig. 5.11), and days of the week (Fig. 5.12) represented by the vertical dimension. The lengths of the dark gray bars are proportional to the numbers of distinct anchoring vessels. The maximal bar length in Fig. 5.11 corresponds to 4 simultaneously anchoring vessels, which is quite few. The pattern of the distribution by the dates and times of the day tells us that there were days when some vessels were anchoring during the whole days but also many days when there were no anchoring vessels at all or a few vessels anchoring for short times. We do not see any pattern regarding busy and less busy hours. However, the histogram in Fig. 5.12 shows us that the number of anchoring vessels tends to decrease starting from the morning of Wednesday (the third row from the bottom of the histogram) till the morning of Thursday (the fourth row), and then it starts increasing again. The weekend (two top rows) and Monday (the bottom row) are the busiest days in terms of vessel anchoring. We looked separately at the temporal distribution of the stops that happened after entering the bay and before going to the port or before entering the bay and found that the patterns observed in Figs. 5.11 and 5.12 are primarily made by these events, which is not surprising since they are the most numerous. The accumulation of the anchoring vessels by the weekend and gradual decrease of their number during the weekdays supports our hypothesis that the stops may be related to the vessels waiting for being served in the port.

5.4.5 Exploration of the Anchoring Events in the Context of the Trajectories

Now we want to look at the movements of the vessels that made stops on their way. We aggregate the vessel trajectories by a set of interactively defined areas, which include the anchoring zones, the port area, the areas at the outer and inner ends of the strait as shown in Fig. 5.9, right, and a few additional regions in the outer sea. The aggregation connects the areas by vectors and computes for each vector the number of moves that happened between its origin and destination areas. The result is shown on a flow map, where the vectors are represented by flow symbols with the widths proportional to the move counts. In our example (Fig. 5.13), the flow symbols are curved lines with the curvature being lower at the vector origins and higher at the destinations.

The aggregation we have applied to the trajectories is dynamic in the sense that it reacts to changes of the filters that are applied to the trajectories. As soon as the subset of trajectories selected by one or more filters changes, the counts of the moves between the areas are automatically re-calculated, and the flow map representing them is immediately updated. The six images in Fig. 5.13 represent different states of the same map display corresponding to different query conditions. The upper left image represents the 1592 trajectories (92.67% of all initially selected trajectories) that did not include stops. This flow map can be considered as showing uninterrupted traffic to and from the port of Brest. The flows between any two areas look symmetric, i.e., the lines have equal widths, which means approximately the same numbers of moves in the two opposite directions.

The remaining images show the flows obtained from the trajectories that included stops. The sizes of the red circles located in the anchoring zones are proportional to the numbers of the stops in these zones. The upper right image represents all 126 trajectories that contained stops. Here, the flows are asymmetric, showing more movements from the outside into the bay than from the bay. Please note that the scales of the widths of the flow symbols differ among the images. The maximal width, which is the same in all images, is proportional to the individual maximal value attained in each image; see the legend in the lower right corner of each image.

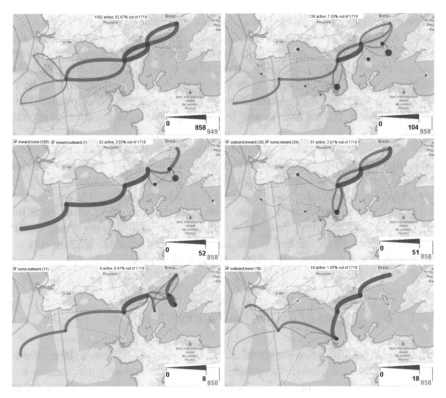

Fig. 5.13 The trajectories under study are represented in an aggregated form on flow map. Top left: trajectories without stops; top right: all 126 trajectories containing stops. The remaining images represent subsets of the trajectories having stops after entering the bay (middle left), before entering the bay (middle right), before exiting the bay (bottom left), and after exiting the bay (bottom right)

Thus, the thickest line represents 858 moves on the top left, 104 moves on the top right, 52 and 51 moves in the two images below, and 8 and 18 moves in the lower two images.

The four images in the second and third rows in Fig. 5.13 represent different subsets of the trajectories containing stops. The images in the second row represent the trajectories that had stops after entering the bay (left, 52 trajectories) and before entering the bay (right, 51 trajectories). Among the latter 51 trajectories, 30 began in the port area, moved through the strait to the outside region, stopped mostly either south of the strait entrance (21 trajectories) or northwest of it (6 trajectories), and then re-entered the bay and moved again into the port. This behavior may mean that the vessels were unloaded in the port and then moved to the outer area for waiting until the loads for their next trips are prepared in the port.

The images in the third row represent the trajectories having stops before (left) and after (right) exiting the bay, 8 and 18 trajectories, respectively. On the left, all 8 trajectories began in the port, 6 of them stopped south of the port and two southwest of the port before going out through the strait. However, three out of these 8 trajectories moved back to the port soon after exiting the bay (they had no stops in the outer area). This behavior may mean that the vessels were waiting for some port services inside the bay and then were going to relocate but changed their intention, perhaps, after being notified that the port is ready to serve them. The image on the right shows that some vessels had stops outside of the bay before going to the outer sea. These vessels, evidently, did not have to wait for port services and stopped for some other reasons.

Concluding the exploration of the stops, we can summarize that most of them are likely to have happened because the vessels had to wait for being served in the port of Brest. The vessels coming from the outer sea were waiting mostly inside the bay (Fig. 5.13, middle left), and the vessels that had been unloaded in the port and had to wait for the next load or another service were waiting mostly outside of the bay (Fig. 5.13, middle right). There were at most 4 simultaneously anchoring vessels (Fig. 5.11). The vessels that had to wait for port services tended to accumulate over the weekend, and their number reduced during the weekend (Fig. 5.12).

5.4.6   Exploration of the Traffic Through the Strait

Although we found out that the vessel anchoring events are unlikely to be related to the traffic in the strait, we are nevertheless interested in exploring the intensity and the temporal patterns of the traffic. From all available data, we select the trajectory fragments contained inside the strait area shown on the left of Fig. 5.9. There are 2,891 trajectory fragments satisfying the spatial query. To guarantee taking into account not only the times of the vessels being inside the strait but also the entering and exiting times, we extend each fragment by adding the preceding and following 45-min parts of the trajectories. We obtain 1,366 trajectories, i.e., some trajectories include two opposite movements through the strait. The trajectories are represented by dashed dark gray lines in Fig. 5.14, top.

We use the two areas shown on the right of Fig. 5.9 for obtaining aggregate flows through the strait in two directions. The total flow magnitudes are represented by the widths of the curved lines in Fig. 5.14, top. Please note that flow symbols, such as the curved lines in our map, only connect the origins and destinations of flows but do not represent the movement paths. Therefore, it should not be thought that the red flow symbol in our map represent movements of vessels through the land.

There were 1,010 inward and 995 outward movements (red and blue, respectively). However, we have calculated not only the totals but have applied spatio-temporal aggregation based on these areas and hourly intervals within the weekly time cycle. Hence, for each direction, we obtained a time series consisting of 168 values of the traffic volume in different hours along the weekly cycle. The aggregation over the weekly cycle puts together the movements that happened in the same hours of different weeks.

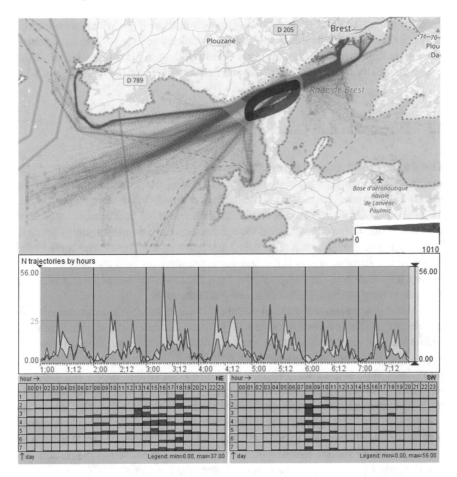

Fig. 5.14 Top: the trajectories going through the strait and the aggregate flows derived from them. Middle and bottom: the temporal distribution of the flow magnitudes over the hours of the week is shown in a line graph and in two 2D histograms where the rows correspond to the days of the week (from 1 for Monday to 7 for Sunday) and the columns to the hours of the day. In all images red is used for the inward flow and blue—for outward

The time series of the traffic volumes in the inward (red) and outward (blue) directions are shown in a line graph and two two-dimensional histograms below the map in Fig. 5.14. We see that the highest intensity of the outward movements was reached in the hour 07:00–8:00 of all days, especially on Wednesday (day 3). The highest peaks of the inward movements happened from 12:00 to 13:00 on Wednesday and early afternoons on Thursday and Friday. The inward traffic was also higher in the hour 18:00–19:00 of the days from Wednesday to Friday and in the hour 17:00–18:00 of the remaining days.

Interestingly, the increase of the inward traffic in the middle of the day and early afternoon from Wednesday to Friday (Fig. 5.14, bottom left) correlates with the decrease of the number of anchoring vessels in these days, which we observed in Fig. 5.12. It is probable that both patterns are related to the port operation schedule. Please note that the histogram rows corresponding to the days of the week are arranged from top to bottom in Fig. 5.14 and from bottom to top in Figs. 5.11 and 5.12.

5.5 Discussion and Conclusion

Visual analytic techniques are meant to support data exploration and analytical reasoning by a human. The human plays the key role in analysis. It is the task of the human to understand what data are available and how to make them appropriate to the analysis goal, what methods can be applied and what their results mean, and how the results differ depending on the methods or the parameter settings applied. It is also the task of the human to gain knowledge of the phenomena represented by the data. Visualization is important for supporting the analysis because it can convey information to the analyst in the most efficient way that enables perceiving and interpreting patterns, abstraction, and generalization.

We have demonstrated the use of visual analytics techniques in examples of analysis scenarios that involved discovery of patterns, interpretation of these patterns, and further analytical reasoning. We applied a number of different visualization techniques in combination with interactive querying and filtering of the data, data transformations, and computational derivation of new data, such as events and temporal relationships between them. Besides data transformations, visual analytics workflows often include methods for computational analysis and modelling, which stem from statistics, data mining, geographic information science, and other areas concerned with data analysis. We want to emphasize that such methods need to be supported by interactive visualizations enabling interpretation and comparison of the results, and that they need to be applied as a part of an iterative procedure in which the human analyst examines the effects of the parameter settings and chooses the most suitable ones.

Due to the interactive character of the visual analytics techniques, they typically require the data under analysis to be loaded in the main memory of the computer, which limits the scalability of these techniques to very large data volumes. Recent research in visual analytics develops approaches to overcoming this problem by combining data processing in a database with interactive operations on data samples, aggregates, or other kinds of derivatives. As an example, paper [4] proposes a scalable approach to clustering of trajectories according to the travelled routes.

Another challenge is application of visual analytics techniques to streaming data such as AIS streams. A possible approach is the use of dynamically updated visual displays in combination with computational analysis techniques specially developed for streaming data. An example is interactive real-time detection and tracing of spatial event clusters [7].

Visual analytics processes need to be performed prior to building computer models. Obtaining a good computer model requires that a human analyst understands what patterns it needs to capture and what modelling methods are suitable for this. The analyst should also understand how the data need to be prepared to the modelling. When a computer model is built, visual analytics techniques can greatly help in investigating its performance and finding ways to improve it. A model developed and tested in this way can be trusted and applied for justifiable decision making.

5.6 Bibliographical Notes

The book by Andrienko et al. [1] gives a detailed presentation of a broad spectrum of visual analytics techniques supporting analysis of movement data. A state of the art survey [5] uses a set of vessel trajectories as a running example to show how different visual analytics techniques can support understanding of various aspects of movement. A more recent survey [2] focuses on the application of visual analytics in transportation studies. The number of visual analytics papers proposing various approaches for analyzing movement data is very large and continues growing. Some of them deal specifically with data describing movements of vessels. Variants of dynamic density maps combined with specialized computations and techniques for interaction [11, 16, 20] support exploration of not only the density but also other characteristics of maritime traffic. Kernel density estimation can be used to compute a volume of the traffic density in space and time [9], which can be represented visually in a space-time cube [10] with two dimensions representing the geographical space and one dimension the time. Tominski et al. [19] apply a 3D view to show similar trajectories as bands stacked on top of a map background. The bands consist of colored segments representing variation of dynamic attributes along the routes. Scheepens et al. [15] propose special glyphs for visualizing maritime data. Lundblad et al. [12] employ visual and interactive techniques for analyzing vessel trajectories together with weather data. Andrienko et al. [6] use vessel movement data to demonstrate the work of an interactive query tool called TimeMask that selects subsets of time intervals in which specified conditions are fulfilled. This technique is especially suited for analyzing movements depending on temporally varying contexts.

Acknowledgements This research was supported by Fraunhofer Cluster of Excellence on "Cognitive Internet Technologies" and by EU in project Track&Know (grant agreement 780754).

References

1. Andrienko, G., Andrienko, N., Bak, P., Keim, D., Wrobel, S.: Visual Analytics of Movement. Springer, Berlin (2013). https://doi.org/10.1007/978-3-642-37583-5
2. Andrienko, G., Andrienko, N., Chen, W., Maciejewski, R., Zhao, Y.: Visual analytics of mobility and transportation: state of the art and further research directions. IEEE Trans. Intell. Transp. Syst. **18**(8), 2232–2249 (2017). https://doi.org/10.1109/TITS.2017.2683539

3. Andrienko, G., Andrienko, N., Fuchs, G.: Understanding movement data quality. J. Locat. Based Serv. **10**(1), 31–46 (2016). https://doi.org/10.1080/17489725.2016.1169322
4. Andrienko, G., Andrienko, N., Rinzivillo, S., Nanni, M., Pedreschi, D., Giannotti, F.: Interactive visual clustering of large collections of trajectories. In: 2009 IEEE Symposium on Visual Analytics Science and Technology, pp. 3–10 (2009). https://doi.org/10.1109/VAST.2009.5332584
5. Andrienko, N., Andrienko, G.: Visual analytics of movement: an overview of methods, tools and procedures. Inf. Vis. **12**(1), 3–24 (2013). https://doi.org/10.1177/1473871612457601
6. Andrienko, N., Andrienko, G., Camossi, E., Claramunt, C., Garcia, J.M.C., Fuchs, G., Hadzagic, M., Jousselme, A.L., Ray, C., Scarlatti, D., Vouros, G.: Visual exploration of movement and event data with interactive time masks. Vis. Inf. **1**(1), 25–39 (2017). https://doi.org/10.1016/j.visinf.2017.01.004
7. Andrienko, N., Andrienko, G., Fuchs, G., Rinzivillo, S., Betz, H.: Detection, tracking, and visualization of spatial event clusters for real time monitoring. In: 2015 IEEE International Conference on Data Science and Advanced Analytics (DSAA), pp. 1–10 (2015). https://doi.org/10.1109/DSAA.2015.7344880
8. Bereta, K., Chatzikokolakis, K., Zissis, D.: Maritime reporting systems. In: Artikis, A., Zissis, D. (eds.) Guide to Maritime Informatics, Chap. 1. Springer, Berlin (2021)
9. Demšar, U., Virrantaus, K.: Space–time density of trajectories: exploring spatio-temporal patterns in movement data. Int. J. Geogr. Inf. Sci. **24**(10), 1527–1542 (2010). https://doi.org/10.1080/13658816.2010.511223
10. Kraak, M.J.: The space-time cube revisited from a geovisualization perspective. In: Proceedings of the 21st International Cartographic Conference, pp. 1988–1996 (2003)
11. Lampe, O.D., Hauser, H.: Interactive visualization of streaming data with kernel density estimation. In: IEEE Pacific Visualization Symposium, PacificVis 2011, Hong Kong, China, 1–4 March, 2011, pp. 171–178 (2011). https://doi.org/10.1109/PacificVis.2011.5742387
12. Lundblad, P., Eurenius, O., Heldring, T.: Interactive visualization of weather and ship data. In: Proceedings of the 13th International Conference on Information Visualization IV2009, pp. 379–386. IEEE Computer Society, Washington, D.C. (2009)
13. Pitsikalis, M., Artikis, A.: Composite maritime event recognition. In: Artikis, A., Zissis, D. (eds.) Guide to Maritime Informatics, Chap. 8. Springer, Berlin (2021)
14. Ray, C., Dreo, R., Camossi, E., Jousselme, A.L.: Heterogeneous Integrated Dataset for Maritime Intelligence, Surveillance, and Reconnaissance (2018). https://doi.org/10.5281/zenodo.1167595
15. Scheepens, R., van de Wetering, H., van Wijk, J.J.: Non-overlapping aggregated multivariate glyphs for moving objects. In: IEEE Pacific Visualization Symposium, PacificVis 2014, Yokohama, Japan, March 4–7, 2014, pp. 17–24 (2014). https://doi.org/10.1109/PacificVis.2014.13
16. Scheepens, R., Willems, N., van de Wetering, H., Andrienko, G.L., Andrienko, N.V., van Wijk, J.J.: Composite density maps for multivariate trajectories. IEEE Trans. Vis. Comput. Graph. **17**(12), 2518–2527 (2011). https://doi.org/10.1109/TVCG.2011.181
17. Tampakis, P., Sideridis, S., Nikitopoulos, P., Pelekis, N., Theodoridis, Y.: Maritime data analytics. In: Artikis, A., Zissis, D. (eds.) Guide to Maritime Informatics, Chap. 4. Springer, Berlin (2021)
18. Thomas, J., Cook, K.: Illuminating the Path: The Research and Development Agenda for Visual Analytics. IEEE, Los Alamitos (2005)
19. Tominski, C., Schumann, H., Andrienko, G., Andrienko, N.: Stacking-based visualization of trajectory attribute data. IEEE Trans. Vis. Comput. Graph. **18**(12), 2565–2574 (2012). https://doi.org/10.1109/TVCG.2012.265
20. Willems, N., van de Wetering, H., van Wijk, J.J.: Visualization of vessel movements. Comput. Graph. Forum **28**(3), 959–966 (2009). https://doi.org/10.1111/j.1467-8659.2009.01440.x

Part III
On-Line Maritime Data Processing

Chapter 6
Online Mobility Tracking Against Evolving Maritime Trajectories

Kostas Patroumpas

Abstract We examine techniques concerning mobility tracking over *trajectories* of vessels monitored over a large maritime area. We focus particularly on maintaining summarized representations of such trajectories in *online* fashion based on surveillance *data streams* of positions relayed from a fleet of numerous vessels using the Automatic Identification System. First, we review generic, state-of-the-art *simplification* algorithms that can offer concise summaries of each trajectory as it evolves. Instead of retaining every incoming position, such methods drop any predictable positions along trajectory segments of "normal" motion characteristics with minimal loss in accuracy. We then discuss online filters that can reduce much of the *noise* inherent in the reported vessel positions. Furthermore, we present a method for deriving *trajectory synopses* designed specifically for the maritime domain. With suitable parametrization, this technique incrementally annotates streaming positions that convey salient *trajectory events* (stop, change in speed or heading, slow motion, etc.) detected when the motion pattern of a given vessel changes significantly. Finally, we discuss a qualitative comparison of maritime-specific synopses along with trajectory approximations obtained from generic simplification algorithms and highlight their pros and cons in terms of approximation error and compression ratio.

6.1 Introduction

Many modern monitoring applications collect and analyze huge amounts of flowing, uncertain and heterogeneous *spatio-temporal* information. Typical examples include location-based services for tourism or advertising, platforms for traffic surveillance or fleet management, notification systems for natural resources and hazards, etc. Our particular interest is on managing spatio-temporal data relayed from vessels sailing at sea, not only in terms of their actual whereabouts, but also regarding their *evolving trajectories* across time. In the past two decades, tracking vessels across the seas has emerged as a precious tool for preventing environmental risks,

K. Patroumpas (✉)
Information Management Systems Institute, Athena Research Center, Marousi, Greece
e-mail: kpatro@athenarc.gr

© Springer Nature Switzerland AG 2021 173
A. Artikis and D. Zissis (eds.), *Guide to Maritime Informatics*,
https://doi.org/10.1007/978-3-030-61852-0_6

through data exchange with other ships nearby, coastal stations, or satellites thanks to the Automatic Identification System [32]. In this chapter, we strictly focus on *dynamic* AIS messages informing about location, heading, and speed of vessels, since this can enable their tracking in real-time. Since AIS data is continuously transmitted from hundreds of thousands of vessels worldwide, maritime surveillance systems must collect, filter, process, and analyze information that presents all four challenges (*Volume, Velocity, Variety, Veracity*) in Big Data management [14]. How such massive, fluctuating, transient, and possibly noisy positional streams of AIS messages can be processed *online* for effective and timely mobility tracking is the topic of this chapter.

In particular, assuming that a large number of ships is being monitored in a given area of interest, we focus on efficient management and processing of their *evolving trajectories* generated by such massive positional updates. The key idea is that incoming trajectory data usually contains redundant information that may waste storage, so *compression* (a.k.a. *simplification*) is necessary. Indeed, except for harsh weather conditions, traffic regulations, local manoeuvres etc., ships normally follow almost straight, predictable routes at open sea. So, a large amount of raw positional updates may be suppressed with minimal losses in accuracy, as they hardly contribute additional knowledge about vessel motion patterns. Such data reduction may also facilitate their timely processing and mining, e.g. producing spatio-temporal analytics (speed, travelled distance, etc.) about individual vessels or entire fleets. In effect, a decision must be made on whether a recently received position report from a vessel should be preserved or not. But, due to the high arrival rate of incoming streaming locations, such trajectory simplification must be performed online. Obviously, the resulting simplified trajectory is only an *approximation* of the original one (i.e., when all reported positions are retained).

In this chapter, we survey several simplification techniques for succinct, lightweight representation of trajectories without harming the quality of the resulting approximation. We focus on online techniques which are able to examine each incoming location only once. Instead of retaining every incoming position for each vessel, such algorithms drop any predictable positions along trajectory segments of "normal" motion characteristics. Thus, they can keep in pace with scalable volumes of streaming positions and achieve low latency, since a position must typically be processed within less than a second after its arrival. Note that all such techniques aim to handle trajectories of moving objects in general (vehicles, devices, humans, etc.), hence they are *agnostic* of the maritime domain. They ignore that the moving objects are actually vessels with specific mobility patterns (manoeuvres when approaching ports, smooth turns due to their large tonnage and length, etc.), and cannot easily cope with the *noise* inherent in AIS data (e.g., off-couse positions due to invalid coordinates, out-of-sequence messages due to unsorted timestamps).

Thus, we present a methodology specifically tailored for vessel trajectories, which not only can achieve their simplification online, but it can also effectively annotate the judiciously retained positions with special *trajectory events* that indicate a stop, a turn, slow motion, change in speed, etc. As illustrated in Fig. 6.1, we may characterize in real-time the current motion of each monitored ship with a particular annotation.

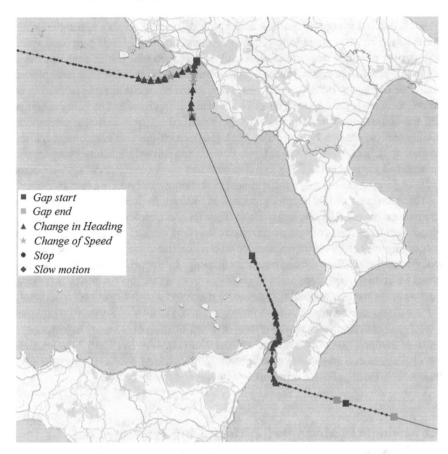

Fig. 6.1 Annotated critical points identified along a vessel trajectory. The sequence of such critical points provides a synopsis of the trajectory (shown as black line segments) with small deviation from the actual one. Original AIS positions not qualifying as critical are depicted in blue dots and generally concern straight segments of the vessel's course. Also note the lack of reported AIS locations (i.e., communication gaps) over large intervals along this vessel's route

Those *critical points* can be detected when the pattern of movement for a given vessel changes significantly and their sequence can actually reconstruct a simplified *synopsis* with small deviation from the original trajectory. Of course, the derived trajectory synopses must keep in pace with the incoming raw streaming data so as to get incrementally annotated with semantically important mobility features once they get detected. This approach attempts to strike a balance between the necessity for real-time monitoring of numerous vessels, while also retaining salient features along their recent motion history. Note that no context (weather, areas of special interest, vessel characteristics, etc.) is taken into account when constructing such trajectory synopses.

Such derived synopses can typically be used for map display and monitoring, but they are mostly valuable in emergency situations, e.g., issuing an alert when a passenger ship has stopped unexpectedly in the open sea, just in case this event required a rescue operation. Further, this summarized information may be exploited online or offline by other applications, such as for recognition and forecasting of complex events (cf. the chapter by Pitsikalis et al. [40]), visual and mobility analytics (cf. the chapter by Andrienko et al. [5]), statistical analysis and spatio-temporal mining (cf. the chapter by Tampakis et al. [50]), maritime data enrichment (cf. the chapter by Santipantakis et al. [46]) etc. Besides, the derived trajectory synopses can be archived for post processing in a database and contextually enriched with other static (e.g., protected areas, fishing zones) or streaming information (e.g., weather).

The remainder of this chapter proceeds as follows. In Sect. 6.2, we discuss how streaming dynamic AIS messages, despite their inherent imperfections, can be used to reconstruct trajectories of vessels. Section 6.3 surveys state-of-the-art research works concerning trajectory simplification, as well as software technologies with spatial extensions for efficient stream-based monitoring. In Sect. 6.4, we examine online filters for reducing inherent noise in positions reported via dynamic AIS messages. Section 6.5 presents a method that detects and annotates mobility events specifically for vessel trajectories and maintains their synopses in real-time. Section 6.6 provides a brief qualitative analysis on the effects of simplification over vessel trajectories in terms of compression efficiency and approximation quality. Section 6.7 summarizes the challenges and objectives of online mobility tracking over evolving maritime trajectories. Finally, Sect. 6.8 provides pointers for further reading.

6.2 Using AIS Messages for Monitoring Trajectories of Vessels

In this Section, we discuss the case of raw surveillance *data streams* with *positional* information in the maritime domain. More specifically, our focus is on dynamic AIS messages arriving in online fashion and concerning timestamped locations of vessels. We next explain how these positions can be used to reconstruct the evolving trajectories of vessels and we further highlight some important issues regarding the quality of this information for effective trajectory data management.

6.2.1 Dynamic AIS Messages as Streaming Data

According to AIS specifications [53], 64 different types of AIS messages can be broadcast by onboard transceivers. As detailed in the chapter by Bereta et al. [6], this data may concern *terrestrial AIS messages* continuously collected by onshore stations or *satellite AIS messages* arriving in bursts when satellites transfer buffered data into a ground station (perhaps several hours after the reported movement).

Irrespective of their collection method, position reports amount to almost 90% of the relayed AIS messages. In the sequel, we only consider such *dynamic* AIS messages that include location-based information, i.e., the current *longitude* and *latitude* of the vessel, as well as some extra *navigational* information: Speed over Ground, Course over Ground, Rate of Turn, navigational status, and true heading. In all such messages, it is assumed that each vessel can be uniquely identified by its 9-digit *Maritime Mobile Service Identity* or its *International Maritime Organisation* number; next, we refer to such a unique vessel identifier as *vid*. However, the *timestamp* of each message is not the one at the time of the measurement on board, but the one assigned at the time when this message was received by the station (terrestrial or satellite). Hence, timestamping might be affected by the accuracy of their clock, occasionally leading to disorder in the collected messages. It should also be noted that navigational information in AIS messages either comes from onboard sensors (e.g., gyroscope) or is manually entered by the crew (e.g., navigational status). So, despite its potential usefulness, it is usually erroneous and perhaps not properly updated. For this reason, navigational information in AIS messages is mostly ignored in online processing, as certain features (e.g., SoG or RoT) can be dynamically deduced with more accuracy from the sequence of locations.

Next, we consider that for a given vessel *vid*, each of its successive positional reports *p* consists of 2-dimensional geographical coordinates (*longitude*, *latitude*) referenced in the World Geodetic System 1984 and observed at discrete, totally ordered timestamps τ (e.g., at the granularity of seconds). This compound is generally referred to as a *timestamped location* (p, τ) reported by each monitored vessel, and it is the only information that will be extracted from AIS messages and used in online monitoring of trajectories. Without loss of generality, we abstract vessels as *point* entities moving across time; hence we ignore their shape, size, tonnage, etc., because our primary concern is to capture their motion features. We consider the case of monitoring a given geographical area (e.g., the Mediterranean Sea), where a fluctuating number of vessels may be sailing at any given time. Formally:

A *Data Stream* of timestamped locations from a fleet of vessels consists of tuples $\langle vid, p, \tau \rangle$, each designating an update from a vessel identified by its *vid* and located at position $p = (Lon, Lat)$ at timestamp τ.

In effect, this constitutes an *append-only* positional data stream, as no deletions or updates are allowed to already received locations.

6.2.2 Trajectories Reconstructed from Dynamic AIS Messages

Even though frequently updated AIS positions are indispensable for monitoring, it is the sequential nature of each vessel's trace that mostly matters for capturing movement patterns *en route* (e.g., a slow turn), as well as spatio-temporal interactions (e.g., ships travelling together). Such a *trajectory* is represented as an evolving sequence of successive point locations, i.e., dynamic AIS messages that locate this vessel at distinct timestamps (e.g., every few seconds). More specifically:

> *Trajectory T_v is a (possibly unbounded) sequence of tuples $\langle vid, p_i, \tau_i \rangle$ concerning a given moving vessel v identified by its unique vid. Each successive position $p_i \in \mathbb{R}^2$ in the Euclidean space has 2-dimensional coordinates (Lon, Lat) recorded at timestamp $\tau_i \in \mathbb{T}$.*

The *time domain* $\mathbb{T}$ is regarded as a set of discrete time instants with a total order $\leq$, which are used for *timestamping* each incoming position (e.g., at the granularity of seconds or milliseconds). Although timestamps may generally have two different interpretations [33], in dynamic AIS messages actually denote the *transaction time*, i.e., when the message was received by the base station. Ideally, each location should report the *valid time* of the actual measurement in the real world, but this is not the case in available AIS datasets. Next, we consider the transaction timestamp assigned to each collected message, and processing is based on them in order to identify the sequence of positions per vessel.

It should be stressed that the collected dynamic AIS messages from a vessel can only provide an *approximation* of its actual course. In the real-world, the trajectory of a vessel is continuous, but each AIS message is a sample of its course at a given time. As illustrated in Fig. 6.2, the trajectory reconstructed from those samples is actually a discretized one, by essentially compiling AIS messages one after the other. Obviously, the reconstructed trajectory deviates from the original continuous course; the magnitude of this deviation varies, depending on the reporting frequency, i.e., how frequently each vessel transmits AIS messages. Note that this frequency is not constant, but it may vary among vessels, whereas a single vessel may also issue messages at a fluctuating frequency (e.g., more frequently when it takes a turn or when it approaches a port). Typically for trajectories [8], linear interpolation is applied between each pair of successive samples (p_i, τ_i) and (p_{i+1}, τ_{i+1}), as shown in Fig. 6.2. For simplicity, we assume that this also holds in the case of vessels. Provided that a vessel transmits AIS messages regularly, its movement between any two consecutive positions practically evolves in a very small area, which can be locally approximated over a Euclidean plane using Haversine distances.

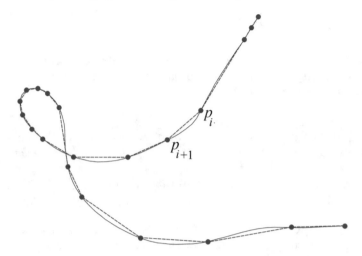

Fig. 6.2 The actual continuous course of a vessel (the solid line) is sampled via discrete AIS position reports (the blue dots) at varying frequency and potential noise. The sequence of these samples can be used to reconstruct the trajectory of this vessel (the dashed blue line). Depending on the sampling rate of positions, the reconstructed trajectory can offer a close approximation of the actual course

As detailed in the chapters by Tampakis et al. [50] and Andrienko et al. [5], dynamic AIS messages are not error-free, due to sea drift, delayed arrival of messages, discrepancies in Global Positioning System signals, and other reasons. Therefore, a monitoring application should account for such stream imperfections, i.e., the *noise* inherent in vessel positions, dealing with issues such as the fluctuating (and sometimes very low) frequency of position reports, duplicate messages, garbled timestamping due to delayed or intermittent messages, erroneous vessel identifiers, etc. As will be discussed in Sect. 6.4, prior to any processing, all incoming AIS positions should be filtered by applying online heuristics.

6.3 Background

As mobility event detection over vessel trajectories effectively also results in their compression, we next survey recent works regarding simplification of trajectories of moving objects. Further, we discuss state-of-the-art monitoring applications that can handle streaming information and their extensions for processing positional streams. Finally, we point to some recent processing frameworks specifically designed to cope with AIS positional information.

6.3.1   Trajectory Simplification

The key idea behind *trajectory simplification* is to discard redundant or less important points from the original raw trajectory, so that it can be suitably approximated by the sequence of the remaining points.

> *Suppose that a trajectory T of a moving object originally consists of n point locations. Its simplification aims to approximate T with a sequence T' of k ≪ n locations, while still preserving its course by minimizing the deviation between original trajectory T and its approximation T'.*

As every data reduction process, effectiveness of trajectory simplification is a trade-off between compression efficiency and approximation accuracy. The former can be measured by the compression ratio achieved by the simplification algorithm, and the latter by estimating the error in the approximate (compressed) trajectory compared to the original one.

Compression ratio. For a given trajectory T, this ratio λ is the percentage (%) of locations dropped from its simplified representation T' over the originally obtained position reports in T, or equivalently:

$$\lambda = 1 - \frac{k}{n},$$

where k is the number of retained locations in the simplified trajectory T', and n is the amount of raw positions available in the original (uncompressed) trajectory T. The higher this ratio, the more compressed the resulting approximation. A compression ratio λ closer to 1 signifies stronger data reduction, as the vast majority of original locations are dropped and few points suffice to represent the trajectory.

Approximation error. Preserving only a portion of the originally reported locations incurs a lossy approximation in trajectory representations. To assess the quality of those compressed trajectories, we typically estimate their deviation from the original ones (i.e., without discarding any raw positions, possibly except for those qualified as noise). Deviation can be computed using the notion of *Synchronous Euclidean Distance*. This is the pairwise distance between *synchronized* locations from the original trajectory T (composed of reported locations) and the simplified trajectory T' (i.e., its synopsis based on retained points only). Suppose that the simplification process evicts as superfluous from T an original point location p_i at timestamp τ_i. As illustrated in Fig. 6.3, its corresponding time-aligned trace p_i' in T' can be estimated using linear interpolation along the simplified path that connects the two retained points A and B along T' before and after τ_i. The error at timestamp τ_i is measured by $SED(p_i, p_i')$, shown as a line segment connecting p_i and p_i'. For a given moving

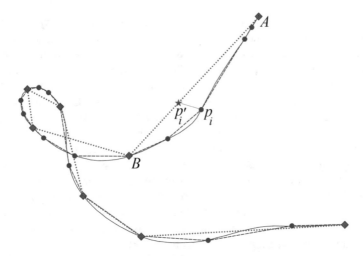

Fig. 6.3 A simplified trajectory representation T' (the dotted red line) consists of the retained points (red diamonds). These are a portion of the originally reported positions (the blue dots) reconstructing the original trajectory T. To assess approximation error, a time-aligned location p_i' (shown with a red star) is taken for each discarded one p_i along the linear path that connects two consecutive retained points A and B

object that has reported n raw positions in total, we can estimate the *Root Mean Square Error* between the original and simplified sequences of its locations as:

$$RMSE = \sqrt{\frac{1}{n} \cdot \sum_{i=1}^{n}(SED(p_i, p_i'))^2}.$$

In the maritime domain, the SED measure employs the Haversine distance formula between geographic coordinates (i.e., longitude and latitude in WGS84) on the surface of the sea, so the estimated $RMSE$ value is in meters.

Next, we outline the main categories of simplification techniques over trajectories of moving objects, particularly focusing on online algorithms that may be applied against streaming positional updates.

6.3.1.1   Taxonomy of Trajectory Simplification Algorithms

Trajectory simplification algorithms can be classified into two major categories, namely those achieving a *lossless compression* and those applying a *lossy compression*. In the first class, methods enable complete reconstruction of the original trajectory without information loss. For example, Trajic [31] employs a novel residual encoding scheme, by storing the initial data point and then the remaining points as a sequence of successive deltas. But lossless compression has the limitation that it generally yields a rather poor compression ratio. In contrast, lossy compression techniques generally provide a far more aggressive data reduction, but they incur approximation error as the compressed trajectories may deviate from the original. For instance, TrajStore [10] offers capabilities for indexing, clustering, and storing trajectory data. Based on an adaptive grid partitioning scheme, it involves a lossy compression scheme to cluster trajectories traveling on nearly identical paths and only store a representative path. However, it is mostly geared towards queries retrieving data about many trajectories passing through a particular location.

Alternatively, simplification techniques can be categorized according to their mode of operation. Under *batch mode*, the complete trajectory is available in storage when the algorithm is applied. Hence, a batch (offline) algorithm may take several passes over the sequence of locations and effectively select which points to keep in the approximation. This can offer a better trade-off between compression ratio and quality, but generally incurs higher computation cost. In contrast, simplification algorithms for streaming applications must work in *online mode*, as trajectories are evolving and the complete history may be neither stored nor indexed. To be affordable, such online techniques consider the data in one pass, i.e., each incoming position must be handled as it arrives. Sometimes, there is also a constraint on the approximation size, so a small number k of points must be kept, especially when simplification is applied locally at each moving object instead of processing all collected messages in a server. In Sect. 6.6, we discuss qualitative results from a first-cut evaluation of several state-of-the-art *online* simplification techniques against a small dataset of vessel trajectories.

6.3.1.2   Batch Trajectory Simplification

The well-known Douglas-Peucker algorithm [11] was the first to achieve *line simplification* by reducing the amount of points (vertices) needed to represent a polyline. Although it is widely used in cartography and computer graphics, this is a batch method that requires knowledge of all points before applying any simplification. In a similar spirit, strategies like [8] also specify an error tolerance for the resulting approximation. Other offline compression techniques aim to preserve the direction of trajectories [25, 26]. Essentially, they address simplification of large polylines, such that the error incurred into the direction of the approximate vector is minimized and the total number of retained points does not exceed the allocated storage. However, temporal information in trajectories is entirely overlooked, and these solutions emphasize on spatial features only. Besides, *map-matching* techniques like [7, 18, 48] leverage knowledge of an underlying network (e.g., roads or railways); instead of dealing with a sequence of point locations, the trajectory consists of a sequence of network segments that the moving object (e.g., a car) has traversed. As this kind of semantic compression necessitates a fixed transportation network, such methods have no straightforward application when dealing with evolving maritime trajectories that may sail freely at open sea.

6.3.1.3 Online Trajectory Simplification

When trajectory simplification is performed in *online mode*, a decision must be taken upon receiving a fresh location from a moving source (i.e., a vessel in our case): either drop or retain this position, depending on the specified criteria. The most straightforward, and also very fast approach is *Uniform sampling*: this algorithm picks samples at a fixed rate, e.g., by 10% it preserves one out of ten incoming positions, eventually yielding $\lambda = 90\%$. When locations are reported frequently, even such a simple approach may provide fair trajectory approximations.

Ideally, samples should keep each compressed trajectory as much closer to its original course, chiefly by minimizing approximation error as in *trajectory fitting* methods [8, 27]. For instance, the sliding window approach in [27] keeps simplifying points along a line until the error exceeds a given threshold. The difference in speed between consecutive sub-trajectories is also used as an error metric; once those two speed values deviate significantly, the common point connecting the corresponding segments should be kept as a sample as well. In contrast, the *Threshold*-guided method in [42] uses the concept of safe areas to generate a simplified trajectory using predefined bounds for allowable deviations on speed dv and orientation $d\varphi$. Thus, a fresh position must be included in the approximate path as long as it reveals significant change in movement which cannot be captured by the current velocity vector and the tolerance bounds.

From another perspective, the memory footprint occupied by the compressed trajectory may also be a constraint in a single-pass evaluation. Under this principle, the *STTrace* algorithm [42] keeps an in-memory buffer with k available points per moving object for approximating its evolving trajectory. Initially, each fresh location is accepted in the buffer. Once its capacity is reached and a fresh position can no longer be accommodated, it discards from the buffer the location that changes the least the existing trajectory approximation, i.e., incurs the minimal error with respect to *Synchronous Euclidean Distance* (SED). For instance, with buffer capacity set to $k = 0.2n$ per vessel, each compressed trajectory consists of 20% of the original locations, hence longer trajectories will not get oversimplified and their approximations will retain more samples compared to shorter ones.

Under a similar error-based principle, *SQUISH* (Spatial QUalIty Simplification Heuristic) [29] drops samples by employing a priority queue. The priority of each retained location is an estimate of the error that will be caused by removing it from the compressed trajectory. Points of lowest priority are removed from the queue in order to achieve a target compression ratio λ. But when λ is set too high (i.e., too many points get dropped), then error estimated with SED may escalate. To alleviate this shortcoming, an adaptive variant called *SQUISH-E* was proposed in [30] and supports two operation modes: *SQUISH-E(λ)* minimizes SED error while targeting a compression ratio λ, and *SQUISH-E(μ)* aims to maximize compression ratio ($\lambda = 1$) while keeping SED-estimated error below an upper bound μ. In empirical tests [30, 58] the latter mode seems to cope better, since it can generally remove more redundant points as long as the increased SED does not exceed the specified error bound μ.

Mainly focusing on savings in communication cost, *dead-reckoning* policies like [52, 55] and *mobility tracking* protocols in [21] may be employed on board and relay positional updates only upon significant deviation from the course already known to a centralized server. Thus, they can guarantee that the approximate trace deviates no more than a certain accuracy bound ϵ from the actual path. To check this, positional batches are collected and filtered at the moving sources (i.e., vessels in maritime surveillance scenarios) before transmitting any updates to the server. However, this can hardly be the case with vessel tracking data. In maritime surveillance, vessels are expected to report regularly their location, at least once every few minutes or even more frequently. Shipping companies wish to be kept aware of the whereabouts of their fleet of vessels so that they can monitor, evaluate, and possibly modify their course. Naval authorities also need to locate vessels as frequently as possible for safety reasons. Hence, communication cost is not the primary concern, since location awareness is far more important for effective maritime surveillance.

Direction-preserving simplification algorithms aim at minimizing the orientation information loss, i.e., they try to capture the direction of movements along a trajectory. *Interval* [17] is one such method that achieves a good trade-off between compression ratio and execution cost by using the concept of interval bound, i.e., an angular range of tolerance ϵ (specified in degrees). Let p_s be the last retained (anchor) point in the compressed trajectory so far, and p_i the currently examined original point. The algorithm also keeps a buffer of uncompressed points arrived after p_s. If the angular difference between line $\overline{p_s p_i}$ and any segment of consecutive points in the buffer exceeds ϵ, then point p_{i-1} is preserved in the compressed trajectory and becomes the new anchor point for the next iteration.

The *Bounded Quadrant System* (BQS) algorithm [23] applies an open window over the recent trajectory portion not yet compressed. Over this subtrajectory, it maintains eight special points and establishes a rectangular bounding box as well as two bounding lines for each of its quadrants. Various compression error bounds can be estimated via convex hulls formed by the box and the bounding lines around all points. Once a new point arrives, it can be readily dropped or retained after being checked against these convex hulls with a given error tolerance ϵ (in distance units). In most cases, there is no need to iterate through the points in the window and calculate the maximum error. A fast, relaxed version (*FBQS*) adds a line segment to the compressed trajectory and starts a new window once the convex hull cannot bound all locations in the current window. Further, an *amnesic* extension (*ABQS*) [24] is geared towards maximizing the trajectory information given a fixed storage space but with varying error tolerance depending on the age of the recorded segments. This aging-aware scheme is similar in spirit to a time-decaying approximation of streaming trajectories developed in [43] by gradually evicting older samples and offer greater precision for the most recent trajectory segments.

Finally, the one-pass, error-bounded algorithm $OPERB$ [22] applies a novel local distance checking method and involves several optimizations in order to achieve higher compression. It approximates the buffered points with a directed line and the distance from a fresh location should be less than a given error bound ζ. If the fresh point does not fit with the directed line, a new segment is added to the approximation and a new starting point for the directed line is considered. A more aggressive variant of this algorithm allows "patches" by interpolating new points when moving objects have sudden changes in their paths or relay updates intermittently.

6.3.2   Geostream-Aware Monitoring Applications

Several platforms and monitoring applications have been proposed for managing and analyzing data streams in order to accomplish their real-time processing demands [49]. Our particular interest is on *geostreaming* data [16] acquired from vessels continuously over time, which must be processed online in order to recognize important phenomena regarding their movement. On the other hand, in the era of Big Data, modern stream processing frameworks like Apache Flink [1, 9], Apache Spark streaming [2], or Apache Storm [3] offer powerful capabilities to ingest, process and aggregate enormous amounts of streaming information from diverse data sources. Note that these are general-purpose platforms, which allow customizations and extensions for specific applications. A review of their pros and cons, along with an experimental study of their current processing capabilities with a particular focus on analytics is available in [19]. Besides, several spatial extensions have emerged for Big Data platforms, like SpatialHadoop [12], GeoSpark [57], or Simba [56]. As examined in recent experimental surveys [13, 35], these modules offer support for representation and indexing of spatial entities, as well as basic topological operators for querying involving point locations and simple spatial features. Unfortunately, they entirely ignore any spatio-temporal notions concerning movement, hence trajectories of moving objects cannot be managed by any of these platforms.

Most recently, Apache Flink has been used as the streaming infrastructure to build custom functionality for event detection and summarization against aircraft trajectories [38], but also applied against vessel trajectories [54]. This algorithm is conceptually similar to the one introduced in [36, 37] and discussed next in Sect. 6.5, as it detects trajectory features from streaming positions, identifies significant changes in movement online, and outputs lightweight synopses of coherent trajectory segments. Apart from its ability to execute on distributed cluster infrastructures, this recent Flink-based method can also issue critical points at operational latency, i.e., within milliseconds after admission to the system so as not to cause delays in subsequent processing. Hence, the derived stream of trajectory synopses keeps in pace with the incoming raw streaming data so as to get incrementally annotated with semantically important mobility events. In contrast, the methodology to be presented in Sect. 6.5 produces trajectory synopses according to a sliding window paradigm [39]. Incoming positions are buffered in a window (e.g., locations received over the past hour) in order to monitor its general course, but new segments are issued once the window slides (e.g., every five minutes). Both approaches are space-efficient in terms of compression ratio and highly accurate with respect to the approximation error, but the Flink-based implementation can natively scale out against increased data volumes. To the best of our knowledge, this is the only streaming framework that has been specifically tailored for surveillance over fluctuating, noisy, intermittent, geostreaming messages from large fleets of vessels.

6.3.3   Processing Maritime Trajectories

CEP-traj [51] is a middleware that processes raw traces of timestamped locations and applies a series of steps, involving data cleaning, segmentation, compression and pattern detection. All this functionality is carried out by means of abstract Complex Event Processing operators that only require a single scan over the incoming data. Although this middleware was evaluated over AIS data, the focus is on complex events [4] like smuggling or illegal fishing as discussed in the chapter by Pitsikalis et al. [40], and not on mobility features as examined here.

In [47] an approach for anomaly detection and classification of vessel interactions is presented. Patterns of interest are expressed as left-to-right Hidden Markov Models and classified using Support Vector Machines, also taking into account contextual information via first-order logic rules. However, this work focuses more on predictive accuracy rather than real-time performance in a streaming scenario. Recently, a hybrid approach adopting a Lambda architecture has been proposed in [45] for processing maritime traffic data coming from both streaming (online) and archived (offline) sources. But this work only suggested interesting design principles towards building such a system and not any concrete processing techniques.

Specifically designed to automatically detect anomalies in AIS locations, *TREAD* (Traffic Route Extraction and Anomaly Detection) [34] can cope with intermittent data, time lags between observations, and also combine data from multiple sources (e.g., both terrestrial and satellite). Yet, this work aims specifically to mine route patterns or anomalies in maritime trajectories rather than online mobility tracking.

None of the generic simplification techniques [58] in Sect. 6.3.1 has ever been applied on trajectories of vessels. Instead, the methodology suggested in [36, 37] and discussed in more detail next, is specifically tailored for streaming trajectories of vessels. Furthermore, it supports mobility-annotated features in the retained samples, and has been extensively tested against diverse dynamic AIS data.

6.4   Online Noise Reduction from Dynamic AIS Messages

Despite its high value in maritime surveillance, AIS data is not error-free. In fact, there are several sources of *noise* that render a portion of this data noisy and sometimes inadequate for monitoring, as discussed in the chapter by Bereta et al. [6]. Next, we briefly discuss certain issues regarding completeness and quality of the incoming AIS data streams.

Most importantly, no precise timestamp value is relayed along with each AIS message. As mentioned in Sect. 6.2, a *transaction timestamp* marking the arrival of

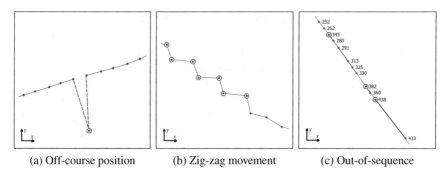

(a) Off-course position   (b) Zig-zag movement   (c) Out-of-sequence

Fig. 6.4   Noise-related situations (depicted as red circles) along a vessel's course (non-noisy original positions are shown as blue dots)

each AIS message at a station has to be used instead. Inevitably, transmission delays may frequently occur between the original message and its arrival. Successive positional messages from a single vessel may often arrive intermingled at a distorted order. Figure 6.4c illustrates such *out-of-sequence* messages, where numbers signify timestamp values since the beginning of this trajectory. If the trajectory gets reconstructed by their order of arrival, the vessel would appear moving back and forth very rapidly at a quite unusual speed. AIS networks do not have synchronized clocks, so if a vessel is within range of two stations, then a broadcasted message may be received by each one and possibly assigned with a different transaction timestamp. It may also happen that the same timestamp is assigned to different locations (maybe of considerable distance) of the same vessel. Therefore, *duplicate* or *contradicting positions* of a vessel may be present in the collected data. Noise might not always be caused by technical issues or the inherent errors and discrepancies in the GPS positions. It may be also due to deliberate, *suspicious actions*, e.g., switching off the transponder or transmitting "spoofed" coordinates in order to avoid surveillance in a sensitive area. To the extent possible, such intermittent or falsified positions should be detected and cleared.

Coping with noisy situations over AIS data is particularly challenging and has attracted significant research interest [15, 28, 34], but all such methods work in offline fashion employing expensive, iterative filters over archived AIS datasets (cf. the chapter by Andrienko et al. [5]). In contrast, for *online* monitoring of vessels it is necessary to apply *single-pass filters* that eliminate noise from the streaming AIS positions. In general, we can afford to lose garbled, out-of-sequence positions and not consider correcting their timestamps. After all, a fresh noise-free location will be soon received from a vessel, effectively compensating for the removal of any erroneous preceding one(s).

In order to effectively and efficiently eliminate noise in timestamped AIS positions, applying a series of simple heuristics [36] can be based on instantaneous velocity vector $\vec{v}_{now}$ of each vessel as computed by its two most recent observations. A noisy situation is identified if at least one of the following conditions apply:

- *Off-course positions* incur an abrupt change both in speed and heading of velocity $\overrightarrow{v}_{now}$. Such an outlier (Fig. 6.4a) can be easily detected since it signifies an abnormal, yet only temporary, deviation from the known course as abstracted by mean velocity $\overrightarrow{v_m}$ of the ship over its previous m positions.

- Vessels (especially larger ones) normally take their turns very smoothly, so a series of AIS locations may be transmitted, each marking a small change in heading as in Fig. 6.4f. However, if the latest position update indicates that a vessel has suddenly made a very abrupt turn (e.g., over 60°) with respect to its known course (even though its speed may not be altered significantly), then this message should better be ignored altogether. In case of adverse weather (e.g., a storm) a vessel's route may appear as a *'zig-zag'* polyline with a series of such abrupt turns as shown in Fig. 6.4b. Dropping those consecutive points as noise is not typically correct; yet, in terms of simplification this is quite desirable, as the vessel does not make any intentional turn and generally follows its planned course.

- When a vessel appears to speed up at a unusual rate, this is another indication of noise. This is typical for *out-of-sequence* messages with twisted timestamps (as the three red spots in Fig. 6.4c). Each such location is along the course of the ship, but due to their late arrival to the base station, the vessel is seen as suddenly retracting backwards at an unrealistically high speed (105 knots in this example).

- If an identical location from the same vessel has been already recorded before, then this might be a sign of error. Note that even if a vessel remains anchored, its successive GPS measurements usually differ by a few meters. In that case, instantaneous velocity $\overrightarrow{v}_{now}$ is infinitesimal, but not exactly zero. Instead, coincidental coordinates in succession should be deemed as almost certain duplicates, possibly concerning position reports that arrived at slightly different times at more than one base stations within range of the vessel's AIS antenna. So, any duplicate locations are dropped.

- A similar problem occurs with conflicts in timestamping, when the same timestamp is assigned to two distinct messages from a given vessel, even though they may be probably reporting different coordinates. In this case, instantaneous velocity $\overrightarrow{v}_{now}$ cannot be computed, signifying that these messages are contradictory (if not violating previous rules, we arbitrarily retain the latest one).

Experimental results in [36, 37] indicate that even 20% of the raw AIS positions may be qualifying as noise, falling in one of the aforementioned cases. Most importantly, accepting noisy positions would drastically distort the resulting trajectory synopsis, as the red dashed line in Fig. 6.4a illustrates. Even worse, noise may affect proper detection of movement events, as discussed next. Hence, although based in empirical heuristics, such noise reduction has proven certainly beneficial in terms of performance without sacrificing accuracy in the resulting trajectory approximations.

6.5   Annotating Mobility Events Along Vessel Trajectories

In [36, 37] a novel trajectory summarization approach was introduced specifically for online maritime surveillance. This framework consumes a geospatial stream of dynamic AIS messages from vessels, and continuously detects important events that characterize their movement. Not surprisingly, detecting trajectory events from positional streams essentially performs a kind of path simplification. In contrast with the techniques presented in Sect. 6.3.1, a major advantage of this scheme is that it annotates the simplified representations according to particular trajectory events (turn, stop, etc.), thus adding rich semantic information all along each compressed trace. Next, we discuss how online detection of such mobility events can be performed along vessel trajectories.

6.5.1   Online Tracking of Moving Vessels

As input, this method accepts dynamic AIS messages with position updates. As discussed in Sect. 6.2, this constitutes an *append-only* positional data stream, as no deletions or updates are allowed to already received locations. A *trajectory* is approximated as an evolving sequence of successive point samples that locate this vessel at distinct timestamps (e.g., every few seconds). Prior to any processing, all incoming AIS positions are filtered for *noise reduction* according to heuristics discussed in Sect. 6.4. Such noise-cleaned data can then be properly used for mobility tracking per vessel, by checking when and how velocity changes with time. Working entirely in main memory and without requiring any index support, this process can detect two kinds of *trajectory events* as shown in Table 6.1:

Table 6.1 Taxonomy of mobility events over vessel trajectories

$$
Trajectory\ Events
\begin{cases}
Instantaneous
\begin{cases}
\text{Pause} \\
\text{Speed change} \\
\text{Turn}
\end{cases} \\[2em]
Long-lasting
\begin{cases}
\text{Gap} \\
\text{Stop} \\
\text{Slow motion} \\
\text{Smooth turn}
\end{cases}
\end{cases}
$$

- *Instantaneous trajectory events* involve individual time points per route, by simply checking potentially important changes with respect to the previously reported location (e.g., a sharp change in heading).
- *Long-lasting trajectory events* are deduced after examining a sequence of instantaneous events over a longer (yet bounded) time period in order to identify evolving motion changes. For example, a few consecutive changes in heading may be very small if each is examined in isolation from the rest, but cumulatively they could signify a notable change in the overall direction.

In order to meet the requirements of the geostreaming paradigm [16], this online process makes use of a *sliding window* [20, 39]. Typically, such a window abstracts the time period of interest by focusing only on positions reported in a recent *range* ω (e.g., during the past 10 min). This window slides forward to keep in pace with newly arrived positions, so it gets refreshed at a specific *slide step* every β units (e.g., each minute).

At each window slide, trajectory events are issued as a sequence of *"critical" points* (such as a stop or turn), which are much fewer compared to the originally relayed positions. Note that such critical points are not reported immediately after detection, but as soon as the window slides forward, which is a suitably small time period (e.g., one minute or less) to ensure the validity of reported events. Accordingly, the current vessel motion can be characterized with particular *annotations* (e.g., stop, turn). Therefore, a *synopsis* consists of a subset of the original positions received by a vessel. Reconstruction of approximate trajectories can be based on such critical points with tolerable deviations from the original path. By taking advantage of those online annotations at critical points along trajectories, lightweight, succinct synopses can be retained per vessel over the recent past. Formally:

> A *synopsis* S_v over trajectory T_v of a moving vessel v identified by its vid consists of a possibly unbounded sequence of *critical points* represented as tuples $\langle vid, p_i, \tau_i, A_i \rangle$, where $\langle vid, p_i, \tau_i \rangle \in T_v$ and A_i specifies its *annotation* regarding each identified critical point.

Once new trajectory events are detected per vessel after each window slide, the annotated critical points can be readily issued and visualized on maps, e.g., as polylines (for trajectories) or point placemarks (for vessel locations) in Keyhole Markup Language or Comma Separated Values files.

6.5.2 Detecting Trajectory Events

In order to identify significant motion changes, this method employs an instantaneous velocity vector $\overrightarrow{v}_{now}$ over the two most recent positions reported by each vessel vid. In addition, it maintains a mean velocity $\overrightarrow{v_m}$ per ship over a buffer containing its pre-

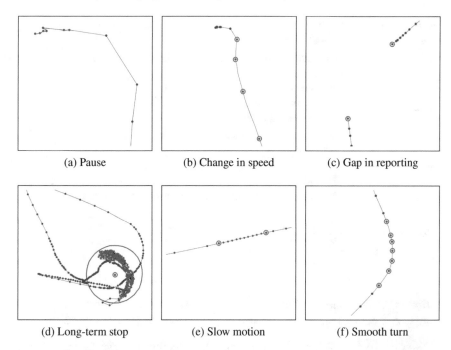

(a) Pause	(b) Change in speed	(c) Gap in reporting
(d) Long-term stop	(e) Slow motion	(f) Smooth turn

Fig. 6.5 Instantaneous and long-lasting trajectory events. Blue dots indicate raw AIS positions, whereas red dots signify positions adhering to a particular event like pause, stop, or slow motion. Red circles mark the critical points issued for each trajectory event

vious m positions (m is a small integer) so as to abstract its short-term course. Using heuristics with a suitable parametrization (indicative parameters listed in Table 6.2), it turns out that a large portion of the raw positional reports can be suppressed with minimal loss in accuracy, as they hardly contribute any additional knowledge. Next, we describe how the sequence of vessel positions can be processed *online* in order to detect trajectory movement events and thus maintain a lightweight synopsis of each vessel's course.

Once potentially noisy positions are cleared, the mobility tracking process can promptly deduce *instantaneous trajectory events* by examining the trace of each vessel alone. In particular:

- *Pause* indicates that a vessel is temporarily halted, once its instantaneous speed v_{now} is below a suitable threshold v_{min}. For example, if v_{now} is currently less than $v_{min} = 1$ knot ($\cong 1.852$ km/h), then the ship is idle, e.g., probably anchored as in the case of Fig. 6.5a, and such small displacements may be recorded due to GPS errors or sea drift.
- *Speed change* occurs once current v_{now} deviates by more than $\alpha\%$ from the previously observed speed v_{prev}. Given a threshold α, the formula $|\frac{v_{now}-v_{prev}}{v_{now}}| > \frac{\alpha}{100}$ indicates whether the vessel has just decelerated or accelerated. This normally happens when approaching to or departing from a port, as depicted in Fig. 6.5b.

Table 6.2 Parameters involved in online detection of trajectory events for vessels

Symbol	Description	Value
v_{min}	Minimum speed for considering a vessel as moving	1 knot
α	Maximum rate of speed change between successive locations	25%
r	Radius to identify long-term stops	250 m
m	Minimal number of positions to identify long-term stop or slow motion	5
$\Delta\theta$	Turn threshold (in degrees) between successive positions	8°
ΔT	Minimum period for asserting a communication gap	10 min
ω	Window range	10 min
β	Window slide	1 min

– *Turn* is spotted when heading in $\overrightarrow{v}_{now}$ has just changed by more than a given angle $\Delta\theta$; e.g., if there is a difference of $\geq 15^o$ from the previous direction.

No critical point gets immediately issued upon detection of any such simple events. An instantaneous pause or turn may be haphazard only; these are not meaningful out of context, because a series of such events may signify that, for example, the ship is stopped for some time. By buffering these instantaneous events within the window, the method can thus detect spatio-temporal phenomena of some duration. Examination of such *long-lasting trajectory events* is carried out in the following order. Note that if a certain long-lasting event has just been detected at a location, then checking if it also qualifies for another event is skipped altogether.

– *Gap* in reporting is examined first. This event is spotted when a vessel has not transmitted a message for a time period ΔT, e.g., over the past 10 min. This may occur when the vessel sails in an area with no AIS receiving station nearby, or because the transmission power of its transponder allows broadcasting in a shorter range. Then, its course is unknown during this period, as it occurs between the two red bullets in Fig. 6.5c. Reporting that contact was lost is important not only for online monitoring, but also for safety reasons, e.g., a suspicious move near maritime boundaries, or a potential intrusion of a tanker into a marine park. A pair of critical points signify when contact was lost (*gapStart* annotates the previously reported location) and when it was restored (*gapEnd* for current location).
– Checking for a *long-term stop* is triggered once the vessel is noticed to move ($v_{now} > v_{min}$) just after a pause. If current location is preceded by at least m consecutive instantaneous pause or turn events in the buffer, and they are all within a predefined radius r (e.g., 250 m), then a long-term stop is identified. In Fig. 6.5d, the red points inside the circle succeed one another and indicate such immobility,

so they are collectively approximated by a single critical point (their centroid) annotated as *stopped* with their total duration.

- *Slow motion* means that a vessel consistently moves at very low speed ($\leq v_{min}$) over its m most recent messages, as in Fig. 6.5e. If those buffered positions have not already qualified as a long-term stop by the previous rule (because they did not fall inside a small circle), then they probably succeed each other slowly along a path. The first and the last of these positions are both reported as critical points, respectively annotated as *lowSpeedStart* and *lowSpeedEnd*.
- *Smooth turns* are examined last. Due to their large size and maritime regulations, vessels normally report a series of locations when they change course. By checking whether the cumulative change in heading over buffered previous positions exceeds a given angle $\Delta\theta$, a series of such critical *turning* points may be issued, as illustrated with the red bullet points in Fig. 6.5f.

Critical points are issued from each detected long-lasting trajectory event, and this also relies on efficient noise reduction (cf. Section 6.4). For instance, an outlier breaking the subsequence of instantaneous pause events could prevent characterization of a long-term stop, and instead yield two successive such stops very close to each other.

This detection process can only lead to a *single annotation* for each critical point as the goal is to achieve a concise trajectory representation by dropping superfluous locations. For instance, if a vessel disappeared for long and is suddenly found anchored somewhere, this event will be spotted either as a gap or a stop, but not both. Further extensions of this method [54] allow multiple characterizations per detected point, while taking into account topological features is planned for future extensions, e.g., the approximation of a vessel trajectory should not cross the land. The aforementioned rules for detecting trajectory events are suitably defined for mobility tracking, which allows fast, in-memory maintenance of mobility features. Note that additional events can be detected by simply enhancing the tracking process with extra rules and conditions. This may take advantage of valuable knowledge from domain experts, but also relies on their particular focus. For instance, such an extension has been suggested in [41] in order to construct maritime activity patterns and thus recognize complex events (like vessel rendez-vous, drifting, loitering, etc.).

The overall performance of this method is more than affordable and detection is near real-time, as empirically verified in [37] and in more detail in [36] against AIS positions in the Aegean Sea. It turned out that online tracking provided quite acceptable accuracy in the resulting approximate trajectories with respect to deviations from the original ones, indicating that it could capture most, if not all, important changes along each vessel's course. Furthermore, it was capable of handling scalable volumes of streaming vessel positions with up to 10,000 locations/sec. Not only has this algorithm shown that it could yield a compression ratio better than 95% and sometimes even 98% over the raw data, but most importantly, it also annotated the identified critical locations with movement characteristics. Hence, point locations along a trajectory can be signaled out as stop, turn, slow motion, etc., thus adding rich semantic information all along each compressed trace.

6.6 Discussion

In this Section, we provide a brief qualitative analysis regarding the effects of simplification over vessel trajectories, primarily in terms of compression efficiency and approximation quality. We simulated several trajectory simplification algorithms working in *online* mode over a sample dataset. This data consists of AIS positions reported by 50 vessels navigating close to Brest, Brittany, France, randomly chosen from a larger, publicly available AIS dataset [44]. In addition, we applied the trajectory event detection method presented in Sect. 6.5, which identifies and annotates critical points online. Since a comprehensive evaluation is beyond the scope of this chapter, we stress that parametrization of each method was not fine-tuned, hence these results can only be seen as indicative. Performance efficiency is also not examined here, since these algorithms are implemented in different programming languages or software frameworks, so a fair comparison would not be possible in terms of execution cost.

Table 6.3 lists the average compression ratio (λ) achieved by each algorithm as well as the average $RMSE$ (in meters) over all 50 trajectories, as discussed in Sect. 6.3.1. Figure 6.6 highlights the different simplification output obtained by each method over a portion of a single trajectory. As detailed in Table 6.3, all methods drop many original locations in order to yield a simplified trajectory, although at a different compression ratio depending on the specifications of each algorithm and its objectives. Therefore, they also differ in approximation quality.

Uniform Sampling is very fast, as it only needs to keep a certain percentage of the original points without any sophisticated criteria. Interestingly, as illustrated in

Table 6.3 Comparison of simplification algorithms on a sample dataset of 50 vessel trajectories

Algorithm	Parametrization	Avg. Compression ratio (λ)	Avg. RMSE (m)
Uniform sampling	Target $\lambda = 0.8$	0.79989726	444.6002394
Dead reckoning [52]	Accuracy bound $\epsilon = 0.0002$	0.84226364	43.6677376
Interval [17]	Angular tolerance $\epsilon = 10^o$	0.49884219	216.1221104
SQUISH-E (μ) [30]	Target $\lambda = 1$, $\mu = 0.0002$	0.82163522	**5.7768162**
OPERB [22]	Error bound $\zeta = 0.0002$	**0.91267798**	127.5501139
FBQS [23]	Error tolerance $\epsilon = 0.0002$	**0.91604916**	91.3804395
Threshold [42]	$dv = 10\,\text{m/s}$, $d\phi = 10\circ$	0.42478510	318.1404981
STTrace [42]	Target $\lambda = 0.8$	0.80002366	11.0573623
Critical points [36, 37]	As in Table 6.2	0.80595616	111.1392395

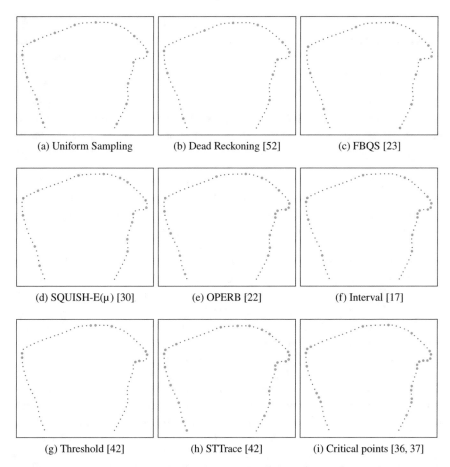

Fig. 6.6 Simplifications resulting from various algorithms over the same vessel trajectory (only a detailed part of the trajectory is shown). The red dots represent locations retained in the compressed trajectory, whereas the blue dots signify original locations that have been discarded

the example in Fig. 6.6a, the method may miss some turning points, but generally provides a decent simplified representation by simply picking samples at a fixed rate (one every five locations in this test). However, this is mainly due to the very frequent arrival of position reports from vessels in this dataset, as well as the rather high sampling rate applied in simplification. Results from this naïve method can be very poor if the reporting frequency widely fluctuates or the sampling rate of the method is low. Therefore, since the method has no intelligence in picking which samples to retain, the approximation error can be very high.

Methods like *Dead Reckoning* (Fig. 6.6b) and *FBQS* (Fig. 6.6c) seem to yield similar, and particularly good approximations. This is just a coincidence for this particular trajectory, as their general operation differs and results in diverse quality. Indeed, as listed in Table 6.3, **Dead Reckoning** can offer better approximations with less error compared to *FBQS*, but it has to keep more points in the resulting compressed trajectories. It may yield more aggressive simplification when its accuracy bound tolerance ϵ is relaxed (i.e., assigning larger ϵ values), but at the expense of deteriorating quality. Instead, **FBQS** is more balanced and yields trajectory approximations with tolerable quality (error is doubled compared to that from *Dead Reckoning*) while also drastically reducing the amount of retained locations (roughly half of those kept by *Dead Reckoning*). **SQUISH-E(μ)** (Fig. 6.6d) offers supreme quality with minimal error in the simplified trajectories. Yet, in order to achieve such accuracy it has to retain more locations compared to *OPERB*, *FBQS*, and *Dead Reckoning* in this test, hence its compression efficiency inevitably drops.

If substantial data reduction is the principal objective, then *OPERB* and *FBQS* should be preferred, as they also offer tolerable approximation error. In effect, **OPERB** (Fig. 6.6e) is competitive to *FBQS* in terms of quality and compression, but it seems to preserve more locations along turns. Apparently, those points exceed the specified error bound ζ with respect to the directed line constructed from the recent motion history, hence they should be kept in the simplified trajectory. Although its framework is different, **Interval** (Fig. 6.6f) essentially has a similar objective and can properly capture the turning points. However, its overall compression ratio is particularly low over all trajectories in this test. It may probably achieve lesser quality with a more relaxed angular range tolerance, as it would only focus on sharper turns so as to retain less information.

The **Threshold**-guided approach (Fig. 6.6g) seems to yield rather poor approximations. It captures only major changes along trajectories by keeping few intermediate points between turns, hence its large approximation error. This may not just be a side effect of inappropriate, ad-hoc parametrization in this test, but also a sign that the algorithm is not suitable for the frequently updated trajectories in this dataset. The applied combination of threshold constraints on speed and orientation may fail to keep important locations, because a point is discarded only when both speed and direction constraints are met. On the contrary, **STTrace** (Fig. 6.6h) accomplishes excellent quality given a target compression ratio λ, as it parsimoniously preserves "good" locations that incur minimal error in the resulting trajectory approximation. Interestingly, it tends to keep more points along turns, so as to closely maintain the shape of the motion path. Note that the method aims to achieve a given compression ratio ($\lambda =80\%$ in this example), so the quality of the approximation largely depends on the available space for storing the compressed trajectories (either on disk or in memory). In these tests, *STTrace* seems to have the most balanced behavior, as it accomplishes a very low approximation error (only *SQUISH-E* fares better in terms of quality) at the expense of a fair compression ratio. In contrast to *Uniform Sampling*, algorithm *STTrace* practically achieves the same compression ratio λ, but with much higher quality in the resulting compressed trajectories.

Mobility tracking of trajectory events (named **Critical Points** in Fig. 6.6i) generally yields high compression efficiency as the amount of retained critical points is comparable with the aforementioned state-of-the-art methods. Since this method (Sect. 6.5) is tailored for maritime trajectories, it purposely detects mobility patterns specific to vessels, like smooth turns or speed changes, so it may preserve the involved locations in the resulting synopses. Most importantly, its added value is that each judiciously retained critical point also carries *annotations*, e.g., turn, stop, gap. etc., which is not available in general-purpose simplification methods. Obviously, such filtering greatly depends on proper choice of parameter values, which is a trade-off between reduction efficiency and approximation accuracy. For a suitable calibration of these parameters, apart from consulting maritime domain experts, exploratory tests should be conducted on randomly chosen data samples. For instance, setting $\Delta\theta = 5^o$ instead of $\Delta\theta = 15^o$ may even double the amount of critical points, because more raw AIS locations would qualify as turning points due to sea drift and discrepancies in GPS signals. With more relaxed parameter values, additional mobility events can be detected, capturing slighter changes along each trajectory.

Overall, it should be emphasized that each algorithm is sensitive to proper choice of parameter values, which also differ amongst algorithms (Table 6.3). Fixed parametrizations applicable to any dataset are rather difficult to specify, as setting suitable parameter values requires careful data exploration. In a preprocessing step, input data should be scrutinized for imperfections such as noise, low or widely fluctuating rate in position reports, invalid coordinates, disorder in timestamps, etc. Map visualization of data samples is particularly helpful, as it can instantly indicate missing data, e.g., areas with few or even no dynamic AIS messages at all, as well as a range of visual analytics over the data as analyzed in the chapter by Andrienko et al. [5]. Clearly, several trial-and-error executions of the chosen simplification method may be needed for fine-tuning its parameters. Extra knowledge from domain experts (naval engineers, port authorities, shipping companies, etc.) is precious since it can guide the analysis to hidden patterns otherwise not identifiable though visualization or statistical analysis. Particularly regarding detection of trajectory events with the method discussed in Sect. 6.5, it would be interesting to try and dynamically adjust the various parameters (Table 6.2) in order to meet different optimization criteria. For instance, if the user wishes a more aggressive compression, then several combinations of parameter values may be tested in order to yield the required compression ratio. A similar approach involving a series of tests against different parametrizations would also apply for other criteria, such as obtaining a target approximation error (RMSE), minimizing the execution cost, etc. In that sense, calibrating these parameters may be carried out even with regard to more advanced conditions, such as minimizing the error incurred by subsequent processing of the resulting critical points, e.g., during complex event recognition (cf. the chapter by Pitsikalis et al. [40]).

6.7 Conclusion

We examined methods that enable mobility tracking over trajectories of vessels. As the amount of incoming streaming AIS messages from vessels reporting their positions can be very high, we focused on techniques that enable filtering out noise and redundant positions along each trajectory. We surveyed state-of-the-art simplification algorithms that can summarize each trajectory online as it evolves, typically applying trade-offs between compression efficiency and approximation error. As none of these generic techniques takes advantage of the mobility patterns of vessels, we presented a specific method that can retain succinct synopses of vessel trajectories, drastically reducing their original paths into few critical points that convey major motion features, while also dealing with inherent noise characteristics in online fashion. Not only this approach offers highly simplified trajectories with minimal error, but it also annotates particular trajectory events (such as stop, changes in speed or heading, slow motion, etc.) detected when the pattern of movement for a given vessel changes significantly. Finally, we briefly analyzed some qualitative results regarding trajectory approximations obtained by various online simplification algorithms against a sample dataset with dynamic AIS messages. Even though all algorithms seem very sensitive to suitable parametrization, it seems that the trajectory event detection method is competitive both in terms of approximation quality and compression ratio.

6.8   Bibliographical Notes

Line simplification methods like the Douglas-Peucker algorithm [11] are mostly applicable in cartography, computer graphics, and spatial databases. In case such paths represent movement of objects (e.g., vessels, cars, merchandise, animals), the temporal dimension should be considered as well, yielding a wide range of *trajectory simplification* techniques. These algorithms can work either in *batch* mode like [8, 25] considering the entire historical trajectory (i.e., no updates) or in *online* mode like [22, 23, 29] where each trajectory evolves in real time. Empirical studies in trajectory compression like [21, 30] offer a nice overview of several methods, along with a comparison of their performance and quality of approximation. The most recent such survey [58] compares about 25 trajectory simplification algorithms running in either processing mode against datasets of different motion patterns and also offers a insightful outlook at future research directions.

Management of vessel trajectories based on AIS positions takes a more practical perspective, although certain research challenges such as timeliness, scalability, quality, and efficiency must be addressed. CEP-traj [51] can consume AIS locations and applies a series of steps, involving data cleaning, segmentation, compression and pattern detection. The architecture outlined in [45] considers hybrid processing of maritime traffic data coming from both online and offline sources. Specifically tailored for streaming trajectories consisting of AIS locations, the methodology suggested in [36, 37] provides high-quality trajectory synopses with mobility-annotated features in the retained critical points, and has been extensively tested against diverse dynamic AIS datasets. Based on the same principles, an advanced framework on top of Apache Flink offers scalable and efficient support for event detection and summarization against vessel trajectories [54], but it is also applicable over aircraft trajectories [38].

References

1. Apache Flink. https://flink.apache.org/. Retrieved 30 Sept 2019
2. Apache Spark Streaming. http://spark.apache.org/streaming/. Retrieved 30 Sept 2019
3. Apache Storm. http://storm.apache.org/. Retrieved 30 Sept 2019
4. Alevizos, E., Artikis, A., Patroumpas, K., Vodas, M., Theodoridis, Y., Pelekis, N.: How not to drown in a sea of information: An event recognition approach. In: IEEE Big Data, pp. 984–990 (2015)
5. Andrienko, N., Andrienko, G.: Visual analytics of vessel movement. In: Artikis, A., Zissis, D. (eds.) Guide to Maritime Informatics, Chap. 5. Springer (2021)
6. Bereta, K., Chatzikokolakis, K., Zissis, D.: Maritime reporting systems. In: Artikis, A., Zissis, D. (eds.) Guide to Maritime Informatics, chap. 1. Springer (2021)
7. Brakatsoulas, S., Pfoser, D., Salas, R., Wenk, C.: On map-matching vehicle tracking data. In: VLDB, pp. 853–864 (2005)
8. Cao, H., Wolfson, O., Trajcevski, G.: Spatio-temporal data reduction with deterministic error bounds. VLDB J. **15**(3), 211–228 (2006)
9. Carbone, P., Katsifodimos, A., Ewen, S., Markl, V., Haridi, S., Tzoumas, K.: Apache Flink: Stream and batch processing in a single engine. IEEE Data Eng. Bull. **38**, 28–38 (2015)
10. Cudré-Mauroux, P., Wu, E., Madden, S.: Trajstore: an adaptive storage system for very large trajectory data sets. In: ICDE, pp. 109–120 (2010)
11. Douglas, D., Peucker, T.: Algorithms for the reduction of the number of points required to represent a digitized line or its caricature. Can. Cartogr. **10**(2), 112–122 (1973)
12. Eldawy, A., Mokbel, M.F.: SpatialHadoop: a MapReduce framework for spatial data. In: ICDE, pp. 1352–1363 (2015)
13. Hagedorn, S., Götze, P., Sattler, K.: Big spatial data processing frameworks: feature and performance evaluation. In: EDBT, pp. 490–493 (2017)
14. Jagadish, H.V., Gehrke, J., Labrinidis, A., Papakonstantinou, Y., Patel, J.M., Ramakrishnan, R., Shahabi, C.: Big data and its technical challenges. Commun. ACM **57**(7), 86–94 (2014)
15. Katsilieris, F., Braca, P., Coraluppi, S.: Detection of malicious AIS position spoofing by exploiting radar information. In: FUSION, pp. 1196–1203 (2013)
16. Kazemitabar, S.J., Demiryurek, U., Ali, M.H., Akdogan, A., Shahabi, C.: Geospatial stream query processing using Microsoft SQL Server Streaminsight. Proc. VLDB Endow. **3**(2), 1537–1540 (2010)
17. Ke, B., Shao, J., Zhang, D.: An efficient online approach for direction-preserving trajectory simplification with interval bounds. In: MDM, pp. 50–55 (2017)
18. Kellaris, G., Pelekis, N., Theodoridis, Y.: Map-matched trajectory compression. J. Syst. Softw. **86**(6), 1566–1579 (2013)
19. Kipf, A., Pandey, V., Böttcher, J., Braun, L., Neumann, T., Kemper, A.: Analytics on fast data: Main-memory database systems versus modern streaming systems. In: EDBT, pp. 49–60 (2017)
20. Krämer, J., Seeger, B.: Semantics and implementation of continuous sliding window queries over data streams. ACM Trans. Database Syst. **34**(1), 4:1–4:49 (2009)
21. Lange, R., Dürr, F., Rothermel, K.: Efficient real-time trajectory tracking. VLDB J. **20**(5), 671–694 (2011)
22. Lin, X., Ma, S., Zhang, H., Wo, T., Huai, J.: One-pass error bounded trajectory simplification. Proc. VLDB Endow. **10**(7), 841–852 (2017)
23. Liu, J., Zhao, K., Sommer, P., Shang, S., Kusy, B., Jurdak, R.: Bounded quadrant system: Error-bounded trajectory compression on the go. In: ICDE, pp. 987–998 (2015)
24. Liu, J., Zhao, K., Sommer, P., Shang, S., Kusy, B., Lee, J., Jurdak, R.: A novel framework for online amnesic trajectory compression in resource-constrained environments. IEEE Trans. Knowl. Data Eng. **28**(11), 2827–2841 (2016)
25. Long, C., Wong, R.C.W., Jagadish, H.V.: Direction-preserving trajectory simplification. Proc. VLDB Endow. **6**(10), 949–960 (2013)

26. Long, C., Wong, R.C.W., Jagadish, H.V.: Trajectory simplification: on minimizing the direction-based error. Proc. VLDB Endow. **8**(1), 49–60 (2014)
27. Meratnia, N., de By, R.: Spatiotemporal compression techniques for moving point objects. In: EDBT, pp. 765–782. Springer (2004)
28. Millefiori, L.M., Braca, P., Bryan, K., Willett, P.: Adaptive filtering of imprecisely time-stamped measurements with application to AIS networks. In: FUSION, pp. 359–365 (2015)
29. Muckell, J., Hwang, J., Patil, V., Lawson, C.T., Ping, F., Ravi, S.S.: SQUISH: an online approach for GPS trajectory compression. In: COM.Geo, pp. 13:1–13:8 (2011)
30. Muckell, J., Jr., P.W.O., Hwang, J.H., Lawson, C., Ravi, S.S.: Compression of trajectory data: a comprehensive evaluation and new approach. Geoinformatica **18**(3), 435–460 (2014)
31. Nibali, A., He, Z.: Trajic: an effective compression system for trajectory data. IEEE Trans. Knowl. Data Eng. **27**(11), 3138–3151 (2015)
32. International Maritime Organization: Automatic Identification Systems. http://www.imo.org/OurWork/Safety/Navigation/Pages/AIS.aspx. Retrieved 30 Sept 2019
33. Ozsoyoglu, G., Snodgrass, R.T.: Temporal and real-time databases: a survey. IEEE Trans. Knowl. Data Eng. **7**(4), 513–532 (1995)
34. Pallotta, G., Vespe, M., Bryan, K.: Vessel pattern knowledge discovery from AIS data: a framework for anomaly detection and route prediction. Entropy **15**(6), 2218–2245 (2013)
35. Pandey, V., Kipf, A., Neumann, T., Kemper, A.: How good are modern spatial analytics systems? Proc. VLDB Endow. **11**(11), 1661–1673 (2018)
36. Patroumpas, K., Alevizos, E., Artikis, A., Vodas, M., Pelekis, N., Theodoridis, Y.: Online event recognition from moving vessel trajectories. GeoInformatica **21**(2), 389–427 (2017)
37. Patroumpas, K., Artikis, A., Katzouris, N., Vodas, M., Theodoridis, Y., Pelekis, N.: Event recognition for maritime surveillance. In: EDBT, pp. 629–640 (2015)
38. Patroumpas, K., Pelekis, N., Theodoridis, Y.: On-the-fly mobility event detection over aircraft trajectories. In: SIGSPATIAL, pp. 259–268 (2018)
39. Patroumpas, K., Sellis, T.: Maintaining consistent results of continuous queries under diverse window specifications. Inf. Syst. **36**(1), 42–61 (2011)
40. Pitsikalis, M., Artikis, A.: Composite maritime event recognition. In: Artikis, A., Zissis, D. (eds.) Guide to Maritime Informatics, chap. 8. Springer (2021)
41. Pitsikalis, M., Artikis, A., Dréo, R., Ray, C., Camossi, E., Jousselme, A.L.: Composite event recognition for maritime monitoring. In: DEBS, pp. 163–174 (2019)
42. Potamias, M., Patroumpas, K., Sellis, T.: Sampling trajectory streams with spatiotemporal criteria. In: SSDBM, pp. 275–284 (2006)
43. Potamias, M., Patroumpas, K., Sellis, T.: Online amnesic summarization of streaming locations. In: SSTD, pp. 148–165. Springer (2007)
44. Ray, C., Dréo, R., Camossi, E., Jousselme, A.L., Iphar, C.: Heterogeneous integrated dataset for maritime intelligence, surveillance, and reconnaissance. Data Brief **25** (2019)
45. Salmon, L., Ray, C.: Design principles of a stream-based framework for mobility analysis. GeoInformatica **21**(2), 237–261 (2017)
46. Santipantakis, G.M., Doulkeridis, C., Vouros, G.A.: Link discovery for maritime monitoring. In: Artikis, A., Zissis, D. (eds.) Guide to Maritime Informatics, chap. 7. Springer (2021)
47. Shahir, H.Y., Glässer, U., Shahir, A.Y., Wehn, H.: Maritime situation analysis framework: Vessel interaction classification and anomaly detection. In: IEEE Big Data, pp. 1279–1289 (2015)
48. Song, R., Sun, W., Zheng, B., Zheng, Y.: PRESS: a novel framework of trajectory compression in road networks. Proc. VLDB Endow. **7**(9), 661–672 (2014)
49. Stonebraker, M., Çetintemel, U., Zdonik, S.: The 8 requirements of real-time stream processing. ACM SIGMOD Rec. **34**(4), 42–47 (2005)
50. Tampakis, P., Sideridis, S., Nikitopoulos, P., Pelekis, N., Theodoridis, Y.: Maritime data analytics. In: Artikis, A., Zissis, D. (eds.) Guide to Maritime Informatics, chap. 4. Springer (2021)
51. Terroso-Saenz, F., Valdés-Vela, M., den Breejen, E., Hanckmann, P., Dekker, R., Skarmeta-Gómez, A.F.: CEP-traj: an event-based solution to process trajectory data. Inf. Syst. **52**, 34–54 (2015)

52. Trajcevski, G., Cao, H., Scheuermann, P., Wolfson, O., Vaccaro, D.: On-line data reduction and the quality of history in moving objects databases. In: MobiDE, pp. 19–26 (2006)
53. International Telecommunication Union: M.1371 : Technical characteristics for an automatic identification system using time-division multiple access in the VHF maritime mobile band. https://www.itu.int/rec/R-REC-M.1371-5-201402-I/en. Retrieved 30 Sept 2019
54. Vouros, G.A., Vlachou, A., Santipantakis, G.M., Doulkeridis, C., Pelekis, N., Georgiou, H.V., Theodoridis, Y., Patroumpas, K., Alevizos, E., Artikis, A., Claramunt, C., Ray, C., Scarlatti, D., Fuchs, G., Andrienko, G.L., Andrienko, N.V., Mock, M., Camossi, E., Jousselme, A., Garcia, J.M.C.: Big data analytics for time critical mobility forecasting: Recent progress and research challenges. In: EDBT, pp. 612–623 (2018)
55. Wolfson, O., Sistla, A., Chamberlain, S., Yesha, Y.: Updating and querying databases that track mobile units. Distrib. Parallel Database 7(3), 257–287 (1999)
56. Xie, D., Li, F., Yao, B., Li, G., Zhou, L., Guo, M.: Simba: Efficient in-memory spatial analytics. In: SIGMOD, pp. 1071–1085 (2016)
57. Yu, J., Zhang, Z., Sarwat, M.: Spatial data management in Apache Spark: the GeoSpark perspective and beyond. GeoInformatica 23(1), 37–78 (2019)
58. Zhang, D., Ding, M., Yang, D., Liu, Y., Fan, J., Shen, H.T.: Trajectory simplification: an experimental study and quality analysis. Proc. VLDB Endow. 11(9), 934–946 (2018)

Chapter 7
Link Discovery for Maritime Monitoring

Georgios M. Santipantakis, Christos Doulkeridis, and George A. Vouros

Abstract Link discovery in the maritime domain is the process of identifying relations—usually of spatial or spatio-temporal nature—between entities that originate from different data sources. Essentially, link discovery is a step towards data integration, which enables interlinking data from disparate sources. As a typical example, vessel trajectories need to be enriched with various types of information: weather conditions, events, contextual data. In turn, this provides enriched data descriptions to data analysis operations, which may lead to the identification of hidden or complex patterns, which would otherwise not be discovered, as they rely on data originating from disparate data sources. This chapter presents the fundamental concepts of link discovery relevant to the maritime domain, focusing on spatial and spatio-temporal data. Due to the processing-intensive nature of the link discovery task over voluminous data, several techniques for efficient processing are presented together with examples on real-world data from the maritime domain.

7.1 Motivation

Analysis of maritime mobility data requires, beyond surveillance data, data coming from disparate and heterogeneous sources, often acquired, stored and curated for special purposes and exploited in isolation, by stakeholders having different interests and operational needs. Data is produced and made available for processing in different modalities, e.g., as streaming or archival data, in varying rate of data generation, and in variable quality. Given the number of vessels and other entities of interest whose data needs to be exploited in conjunction to that of vessels, data is voluminous, even if only a part of the globe is considered.

G. M. Santipantakis (✉) · C. Doulkeridis · G. A. Vouros
Department of Digital Systems, University of Piraeus, Karaoli and Dimitriou 80, Piraeus, Greece
e-mail: gsant@unipi.gr

C. Doulkeridis
e-mail: cdoulk@unipi.gr

G. A. Vouros
e-mail: georgev@unipi.gr

© Springer Nature Switzerland AG 2021
A. Artikis and D. Zissis (eds.), *Guide to Maritime Informatics*,
https://doi.org/10.1007/978-3-030-61852-0_7

Integration of surveillance data for large number of entities in large geographical areas with other data coming from different sources presents opportunities and challenges [24], especially when the integration needs to be performed in an online way (i.e., in near real-time). Also, providing integrated views of data enhances our abilities: (a) to reveal patterns of movement and understand the rationale behind patterns of behaviour, thus increasing our abilities to make accurate mobility predictions (e.g., predicting future locations of vessels, or their trajectories in specific time horizons), (b) to detect and forecast important events that may affect operations, safety in sea, or may affect the environment and life in sea, or (c) to reveal important behaviour that needs to be monitored. Situation awareness and monitoring can be done effectively, as long as the necessary information (i.e., the information that operators need) is provided in integrated ways, at multiple levels of detail, and in appropriate forms.

Challenges include reducing the computational complexity of the integration task without reducing the quality of the integrated data set, and achieving scalable processing in highly distributed operational and computational settings.

In this chapter, we explain how maritime data can be integrated, focusing on mobility data related to events and trajectories. An event is any happening that is related to the trajectories of vessels, and may concern other entities beyond the moving vessel. A trajectory is a time series of positional data, i.e., a series of 2D spatio-temporal entities, but it can also be considered as a mere line (i.e., a geometry), or a sequence of geometries (e.g., sub-trajectories). It should be mentioned that there are other types of trajectory abstractions that are out of the scope of this chapter.

Going from positional information of vessels to trajectories with a given start (e.g., port of departure) and destination (e.g., point of arrival) is a relevant Link Discovery task. Beyond trajectories in isolation from the context in which they occur, of particular interest are also events concerning the relation of a trajectory to other trajectories, which may be translated to relations between spatio-temporal geometries (points or lines), or relations between trajectories and static points, or areas of special interest. Generally, all these events require the detection of relations between spatio-temporal entities, represented as points, polylines, or polygons (areas).

The main sources to be exploited in integration concern data which need to be associated with a trajectory or with an event. Consider the following:

- Surveillance data sources (e.g., AIS streams), which provide positional data gathered from terrestrial and/or satellite stations.
- Weather data in the form of forecasts for specific areas. Weather-related variables of interest may include wind, sea state (e.g., wave height and direction) and precipitation at specific positions. Severe phenomena that may affect operations and safety at sea are also of major interest. These phenomena may be associated to areas, and thus represented as spatio-temporal entities, given their temporal and spatial extent and evolution.
- Contextual data from different sources, including areas of interest (e.g., areas protected from fishing, areas of fragile environmental nature or of increased fishing pressure/traffic, traffic separation schemes, etc.).

- Other data sources concerning vessels' type, dimensions, nominal speed, identification, type of cargo, etc., that are of interest for correctly detecting trajectories and performing any kind of trajectory analysis.

In this chapter we describe basic and advanced techniques towards integrating data sources, focusing on "discovering" relations among spatio-temporal entities: This task is called *spatio-temporal Link Discovery*.

Formally, given two data sources providing information about spatio-temporal entities, one called *target*, denoted by $\mathcal{T}$, and the other *source*, denoted by $\mathcal{S}$, and a set of relations $\mathcal{R}$, the objective of spatio-temporal Link Discovery is to compute tuples of spatio-temporal entities $\langle s, t \rangle$, such that $\langle s, t \rangle$ satisfies r, where $s \in \mathcal{S}$, $t \in \mathcal{T}$ and $r \in \mathcal{R}$. The computational complexity of this task, if done in a naive way, is $|\mathcal{S}| \cdot |\mathcal{T}| \cdot |\mathcal{R}|$. This chapter presents techniques towards reducing this complexity, without missing to identify relations or sacrificing correctness, although approximations are in service.

The structure of this chapter is as follows: Sect. 7.2 briefly presents the relations typically involved in the spatio-temporal Link Discovery tasks, Sect. 7.3 introduces the basic Link Discovery concepts and algorithm, while Sect. 7.4 outlines extensions and generalizations of the algorithm. Section 7.5 presents the most common issues that can be found in Link Discovery tasks with real-world data sets and Sect. 7.6 demonstrates Link Discovery using Open Data available online. Section 7.7 concludes the chapter and provides resources for related work and further reading.

7.2 Spatial and Temporal Relations

Spatial data is typically associated with a location or region. Two basic representations are used for spatial data: *vector approximation* and *raster approximation*. In the case of vector approximation, spatial data is represented using geometrical concepts: points, polylines and polygons (Fig. 7.1). Formally, a point is defined by exactly one pair of values (or triple of values for three dimensions) representing the projection of the point to the corresponding axis of each dimension. A polyline is constructed by a sequence of points, and a polygon is constructed by a polyline that is starting and closing at the same point, and the area defined in it. For example, a city can be represented by a polygon, whereas a vessel trajectory is represented as a polyline. More complex polygons can be constructed, where polylines are also used to exclude parts of the internal region of the polygon (i.e., they create "holes"). In the case of raster approximation, an object is approximated by a set of pixels, with non-zero values in any of these pixels indicating the presence of the spatial object.

Spatial relationships often indicate the distance and angle between objects represented by spatial data. Spatial relationships between spatial data objects can be classified as *topological*, *directional*, and *distance* relationships [10]. Topological relationships are defined in the Dimensionally Extended nine-Intersection Model (DE-9IM). Typical relations include *equals*, *disjoint*, *touches*, *contains* and *covers*, and they are defined on the geometric representations of objects. Directional relation-

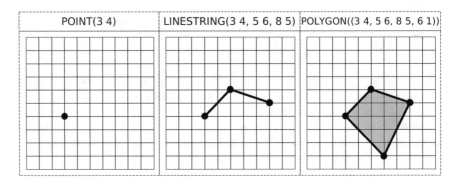

Fig. 7.1 The primitive geometry types in 2D space. The origin (0, 0) is at the bottom left

ships (such as north/south or left/right) compare relative locations of objects with respect to some reference system. Finally, distance relationships express distance information between objects (such as "nearby").

Temporal data can be instantaneous (e.g., associated with a specific timestamp) or durative (e.g., associated with a time interval). Allen's interval algebra [2], a calculus for temporal reasoning that defines relations between time intervals (*before*, *meets*, *overlaps*, *starts*, *during*, *finishes* and *equals*), is widely used to express temporal predicates.

7.3 Basic Concepts of Link Discovery

The objective of a Link Discovery (LD) task is to detect relations (links) between representations of entities[1] reported in given data sets. Link Discovery is a two-step *filter-and-refine* process, where the first step creates candidate pairs of entities (for linking), and the second step evaluates whether each candidate pair satisfies the relation(s) of interest. This section focuses on a special case of Link Discovery, namely spatial Link Discovery, which involves spatial data and the relations of interest are of spatial nature. Although examples in this chapter are mainly 2D, we note that the techniques presented in the following paragraphs are also applicable for multidimensional (spatio-temporal) data.

Link Discovery tasks typically involve the combination of two data sets. However, Link Discovery can be also applied on a single data set to detect links between its own entities. This can be useful as a part of data cleaning and de-duplication process [5], with respect to the set of relations to be discovered. For example, a Link Discovery task for "sameAs" relations can identify and remove duplicate ports by their position, in data sets that may have been acquired from different sources, since at each position there can be at most one port.

---

[1] In this chapter, entities correspond to spatial objects, so these terms are used interchangeably.

Let S and T denote the *source* and *target* data sets respectively. Then, entities from the source data set need to be linked to entities of the target data set. Furthermore, let $\mathcal{R}$ denote a set of formally defined relations as discussed in Sect. 7.2. The Link Discovery task computes a possibly empty set r of pairs $(t, s) \in r$, such that $t \in T$, $s \in S$ and $r \in \mathcal{R}$.

In the maritime domain, we are often interested to detect if a moving object has crossed or was within a region of interest for any time interval. To simplify the presentation, we focus on the special case of discovering the relation $\langle s, t \rangle \in within$ between any *point* $s \in S$ and any *polygon* $t \in T$. Later, we extend the discussion to the more general case of multiple relations and other spatial representations. Obviously, a brute-force approach for discovering all the relations between the given data sets would have to evaluate all pairs of entities in S, T with complexity $O(|S| \cdot |T|)$. In the following, we outline the basic technique for reducing this cost in the filtering step, which is known as *blocking*.

The key idea is that blocking identifies pairs of entities, from the source and target data sets respectively, that are considered as candidates to be linked. Next, the refinement step evaluates each pair of entities, i.e., it compares the entities of a pair under the scope of the relations of interest. For this reason, the refinement step has usually the principal cost factor that determines the overall performance of Link Discovery, since it requires the evaluation of distance functions on complex geometrical entities. Therefore, blocking techniques aim at reducing the number of candidate entities that need to be examined in the refinement step.

7.3.1 Blocking

Blocking is the process that organizes entities from the two data sets S and T in groups of entities (blocks), which can be processed independently and still discover all valid relations. Ideally, blocking should group entities in such a way that guarantees that entities in different blocks cannot be related, thereby avoiding many comparisons of pairs of entities.

In the case of spatial Link Discovery, the prevalent blocking mechanism is to apply *grid partitioning* of the 2D space, also known as *space tiling*. Essentially, the space is partitioned to adjacent, non-overlapping cells of equal size, and each entity is assigned to cell(s) given that its spatial representation and the spatial representation of any of these cells overlap. For example, in the case of entities represented by points, we need to consider the cell that includes the point, whereas in case of areas, we need to consider the set of cells overlapping with the area.

7.3.1.1 Pre-processing: Grid Construction

For the construction of the grid and its constituent cells, the first step is to derive the bounding box of the two data sets S and T. Essentially, this bounding box is the 2D data space, which is of interest for the current Link Discovery task. The bounding box is defined by two points: its lower left corner $(x_L, y_L) = (\min_{\forall i} x_i, \min_{\forall i} y_i)$ and its upper right corner $(x_U, y_U) = (\max_{\forall i} x_i, \max_{\forall i} y_i)$, where x_i and y_i represent the X- and Y-coordinate value of the ith point of a spatial object. Let a function $MBR(.)$ such that the bounding box of a given geometry g is denoted as $MBR(g) = \langle (x_L, y_L), (x_U, y_U) \rangle$.

The second step is to construct the grid cells. For this purpose, two parameters m_x and m_y need to be set, corresponding to the number of splits in the X-axis and Y-axis respectively. As a result, the number of grid cells is $m_x \cdot m_y$. These parameters also define the *granularity* of the grid, since the sides of a cell are $\frac{x_U - x_L}{m_x}$ and $\frac{y_U - y_L}{m_y}$ respectively. For example, a 2D grid in World Geodetic System (WGS84) reference system, with 0.5×1.5 granularity, is built by cells of size 0.5 by 1.5 degrees.[2]

The last step is to organize the entities of the given data sets in grid cells. This step is performed using a function that assigns each entity to one or more cells of the grid.

7.3.1.2   Filter by Blocking

The formal presentation of the blocking mechanism is presented in this section. An example is also provided in Sect. 7.3.1.3 to illustrate the process.

Given a set of cells forming a grid $\mathcal{G}$, let us consider the target data set $\mathcal{T}$ first. A basic functionality is how to derive the enclosing cell for a given spatial point (x, y) of an entity $t \in \mathcal{T}$. Let $c(i, j)$ denote a cell, where $i \in [0, m_x)$ and $j \in [0, m_y)$. For a point (x, y), its enclosing cell $c(i, j)$ is determined as follows:

$$i = \lfloor \frac{x - x_L}{\delta x} \rfloor, \quad j = \lfloor \frac{y - y_L}{\delta y} \rfloor$$

where $\delta x = \frac{x_U - x_L}{m_x}$ and $\delta y = \frac{y_U - y_L}{m_y}$.

Let $b(x, y)$ denote a function that computes the enclosing cell $c(i, j)$ for a given point with coordinates (x, y), i.e., $b(x, y) = c(i, j)$. For a given spatial region g defined by a set of points $\{(x_0, y_0), \ldots, (x_k, y_k)\}$ of an entity $t \in \mathcal{T}$, the set of enclosing cells $\mathbf{C}$ can be computed by the minimum bounding rectangle (MBR) of the geometry $\langle (x_L, y_L), (x_U, y_U) \rangle = MBR(g)$ as follows:

$$\mathbf{C} = \{c(i, j) | i \in [i_{min}, i_{max}], j \in [j_{min}, j_{max}]\}, \text{ where}$$
$$c(i_{min}, j_{min}) = b(x_L, y_L) \text{ and } c(i_{max}, j_{max}) = b(x_U, y_U)$$

Blocking the $MBR(g)$ of a geometry g can often include irrelevant cells. More sophisticated blocking techniques can improve the selection of cells, by processing each point $\{(x_0, y_0), \ldots, (x_k, y_k)\}$ of g and interpolating between successive points. Figure 7.2 illustrates the result of blocking Italy as a single entity (one multipolygon)

---

[2]Other Coordination Reference Systems (CRS) can be also used for the construction of the grid.

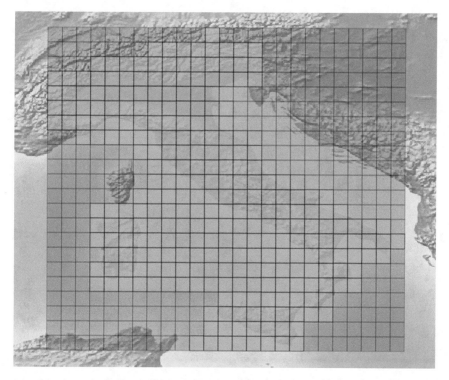

Fig. 7.2 Example of blocking Italy as a single target entity in a grid of 0.5×0.5 granularity. Red cells indicate blocking using MBR, while green cells illustrate blocking by interpolation

in a grid of 0.5×0.5 granularity, using the MBR of the geometry (illustrated as red cells), and point-by-point with interpolation (illustrated by green cells). Other examples of blocked polygons (highlighted with yellow), are illustrated in Fig. 7.3 where an entity can be a multipolygon extended in two cells, containing holes and overlapping with other geometries in the cell.

It follows that given any geometry g, we can always determine the non-empty set of cells $\mathbf{C}$, such that each $c \in \mathbf{C}$ either encloses the geometry g or overlaps with any part of g. We denote this process on a given geometry g, as $\mathbf{C} = block(g)$ and we say that the geometry g is *assigned* to cells $\mathbf{C}$, and $\mathbf{C}$ *block* the geometry g.

The filtering part of the filter-and-refine algorithm is based on the blocking function. A common strategy is to block each entity $t \in \mathcal{T}$ to populate the grid and then block entities of $\mathcal{S}$. Given that each relation holds between exactly one entity of $\mathcal{S}$ and exactly one entity of $\mathcal{T}$, only one data set needs to be preserved in memory. We often decide to preserve the smallest data set, or the data set that is less likely to be modified in the future (e.g., protected areas).

When all the entities $t \in \mathcal{T}$ have been assigned to cells, each cell $c(i, j)$ of the grid will block a possibly empty set of entities $\mathcal{T}_{(i,j)} \subseteq \mathcal{T}$, denoted as $\mathcal{T}_{(i,j)} = \{t | c(i, j) \in block(t), \forall t \in \mathcal{T}\}$.

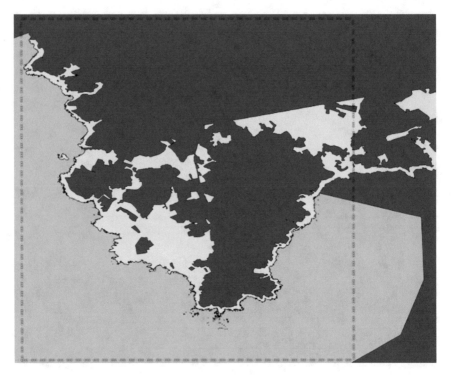

Fig. 7.3 Detail of blocking areas, red dashed lines indicate the borders of a cell, and yellow polygons highlight complex spatial representations, such as detailed multipolygons with holes, overlapping with other geometries and spanning across multiple cells

The blocking function that organizes all the entities $t \in \mathcal{T}$, is also applied for blocking each entity $s \in \mathcal{S}$. Thus, for each entity $s \in \mathcal{S}$ we obtain a set of entities $t \in T \subseteq \mathcal{T}$, that have been assigned to the same cells as s. For each entity $s \in \mathcal{S}$ blocking will result to a pair of entities (named as *candidate pairs*). Formally:

$$\mathbf{C} = block(s),$$
$$\mathcal{T}_C = \bigcup \mathcal{T}_{(i,j)}, \ \forall c(i,j) \in \mathbf{C},$$
$$\mathcal{Y}_s = \{\langle s, t \rangle | \forall t \in \mathcal{T}_C\}$$

where $\mathcal{Y}_s$ is the set of candidate pairs constructed for some $s \in \mathcal{S}$. It follows that $\mathcal{Y} = \bigcup \mathcal{Y}_s, \forall s \in \mathcal{S}$, i.e., $\mathcal{Y}$ denotes the set of all candidate pairs computed during filtering.

The final step of Link Discovery is the refinement, where each candidate pair of entities is processed in order to compute the set $\mathcal{H}_r$ of pairs satisfying the given relation. Formally, $\mathcal{H}_r = \{\langle s, t \rangle | \langle s, t \rangle \in r, r \in \mathcal{R}, \forall s \in \mathcal{S}, \forall t \in \mathcal{T}\}$. The complete set of pairs satisfying at least one relation in $\mathcal{R}$, is denoted as $\mathcal{H} = \bigcup \mathcal{H}_r, \forall r \in \mathcal{R}$. The

Fig. 7.4 Blocking example on Natura 2000 protected areas, in a grid of 0.5 × 0.5 degree granularity. Dashed lines indicate the borders of cells, while green polygons are the protected areas in the data set

blocking function has to guarantee that $\mathcal{H} \subseteq \mathcal{Y} \subseteq \mathcal{T} \times \mathcal{S}$. An ideal grid configuration and blocking function would result to $\mathcal{H} \equiv \mathcal{Y}$, i.e., the computed pairs of candidates are exactly the pairs satisfying the relation under consideration.

7.3.1.3 Example

Let $\mathcal{T}$ be a data set of Natura 2000 protected areas,[3] where each area is represented as a polygon. Let $\mathcal{S}$ a surveillance data set, reporting the positions of moving vessels. The Link Discovery task can be employed to detect topological relations between entities in $\mathcal{S}$ and $\mathcal{T}$. Specifically, the topological relations for the point-to-polygon combination of entities, are: $\mathcal{R} = \{within, \ touches, \ disjoint\}$. Two entities related by *within* or *touches* cannot be also related by *disjoint*, and vice versa. Figure 7.4 illustrates the result of blocking the target data set in a grid of 0.5 × 0.5 degree granularity, where the dashed lines indicate borders of cells, comprising 4413 Natura 2000 protected areas at sea. Figure 7.5 illustrates that the maximum number of target entities in a cell for this grid granularity is 39. Put differently, if a vessel position is enclosed in that cell, there will be at most 39 pairs to be evaluated for each relation in $\mathcal{R}$. Also, it appears that most of the cells contain exactly one Natura 2000 area, i.e., if a vessel position is assigned to any of these cells, exactly one pair (for each relation) will be evaluated at the refinement step.

---

[3] Available online https://zenodo.org/record/1167595 and https://zenodo.org/record/2576584.

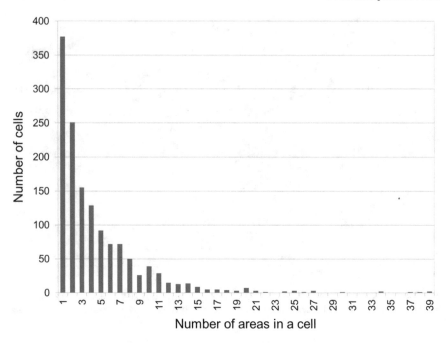

Fig. 7.5 Frequency of number of areas of target data set in cells of 0.5×0.5 grid granularity

Although in this example we have assumed that S reports positions of moving vessels (as points), more complex geometries can also be used (e.g., sequence of positions as trajectories, aggregates of trajectories as flows, etc.) in both $\mathcal{T}$ and S input data sets. In this case, it is expected that the block function will assign each $s \in S$ to more than one cells. This typically increases the candidate pairs of entities for s (and the comparisons during refinement), since it always holds that $|\mathcal{T}_{(i,j)}| \leq |\mathcal{T}_C|$, where $\mathcal{T}_{(i,j)}$ is the set of $t \in \mathcal{T}$ assigned to a single cell $c(i,j)$, when s is a point and $c(i,j)$ is the cell it is located in, and $\mathcal{T}_C$ the set of $t \in \mathcal{T}$ assigned to cell $c(i,j)$ and all the adjacent cells that overlap with s when s is a polygon. This can become more clear, when considering the case of blocking a point entity s, thus $|\mathbf{C}_p| = |block(s)| = 1$, and the case of blocking the buffered point $s' = buffer(s, k)$, by a factor k, such that $|\mathbf{C}_b| = |block(s')| > 1$. Then, it is expected that the total number of entities $t \in \mathcal{T}$ assigned to $\mathbf{C}_b$ are more than those assigned to $\mathbf{C}_p$ (i.e., by construction, $\mathbf{C}_b$ overlaps with $\mathbf{C}_p$ and at least one of its adjacent cells).

7.3.2 Generalization of Blocking

The aforementioned basic blocking technique demonstrates the use of Link Discovery on point-to-polygon topological relations. In the general case, the source and target data sets can contain any combination of geometry types, allowing for the detection of any type of spatio-temporal relation between the entities of the data sets. In the next paragraphs we consider the following generalizations of the aforementioned basic blocking technique: (a) the case of proximity relations, and (b) having multiple relations that need to be evaluated, instead of a single relation.

Proximity relations. In proximity-based Link Discovery, the task is to discover pairs of entities $s \in S$ and $t \in T$, whose distance is below an application-defined threshold θ. This proximity relation can be represented as $\langle s, t \rangle \in nearby_\theta$. We consider again the case where S is a data set of points, such as AIS position signals, and T a data set of polygons, expressing, for example, Natura 2000 areas. The blocking technique presented above needs to be modified.

The main difference is in the processing of the source data set. Namely, the blocking function may now return multiple cells for a given point (x, y) of an entity in S. This is due to the fact that such a point needs to be assigned to all cells that intersect with a circle centered at (x, y) and having radius θ, in order to ensure *correctness*. Given a distance function $d(s, t)$ that computes the distance between geometries s and t, correctness is ensured when: $\langle s, t \rangle \in nearby_\theta$ if and only if $d(s, t) \leq \theta$. Thus, the mapping of a point from S may result in more than one cells. A second difference concerns the refinement step, where the evaluation of a candidate pair should now involve a distance computation; however this is a straightforward modification.

Multiple relations. In the general case, a Link Discovery task needs to evaluate multiple relations R instead of a single relation. Given a source and a target data set, when multiple relations need to be discovered, changes are necessary both in the blocking as well as in the refinement phase.

In blocking, the assignment of entities of the source data set to cells should be guided by the most general relation in set R. For instance, consider the Link Discovery task in which we want to discover both $\langle s, t \rangle \in within$ and $\langle s, t \rangle \in nearby_\theta$, i.e., $R = \{within, nearby_\theta\}$. In this case, the blocking of entities of the source data set should be performed as dictated by relation $\langle s, t \rangle \in nearby_\theta$, which is more general than $\langle s, t \rangle \in within$, in the sense that it results in assigning blocks comprising cells enclosing the source entity (which is adequate for $\langle s, t \rangle \in within$), but also other cells based on the distance threshold θ, which is necessary for guaranteeing the correctness of $\langle s, t \rangle \in nearby_\theta$.

Furthermore, the refinement process that evaluates the pairs of candidates, can take into consideration the semantics and properties of spatial relations. For example, when considering the set of topological relations, a well-known property states that if two entities s, t are proved to be disjoint (i.e., there is no shared point on the perimeter or the internal areas of these geometries), then no other topological relation holds (recall that $nearby_\theta$ is a proximity relation) between s and t. In turn, this means that it is beneficial to process the *disjoint* relation first, and when $\langle s, t \rangle \in disjoint$, we can safely avoid evaluating any other topological relation between s and t, thereby saving significant computational cost. As another example, if two entities t_a, t_b of the target data set T are known to satisfy the relation $\langle t_a, t_b \rangle \in contains$, then for any $s \in S$ such that $\langle t_b, s \rangle \in contains$, it follows directly that $\langle t_a, s \rangle \in contains$. In this case, refinement takes into consideration the transitivity of relation *contains*, thus avoiding the evaluation of the pair $\langle t_a, s \rangle$.

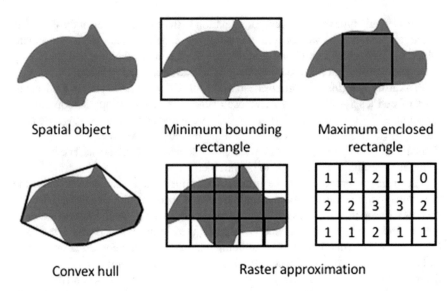

Fig. 7.6 Approximations of a spatial object

7.4 Special Topics, Extensions and Generalizations

7.4.1 Use of Approximations

Processing geometry data by means of geometric operations is typically costly, in particular in the case of large data sets, where such operations need to be performed for many objects. To reduce the associated processing cost of complex geometries, different types of approximations of spatial objects can be used: minimum bounding rectangle (MBR), convex hull, maximum enclosed rectangle (MER), and raster approximation.

Figure 7.6 depicts these approximations for an exemplary spatial object. A minimum bounding rectangle is defined as the minimum rectangle that contains the spatial object. It is represented with two points: the lower left corner $(x_L, y_L) = (\min_{\forall i} x_i, \min_{\forall i} y_i)$ and the upper right corner $(x_U, y_U) = (\max_{\forall i} x_i, \max_{\forall i} y_i)$, where x_i and y_i represent the X- and Y-coordinate value of the ith point of the spatial object. The maximum enclosed rectangle is defined as the largest rectangle enclosed in the spatial object. The convex hull of the set of points P representing the spatial object is the smallest convex set that contains P. Last, a raster approximation can form a signature of a spatial object, which is defined as a 4-valued grid. A cell with value 0 has no overlap with the spatial object, whereas value 3 indicates total coverage. Values 1 and 2 correspond to the cases of less than and more than 50% overlap respectively.

Object approximations can be used to determine fast if certain spatial relations hold between the respective spatial objects. We state the following main properties:

Property 1: *If two MBRs have no overlap, then the spatial objects cannot overlap.*

Property 2: *If two MERs have overlap, then the spatial objects also overlap.*

Property 3: *If a pair of common grid cells with values 2 or 3 are found, then the spatial objects overlap.*

Property 4: *If all common grid cells have value 0, then the spatial objects cannot overlap.*

The above properties can be used during the filtering phase, in order to discard pairs of spatial objects or report them as results, *without processing* their geometries. Let us consider the case of finding spatial objects that have overlap. When Property 1 holds, the spatial objects can be safely discarded, whereas when Property 2 holds, the pair of spatial objects can be immediately reported as result.

7.4.2 Handling Data Skew

As described so far, the blocking technique is based on the fact that all cells have equal size, thus forming an "equi-grid". From the overall performance perspective, the ideal distribution of data in such a grid would be a uniform distribution, i.e., each cell would be assigned exactly the same number of entities, such as AIS position signals, thus the ideal distribution of workload across the cells would be achieved. However, real-life data sets follow data distributions that only seldom resemble a uniform data distribution, for diverse reasons. In the maritime domain, the movement of vessels and their respective trajectories is determined by various external factors, such as location of origin and destination ports, obstacles (e.g., islands), bathymetric data, sea currents, etc. As a result, when the movement of vessels is visualized on a map, the depicted vessel positions do not resemble a uniform distribution of points in the 2D space, but they typically correspond to a skewed distribution (i.e., positions are not reported evenly in space, but only near to specific areas and routes). As an example, it is expected that the area around a port is a source of significantly more position reports, compared to any other area. This case is also illustrated in Fig. 7.7, where the total number of reported positions over the period of one month is illustrated in a 0.5×0.5 degree grid. This effect is also encountered in other maritime geographical data, e.g., protected areas, fishing areas, etc.

The presence of skew in the data complicates the Link Discovery task considerably. The main problem is that the underlying blocking mechanism cannot be an equi-sized grid anymore, as this would lead to grid cells with high variance with respect to the number of enclosed objects. In turn, for a cell that encloses a high number of objects, this results in many computations that are necessary in order to process a Link Discovery relation in that given cell, thus higher execution time. Therefore, this calls for alternative techniques that partition the data in cells of uneven size, aiming at minimizing the variance of the number of enclosed objects in cells.

Many approaches exist for handling the problem of *load balancing* spatial objects to cells, and they can be classified based on whether the cells are overlapping or disjoint:

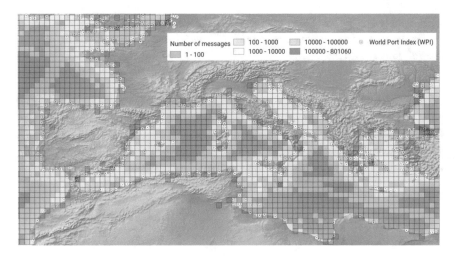

Fig. 7.7 Total number of reported positions during January 2016. White dots indicate positions of ports

- *Disjoint cells:* One approach is to create a grid of finer granularity and then solve an optimization problem of grouping cells in order to produce K groups that minimize the variance in terms of enclosed spatial objects. Another approach is to build a QuadTree [16], a data structure that recursively partitions the data space into four quadrants (cells), based on a user-specified capacity value for the maximum number of objects in a quadrant.
- *Overlapping cells:* Typical approaches include spatial indexing methods, such as R-trees [6], which result in potentially overlapping leaf nodes that contain the spatial objects. The leaf nodes of an R-tree can be used as cells, and by construction leaf nodes have a maximum capacity.

The disadvantage of having overlapping cells is that multiple cells may need to be examined in order to detect the link for a given spatial object; in fact, all overlapping cells with the given spatial object should be examined. On the other hand, the disadvantage of disjoint cells is that they typically partition the entire space, regardless of the presence of spatial objects, thus resulting in a higher number of cells that are not necessary.

7.4.3 Handling Empty Space

Let us consider the case of point-to-polygon Link Discovery using the relation: *within*. In several applications of spatial Link Discovery, such as the maritime domain, grid cells contain a significant amount of "empty space", namely the space of the cell that overlaps with no spatial objects. Then, if a point is located in this empty space, it

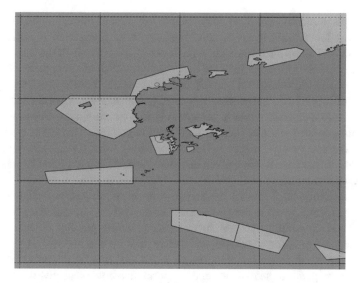

Fig. 7.8 Example of masks (in dark blue) of cells with polygons (in green), and empty cell (light blue)

cannot produce a link to any of the (say) n polygons of the same cell. Thus, if one could detect this case, the associated cost of n computations could be avoided.

The *MaskLink* technique [17, 20] is inspired by this observation. Its main idea is to explicitly model the empty area of each cell (called *mask*) as a separate polygon. This can be achieved by a geometrical operation (difference) between the rectangle representing the cell and the union of the intersections of each polygon in that cell with the cell. Then, the cell contains $n + 1$ polygons, instead of n. During the filtering phase of Link Discovery, the point at hand is first compared against the empty area, and if it is located there, then it is immediately discarded, saving $n - 1$ computations. If it is not located in the empty area, then the filtering continues as usual, comparing it to all polygons, thus resulting in $n + 1$ computations. Figure 7.8 illustrates grid cells that contain polygons (in green color) whereas the dark blue color represents the mask of each cell.

The benefit of using MaskLink depends on the size of the empty area, and it is expected that the gain is larger for cells with large-sized empty areas. Interestingly, the above technique can be generalized for more spatial relations and types of spatial objects, other than the ones used in the example above.

7.4.4 Handling the Temporal Dimension

So far, the Link Discovery tasks presented were related to purely spatial relations, thus ignoring the temporal dimension. However, when dealing with vessel trajectories, it is important to identify relations between moving vessels. For example, identifying vessels that move close in space and time may be of interest as an indication of possible collision. As another example, two vessels that remain at the same position for a long period of time can indicate illegal activity, such as smuggling.

The case of spatio-temporal Link Discovery addresses these topics. The objective is to identify pairs of entities that satisfy a spatial and a temporal relation. The most common data source that is subject to spatio-temporal Link Discovery is the surveillance data of vessels, such as AIS position streams. Therefore, we turn our attention to a special case proximity relation $nearby_{\theta,\tau}$, identifying pairs of positions of vessels with distance below a given distance threshold and temporal interval lower than a temporal threshold.

More formally, let S denote the source data set of surveillance data. Then, an object $s \in S$ corresponds to the position of a vessel and is represented by its location $(s.x, s.y)$ and timestamp $s.t$. The objective is to identify pairs of objects $s, s' \in S$ that correspond to different vessels, and satisfy: $d(s, s') \leq \theta$ and $t(s, s') \leq \tau$, where $d(s, s')$ denotes a distance function, $t(s, s') = |s.t - s'.t|$ is the temporal difference, and θ, τ correspond to the distance and temporal threshold respectively.

The solution to this spatio-temporal Link Discovery task can again exploit the grid-based approach outlined earlier for spatial Link Discovery. However, it may not be feasible to store the entire data set S in memory, since this is typically a stream of positions of moving vessels. Apart from that, the temporal threshold τ indicates that for any given timestamped object s, we only need to check against objects in the interval $[s.t - \tau, s.t + \tau]$ for potential links.

Therefore, the grid structure is used in an incremental way, by inserting (one-by-one) and maintaining the objects of S. When a new vessel position s is processed, in the filtering step, we examine the cells that intersect with a circle centered at $(s.x, s.y)$ with radius θ, and retrieve all vessel positions s'_i that satisfy the spatial constraint, i.e., $d(s, s'_i) \leq \theta$. Then, in the refinement step, we identify the subset of vessel positions $\{s'_i\}$ that in addition satisfy the temporal constraint, i.e., $t(s, s'_i) \leq \tau$. This solution avoids the exhaustive comparison of s to all other positions that have arrived before s.

However, in order to manage the available memory effectively, we need to find a suitable way to clean up the grid, since vessel positions that have arrived before more than τ time units will never produce a link to a new vessel position. A second reason to perform this cleaning operation is efficiency. If no cleaning were performed, then too many (old) vessel positions would be retrieved that satisfy the spatial constraint, but would be eliminated due to the temporal constraint, leading to wasteful processing.

A naive approach for cleaning would be to scan all grid cells (set at time t_{now}) and delete vessel positions s whose timestamp $s.t$ is more than τ units ago, i.e., $t_{now} - s.t > \tau$. However, this operation has linear complexity with respect to the number of positions in the grid. To address this problem, one solution is to introduce an auxiliary, bookkeeping data structure that efficiently detects the vessel positions that need to be deleted. For this purpose, a list of pointers to the vessel positions in the grid is maintained. The list is in temporal order, since vessel positions are inserted in the list in the order in which they arrive. Typically, vessel positions are processed ordered in time as they are recorded by surveillance/monitoring systems. When this is violated (due to technical constraints or noise), a buffer is used that keeps a pre-defined set of positions and performs temporal sorting, prior to assignment to grid cells. Then, cleaning can be performed efficiently, by traversing the list and deleting vessel positions (from the grid and list) until a position s is found with timestamp $t_{now} - s.t \leq \tau$. The maintenance cost of this list is small; insertions have $O(1)$ cost, whereas deletions have linear cost to the number of vessel positions that need to be deleted.

7.5 Common Issues in Link Discovery

In this section we highlight some of the common issues encountered when dealing with Link Discovery for spatio-temporal data. Although there is no silver bullet to solve every data related problem, we outline common tactics to analyze and handle each case separately.

One common problem is the Link Discovery between data sets that are not com-pletely spatially and/or temporally aligned. Link Discovery detects relations between entities of a source and a target data set. It follows that Link Discovery can detect spatio-temporal relations only between data sets that their bounding boxes overlap. A common tactic in this case is to compute the bounding boxes of each data set to verify that they overlap, thus there is a potential for discovering links. It also follows that Link Discovery can detect relations only in the intersection of the MBRs of the source and target data sets. Specifically, a relation can hold between exactly one entity of source and one entity of target data set. Thus, entities of source (respec-tively target) data set that are not in the intersection of the MBRs of the data sets can be safely ignored, i.e., there is no entity of target (respectively source) data set to create a candidate pair of entities. Therefore, Link Discovery can process only the entities in the intersection of MBRs of source and target data sets. Given that other data related issues also may need to be tackled, reducing the size of data sets also will preserve from any waste of resources on curating data that will never be used. Regarding the temporal dimension of data sets, verifying that the overall temporal intervals of data sets overlap, is not always sufficient. It is often necessary to verify that the intersection of these intervals is not empty, thus they can potentially lead to the discovery of links. These data analysis tasks can lead to the decision of which data sets (and subsets) to be used for Link Discovery and which relations can be expected in the results.

An important issue is the quality of the block function to be applied. For most of the cases blocking a spatial object by its bounding box can be a fast and sufficient method, i.e., it is easy to find the cells enclosing the bounding box of a geometry (thus the cells enclosing the blocked geometry). However, in some cases this method can assign the geometry to several unrelated cells, increasing the number of pairs of candidates that need to be validated during refinement. For example, consider the case of Link Discovery for proximity relations between vessels and a moving

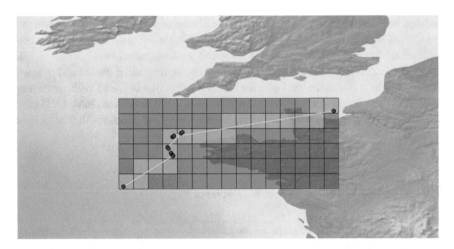

Fig. 7.9 Example of blocking a trajectory reconstructed from reported positions of a moving vessel. Green cells indicate blocking by geometry, while red cells indicate blocking by bounding box. Granularity of grid is 0.5×0.5

vessel in a given time period. Figure 7.9 illustrates such a case, where the trajectory of the vessel is reconstructed from its reported positions, red cells indicate the result of blocking the trajectory by its bounding box, while green cells indicate blocking by its geometry. If we consider that some of the red cells may contain ports (thus a few hundreds of vessels are also blocked at any time in these cells), the processing overhead for the refinement part of Link Discovery when blocking by bounding box is unnecessarily high. Any pair constructed by any position of the moving vessel and the positions of vessels blocked in red cells is wasteful processing, since it will not produce any links. However, blocking by geometry can become an expensive task with respect to the grid granularity, as the complexity of the block function is higher compared to blocking by bounding box. In addition to that, if the relations evaluated during the refinement part have low complexity, any additional effort to reduce the set of candidate pairs can affect the overall performance. In other words, more computational resources are needed to improve filtering, compared to the resources that would have been used for validating the unfiltered pairs. This example indicates that the type of blocking function to be applied is one of the first important decisions that needs to be taken, while also taking into consideration the nature of data sets, the complexity of relations, the bounding boxes of data sets, and the granularity of grid.

Also the origin of the data sets can trigger some issues. Data sets providing spatial information always impose precision and scale limitations, i.e., geometries in a spatial data set, represent real world entities in a predefined scale and precision, subject to data acquisition methods applied and preprocessing. When combining different data sets in a Link Discovery task, it is necessary to verify that the precision and scale of the data sets match, or they should be converted accordingly. Some data sets that have been modified by application of geoprocessing tools often carry more decimal digits

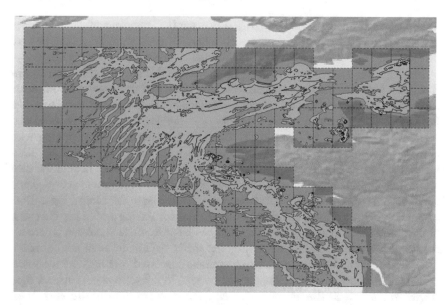

Fig. 7.10 The European Commission Fishing Areas data set, blocked in a grid of 0.5 × 0.5 granularity. The dashed lines indicate the borders of cells

than in the original data sets. This can lead to a waste of computational resources, since the original data were not as precise, as the computed results. For instance, computing the mask of a cell with a precision of a few centimeters on the surface of the earth is pointless, if the original data set has a rough representation of the entity in the cell. Similarly, when combining data sets of different scale in Link Discovery, the computed results can be inaccurate. For example, the Link Discovery task may detect a relation between pairs of entities, that are not actually related in the real world. Unfortunately, in this case reconstructing the data to the desired scale (if possible, e.g., from raw satellite images) is the safest solution.

Additionally, spatial vector data that are derived from data analysis, may result to large complex multipolygon geometries. For example, European Commission Fishing Areas[4] as identified by fishing vessel activity and illustrated in Fig. 7.10, is a relatively small data set, which contains 502 geometries for the MBR defined by $(x_L, y_L) = (45, -10)$, $(x_U, y_U) = (51, 0)$ degrees (WGS84). However, the largest geometry (covering approximately the 1/3 of the region) is formed by 247,161 points, and overlaps with most of the rest of the geometries in the data set. This means that the complex geometry will be assigned to several cells, thus it will frequently participate in pairs to be evaluated during the refinement process. Complex geometries are more expensive to process than simple geometries. In addition to that, a complex geometry that spans more than one cells increases the probability that it will be evaluated, and this also increases the overall processing time. Also, when converting raster to

---

[4] Available online at https://zenodo.org/record/1167595.

vector data, self-intersecting geometries may occur (i.e., segments of the boundaries of a geometry, crossing or touching each other). In general, these geometries are considered to be invalid and any validity routine can be used to detect them.

Finally, several issues are encountered in data sets that do not follow well known standards. For example, data sets collected by hand-held devices or custom made hardware/software may not follow common standards, and they can be in raw data formats or even binary file-dumps. In this case, the data need to be decoded and converted in a standard format, so as to enable data reuse and interlinking with other data sets. Decoding the data is most times a hard and time consuming task, which requires good perception of the domain and the semantics behind the data set. Reusability and integration of a decoded data set into a commonly accepted and well known schema will render the data set to be visible to the Linked Open Data community. This provides a wide range of options for linking with other data sets, or integrating and expanding the data set in hand. For example, there can be several regions recorded by surveillance devices (recording data in different raw formats), that can be integrated under a common schema and processed in a Link Discovery task as a single data set. Another example can be seen in the various port indices (i.e., *World Port Index*, *SeaDataNet Fishing Port Index*, *NaturalEarthData Port Index* to name just a few), each providing additional attributes to the reported ports. A conversion of each data set under a common schema will enable Link Discovery tasks on relations that require a combination of attributes provided in different data sets.

7.6 Application on Real-Life Data

In this section, we discuss Link Discovery tasks on topological and proximity relations between real world data sets from the area of Brest, France, accessible online at [14, 15]. Specifically, we demonstrate Link Discovery on topological relations between surveillance and contextual data sets, to detect reported positions of moving vessels within protected areas. We also compute proximity relations with a distance threshold of 0.003 degrees (approximately 100 m on the surface of the earth), to identify reported positions of vessels close to ports. The position of each port in this data set is reported as a point; however any other geometry type can be also supported, provided that a distance function between the position of a moving vessel and the port may be defined.

The next paragraphs describe the surveillance and contextual data sets used and justify the preprocessing tasks applied. The section concludes with a brief discussion on the findings.

7.6.1 Data Set Analysis

The contextual data sets with spatial features are provided as ESRI (Environmental Systems Research Institute) shapefiles with different spatial coverage; the coastline of Brittany (as in "C1 Ports of Brittany"), the European region (as in "C2 European Coastline" and "C5 Marine Protected Areas Natura 2000"), as well as worldwide coverage (as in "C1 World Port Index" and "C2 World EEZ"). Details on the spatial data sets are provided in Table 7.1. The first column provides the data set name, the second column reports the number of geometries in the data set, and the third column reports the type of geometries as one of {point, line, polygon}. The fourth column reports the number of attributes for each geometry (e.g., for World Port Index these include the name, the type, and other characteristics of each port) in the data set, and the fifth column reports the number of decimal places used in the coordinates of the geometries. Finally, the sixth, seventh and eighth columns, report the minimum, maximum and average number of points used for the geometries in each data set. We observe that the European coastline data set contains only one line geometry to describe the entire continent. We also observe that among the data sets that provide polygon geometry types, the data set "Fishing Areas" contains the highest number of maximum points in geometries, for the smallest MBR, i.e., the region of Brittany. This indicates that this data set contains large geometries compared to the MBR of the data set. This is good evidence that deciding a cell size such that it will partition the larger geometries can improve the performance of Link Discovery, with respect to the data sets used and the relations to be discovered.

We demonstrate the Link Discovery task using as source data set a surveillance data set available online [14]. This data set reports the positions of moving vessels in the region around Brittany in a comma separated (CSV) file. Specifically, it reports 4,637,652 positions of 5055 distinct moving vessels in the Brittany region, from 30th of September, 2015 22:00:00 to 31st of March, 2016 21:59:45. The coverage of the source data set shows that target data sets can be trimmed to the Brittany region before used in Link Discovery, in order to reduce resource requirements and improve performance, i.e., links between surveillance and contextual data sets can be discovered only in the region of Brittany.

The relations that will be discovered between the surveillance and contextual data sets are topological and proximity relations. For demonstrating the Link Discovery task, we focus on the *within* relation between positions of vessels and polygons of contextual data sets, and the proximity relation between positions of vessels and ports represented as points in the contextual data sets. Given the data set analysis of Table 7.1, we demonstrate the topological relation *within* using the contextual data sets "Fishing Areas", "Fishing constraints" and "Natura 2000". We demonstrate the proximity relation using the point data sets "Ports of Brittany", "SeaDataNet Port Index" and "World Port Index". The positions that are detected to be close to a port, can serve as departure or destination of a moving vessel, and thus they can be used to define the starting and ending point of a trajectory in the data set.

Table 7.1 Statistics on maritime data sets

Data set	Total geome-tries	Type	Attributes	Precision	MinPoints	MaxPoints	AvgPoints
[C1] Ports of Brittany	222	Point	7	16	1	1	1
[C1] Sea-DataNet Port Index	4,896	Point	9	5	1	1	1
[C1] World Port Index	3,684	Point	79	15	1	1	1
[C2] European Coastline	1	Line	2	11	2,912,707	2,912,707	2,912,707
[C2] IHO World Seas	101	Polygon	5	15	986	824,576	91,768.37
[C2] World EEZ	281	Polygon	24	15	4	1,693,733	36,430.56
[C4] FAO Maritime Areas	322	Polygon	12	13	5	37,346	1,639.48
[C4] Fishing Areas (European commis-sion)	502	Polygon	8	16	7	247,161	883
[C5] Fishing Con-straints	2	Polygon	2	16	797	993	895
[C5] Marine Protected Areas (EEA Natura 2000)	27,295	Polygon	6	15	4	862,147	2,395.57

7.6.2 Applying Link Discovery

Given the brief data analysis on the selected data sets, a set of data preparation tasks needs to be performed, before Link Discovery is initiated. Specifically, a common Coordination Reference System (CRS) needs to be selected and all the data sets should be converted to that. Most of the data sets in Table 7.1 (including the surveillance data) are provided in WGS84/SRID 4326. Any data set using different CRS has to be converted before any topological or proximity relations are computed to a common reference system (e.g., "[C5] maritime Natura 2000" is under ETRS89/SRID 3035).

Moreover, we have converted the selected data sets into RDF triples to enable linking with other Open Data on the Web in the future. Specifically, we have employed the datAcron ontology [21], which describes trajectories of moving objects and enables data transformations for analytics [7]. We have used RDF-Gen [19] for converting the data, populating the datAcron ontology and GeoSPARQL [1] for the geometries.[5]

The Link Discovery framework that we have used for the detection of relations performs automatically a set of processing tasks. It first computes the region that can provide topological and proximity relations. When both source and target data sets are archival, it scans the data to compute the intersection of the MBRs of the data sets. Any geometry beyond the boundaries of the intersection can be safely ignored, since no topological relations (other than disjoint) can be detected. In the case of online processing the boundaries can be either defined manually or by the target data set. The same holds for the proximity relation, if the boundaries of the intersection are buffered by the distance threshold applied.

The Link Discovery framework also computes the mask, i.e., the geometry representing the empty space of each cell in the grid and crops the geometries of data sets into cells of the grid. Both computations require valid spatial data, i.e., no self-intersecting geometries should be present, otherwise the computations are not possible. The mask of a cell as well as geometry cropping are computed only for the target data set. The computation of masks in cells applies for both topological and proximity relations; in the latter case the geometry is buffered by the distance threshold before the mask of the cell is computed. Thus, any geometry falling in the mask of the cell (and not within the buffer of any geometry), can safely be ignored as it will not be nearby to any geometry with respect to the distance threshold. Geometry cropping for data sets reporting point geometries (e.g., World Port Index) has no impact when only topological relations are to be detected.

Finally, a minor task in the preprocessing phase is to integrate data sets that share common properties. Given the analysis displayed in Table 7.1, it follows that the RDF triple sets generated from the data sets "[C1] Ports of Brittany", "[C1] SeaDataNet Port Index", "[C1] World Port Index", can be integrated since they report ports using the same geometry type (point). The integration of these three data sets appends the attributes of the ports. For example the data set "Ports of Brittany" does not

---

[5]The generated RDF triples from contextual data sets are available online at https://zenodo.org/deposit/2576584 and the surveillance data set at https://zenodo.org/deposit/2576152.

provide the type of shelter or harbor size of each port, which are described in the "World Port Index" data set. Also, some ports may not be reported in some of these data sets, as they may not be complete. This task is a simple text concatenation of the corresponding triple sets, since all the data sets have been converted to RDF triples under the same schema (ontology). The properties of the ports involved in Link Discovery, are mapped to elements named by the same terms in the schema. Thus, the integrated data set will be used as the target data set for the computation of relations between surveillance data and sea ports, instead of repeating the Link Discovery task for each one of the above three data sets, in which case the source data set would have also been processed three times.

The execution of Link Discovery has produced four data sets, one for each of the following cases:

LD1: proximity relations between surveillance data and ports,

LD2: topological relation *within*, between surveillance data and fishing areas,

LD3: topological relation *within*, between surveillance data and fishing constraints (represented as polygons),

LD4: topological relation *within*, between surveillance data and Natura 2000 protected areas.

The results of Link Discovery for proximity and topological relations are available at https://zenodo.org/record/2597641.

The LD1 case results in 167,209 links between reported positions and ports in the integrated target data set. This set of links involves 376 distinct vessels and only 37 distinct ports. This set of results also enables further useful analytics, such as the relations between the type of vessel and the duration being close to the port, the frequency of vessel types for each port, etc.

The LD2 case results in 984,880 links involving 4,246 distinct vessels and only 158 distinct fishing areas. This result set can also support further analytics and detection of critical events. For example, it may be dangerous for a non-fishing vessel to sail through a fishing area. The computed links are illustrated in Fig. 7.11.

The LD3 case results in 322,430 links, involving 519 distinct vessels and 2 fishing areas (representing fishing interdictions). This result set can guide the detection of fishing violations, i.e., any fishing vessel that is found to be stopped for some time period within the predefined regions should be further investigated for potential violation of a fishing interdiction. The results are illustrated in Fig. 7.12.

The LD4 case results in 891,215 links, involving 1,289 distinct vessels and only 60 Natura 2000 protected regions. Further analytics can be also supported using this result set, i.e., vessels carrying dangerous cargo within or crossing these protected areas may pose an environmental threat. The results are illustrated in Fig. 7.13.

7.6.3 Summary

The presentation above points to three fundamental aspects affecting Link Discovery:

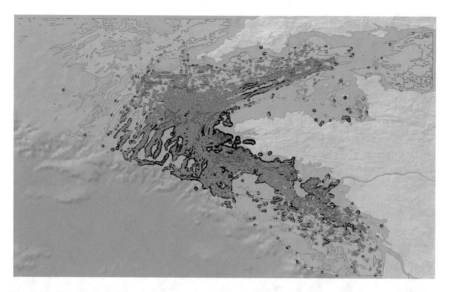

Fig. 7.11 Reported positions of vessels that are detected within fishing areas. Red dots indicate positions and the overlaying green regions illustrate fishing areas

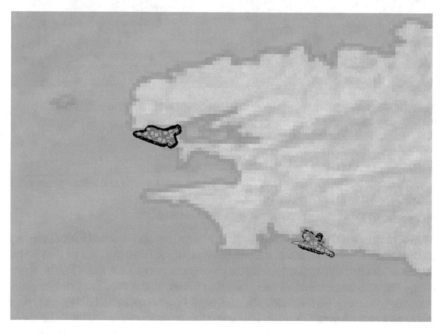

Fig. 7.12 Reported positions within areas of fishing interdictions. Dots indicate positions and the overlaying semitransparent regions illustrate fishing areas

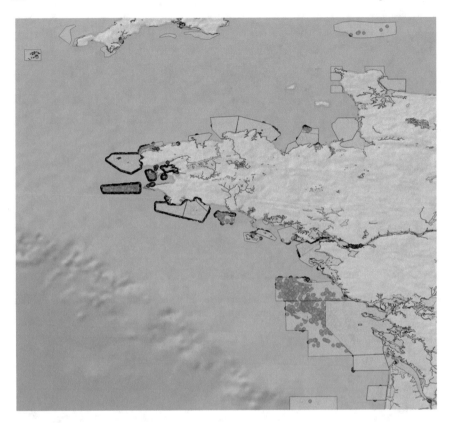

Fig. 7.13 Reported positions (represented as red dots), within Natura 2000 protected areas (represented as green polygons)

- data preperation,
- identification of properties of the data, such as geometry type and size of geometries with respect to grid granularity,
- knowledge about the semantics of data and relations.

In several cases, data sets need to be preprocessed before the actual Link Discovery task is initialized. For example, the computation of the bounding box of data sets is important as a first step to validate that the process can provide any links. Furthermore, computing the intersection of the bounding boxes of the data sets, can improve the Link Discovery performance with respect to the relations to be detected. As a minor preprocessing task, we can also consider the data conversion into a common projection, measurement units and schema (to allow the integration of data sets, as we have done for the case of ports).

Additionally, analyzing the data in the target and source data sets, leads to the identification of properties of the data (e.g., the number of geometries, the average or maximum size, the type of geometry, etc.) and supports the decision on which techniques should be applied for improving the performance of Link Discovery. For example, data sets providing large geometries with respect to grid granularity and the desired relations, are often suggested to be preprocessed, such that the geometries are cropped to the size of the cells. However, when used for relations between polygons, it may not always improve performance. For instance, to decide that a polygon of a source data set is covered by a cropped polygon of the target data set, it is necessary to compute the union of all parts of the target data set polygon that cover the MBR of the source polygon. In some cases this can be an expensive task, with respect to the granularity of the grid and the size and complexity of the polygons.

Another important property of the data is sparseness of geometries. In the data sets used for the demonstration presented above, we observed that for most of the cases, polygons are not adjacent (i.e., as it would have been in a data set about administrative regions, or country borders). This allows the use of mask, i.e., the computation of empty space in cells. On the other hand, for data sets providing adjacent geometries, this computation is pointless, since it will barely prune any candidate pairs of entities. Furthermore, the distribution of geometries in each data set can support the decision for grid granularity, in such a way that the empty space computed in cells prunes candidate pairs of entities, while the target entities assigned to cells remain relatively small. The grid granularity, can also affect the decision on the block function to be used, i.e., a fine-grained grid granularity, may increase the computational cost of a point-by-point block function.

Finally, the semantics between the relations to be detected need to be identified and exploited during the refinement process. For example, transitive relations, such as *contains*, or reflexive ones, such as *touches*, can prune candidate pairs of entities, since some pairs satisfying the relation can be implied by others, and thus they do not need to be evaluated.

7.7   Bibliographical Notes

Link Discovery is a challenging topic which relates to entity resolution and record linkage [4], deduplication [5], and data fusion [3]. A recent survey can be found in [11]. However, only limited work exists on spatio-temporal Link Discovery or on Link Discovery over streaming data.

Generic Link Discovery frameworks include LIMES [13] and SILK [8]. LIMES is a Link Discovery framework for metric spaces that uses the triangular inequality in order to avoid processing all pairs of objects. LIMES employs the concept of exemplars, which are used to represent areas in the multidimensional space, and prunes entire areas (and the respective enclosed entities) during Link Discovery. SILK [8] is a Link Discovery framework proposing a novel blocking method called MultiBlock, which uses a multidimensional index. The only work targeting a generic spatio-temporal Link Discovery framework is [18]. This framework supports all combinations of spatial relations and exploits the MaskLink technique to achieve high performance. Innovative features of this framework include the ability to operate on streaming data sources, supporting both 2D and 3D geometries, as well as supporting proximity-based LD tasks.

Spatial and spatio-temporal Link Discovery methods, such as ORCHID [12] and RADON [22], apply grid partitioning (a.k.a. space tiling), in order to perform efficiently the filtering step. Then, in the refinement step, different optimizations are employed in order to minimize the number of computations necessary to produce the correct result set. In [23], spatio-temporal Link Discovery is studied focusing on topological relations and specific data types (polygons). More recently, Santipantakis et al. [17, 20] have worked on spatio-temporal Link Discovery problems that are useful in the maritime domain.

Finally, spatial join techniques [9] are tightly related to the problem of spatial Link Discovery, since the underlying problem can be seen as a spatial join operation between two data sets.

Acknowledgements The research work was supported by the Hellenic Foundation for Research and Innovation (H.F.R.I.) under the "First Call for H.F.R.I. Research Projects to support Faculty members and Researchers and the procurement of high-cost research equipment grant" (Project Number: HFRI-FM17-81).

References

1. Open Geospatial Consortium. GeoSPARQL - A geographic query language for RDF data. OpenGIS Implementation Standard (2012). https://portal.opengeospatial.org/files/?artifact_id=47664
2. Allen, J.F.: Maintaining knowledge about temporal intervals. Commun. ACM **26**(11), 832–843 (1983)
3. Bleiholder, J., Naumann, F.: Data fusion. ACM Comput. Surv. **41**(1), 1:0–1:41 (2008)
4. Christen, P.: A survey of indexing techniques for scalable record linkage and deduplication. IEEE Trans. Knowl. Data Eng. **24**(9), 1537–1555 (2012)
5. Elmagarmid, A.K., Ipeirotis, P.G., Verykios, V.S.: Duplicate record detection: a survey. IEEE Trans. Knowl. Data Eng. **19**(1), 1–16 (2007)
6. Guttman, A.: R-trees: a dynamic index structure for spatial searching. In: SIGMOD'84, Proceedings of Annual Meeting, Boston, Massachusetts, USA, June 18–21, pp. 47–57 (1984)
7. Vouros, G., Santipantakis, G., Doulkeridis, C., Vlachou, A., Andrienko, G., Andrienko, N., Fuchs, G., Martinez, M.G., Cordero, J.M.G.: The datacron ontology for the specification of semantic trajectories: specification of semantic trajectories for data transformations supporting visual analytics. J. Data Semant. (In Press)
8. Isele, R., Jentzsch, A., Bizer, C.: Efficient multidimensional blocking for link discovery without losing recall. In: Proceedings of WebDB (2011)
9. Jacox, E.H., Samet, H.: Spatial join techniques. ACM Trans. Database Syst. **32**(1), 7 (2007)
10. Mamoulis, N.: Spatial data management. Synthesis Lectures on Data Management. Morgan & Claypool Publishers, San Rafael (2011)
11. Nentwig, M., Hartung, M., Ngomo, A.N., Rahm, E.: A survey of current link discovery frameworks. Semant. Web **8**(3), 419–436 (2017)
12. Ngomo, A.N.: ORCHID - reduction-ratio-optimal computation of geo-spatial distances for link discovery. In: Proceedings of ISWC, pp. 395–410 (2013)
13. Ngomo, A.N., Auer, S.: LIMES - a time-efficient approach for large-scale link discovery on the web of data. In: Proceedings of IJCAI, pp. 2312–2317 (2011)

14. Patroumpas, K., Spirelis, D., Chondrodima, E., Georgiou, H., Petrou, P., Tampakis, P., Sideridis, S., Pelekis, N., Theodoridis, Y.: Final dataset of trajectory synopses over AIS kinematic messages in Brest area (ver. 0.8) (2018). https://doi.org/10.5281/zenodo.2563256

15. Ray, C., Dreo, R., Camossi, E., Jousselme, A.L.: Heterogeneous Integrated Dataset for Maritime Intelligence, Surveillance, and Reconnaissance (2018). https://doi.org/10.5281/zenodo.1167595

16. Samet, H.: The quadtree and related hierarchical data structures. ACM Comput. Surv. **16**(2), 187–260 (1984)

17. Santipantakis, G.M., Doulkeridis, C., Vouros, G.A., Vlachou, A.: Masklink: efficient link discovery for spatial relations via masking areas (2018). arXiv:1803.01135

18. Santipantakis, G.M., Glenis, A., Doulkeridis, C., Vlachou, A., Vouros, G.A.: stLD: towards a spatio-temporal link discovery framework. In: SBD@SIGMOD 2019 (2019)

19. Santipantakis, G.M., Kotis, K.I., Vouros, G.A., Doulkeridis, C.: Rdf-gen: Generating RDF from streaming and archival data. In: WIMS, pp. 28:1–28:10. ACM (2018)

20. Santipantakis, G.M., Vlachou, A., Doulkeridis, C., Artikis, A., Kontopoulos, I., Vouros, G.A.: A stream reasoning system for maritime monitoring. In: 25th International Symposium on Temporal Representation and Reasoning, TIME 2018, Warsaw, Poland, October 15–17, pp. 20:1–20:17 (2018)

21. Santipantakis, G.M., Vouros, G.A., Doulkeridis, C., Vlachou, A., Andrienko, G.L., Andrienko, N.V., Fuchs, G., Garcia, J.M.C., Martinez, M.G.: Specification of semantic trajectories supporting data transformations for analytics: the datacron ontology. In: Proceedings of the 13th International Conference on Semantic Systems, SEMANTICS 2017, Amsterdam, The Netherlands, September 11–14, pp. 17–24 (2017). https://doi.org/10.1145/3132218.3132225

22. Sherif, M.A., Dreßler, K., Smeros, P., Ngomo, A.N.: Radon - rapid discovery of topological relations. In: Proceedings of AAAI, pp. 175–181 (2017)

23. Smeros, P., Koubarakis, M.: Discovering spatial and temporal links among RDF data. In: Proceedings of LDOW (2016)

24. Vouros, G.A., Vlachou, A., Santipantakis, G.M., Doulkeridis, C., Pelekis, N., Georgiou, H.V., Theodoridis, Y., Patroumpas, K., Alevizos, E., Artikis, A., Claramunt, C., Ray, C., Scarlatti, D., Fuchs, G., Andrienko, G.L., Andrienko, N.V., Mock, M., Camossi, E., Jousselme, A., Garcia, J.M.C.: Big data analytics for time critical mobility forecasting: recent progress and research challenges. In: Proceedings of the 21th International Conference on Extending Database Technology, EDBT 2018, Vienna, Austria, March 26–29, pp. 612–623 (2018)

Chapter 8
Composite Maritime Event Recognition

Manolis Pitsikalis and Alexander Artikis

Abstract Composite maritime event recognition systems support maritime situational awareness as they allow for the real-time detection of dangerous, suspicious and illegal vessel activities. To illustrate the use of such systems, we motivate and present a series of composite maritime event patterns in a formal language. For effective recognition, the presented maritime patterns have been developed in close collaboration with domain experts, and evaluated with the use of real-world Automatic Identification System (AIS) datasets.

8.1 Introduction

Systems for composite event recognition—'event pattern matching'—accept as input a stream of time-stamped simple, derived events [7, 16, 24]. A simple or 'low-level event' is the result of applying a computational derivation process to some other event, such as an event coming from a sensor. Using simple events as input, event recognition systems identify composite events of interest—collections of events that satisfy some pattern. The pattern of a composite event ('high-level event') imposes temporal and, possibly, atemporal constraints on its sub-events, i.e. simple events or other composite events. Consider, as an example, the recognition of fishing activity, such as trawling, given, among others, streams of Automatic Identification System (AIS) position signals and polygons expressing fishing areas.

Systems for composite *maritime* event recognition have been attracting considerable attention for economic as well as environmental reasons. Terosso-Saenz et al., for example, presented a system detecting when two vessels are in danger of colliding

M. Pitsikalis
Department of Computer Science, University of Liverpool, Liverpool, UK
e-mail: E.pitsikalis@liverpool.ac.uk

M. Pitsikalis · A. Artikis (✉)
Institute of Informatics & Telecommunications, NCSR Demokritos, Athens, Greece
e-mail: a.artikis@unipi.gr

A. Artikis
Department of Maritime Studies, University of Piraeus, Piraeus, Greece

© Springer Nature Switzerland AG 2021
A. Artikis and D. Zissis (eds.), *Guide to Maritime Informatics*,
https://doi.org/10.1007/978-3-030-61852-0_8

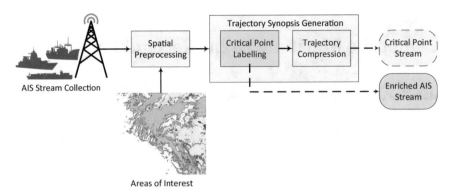

Fig. 8.1 Steps prior to composite maritime event recognition

[43]. SUMO is an open-source system combining AIS streams with synthetic aperture radar images for detecting illegal oil dumping, piracy and unsustainable fishing [19]. Mills et al. presented a method for identifying trawling using speed and directionality rules, thus helping in studies of how trawling impacts on species, habitats and the ecosystem [27]. In [31], we presented a maritime monitoring system with a component for trajectory simplification, allowing for efficient maritime stream analytics, and a composite event recognition component, combining vessel position streams with contextual (geographical) knowledge, for real-time vessel activity recognition.

Prior to composite maritime event recognition, two online tasks/steps must be performed:

- Computation of a set of spatial relations among vessels, such as proximity, and among vessels and areas of interest, such as when a vessel sails within a protected area.
- Labelling position signals of interest as 'critical'—for example, when a vessel changes its speed, turns, stops, moves slowly or stops transmitting its position.

Figure 8.1 illustrates these steps. AIS position signals are streamed into the system and go through a spatial preprocessing step, for the computation of the spatial relations required by the composite event patterns. The details of this step are presented in the chapter of Santipantakis et al. [37].

Then, critical point labelling is performed, as part of trajectory synopsis generation, whereby major changes along each vessel's movement are tracked. This process can instantly identify critical points along each trajectory, such as a stop, a turn or slow motion. Using these critical points, it is possible to reconstruct a vessel trajectory with small acceptable deviations from the original one. The details of this process are presented in the chapter of Patroumpas [30]. Subsequently, the position signals may be consumed for composite maritime event recognition either directly (see 'enriched AIS stream' in Fig. 8.1), or after being compressed, i.e. after removing all signals that have not been labelled as critical (see 'critical point stream' in Fig. 8.1).

Composite maritime event recognition systems rely on a library of patterns expressing various types of activity of interest, such as patterns for ship-to-ship transfer, (illegal) fishing, loitering, drifting, and so on. van Laere et al., for instance, evaluated a workshop aiming at the identification of potential vessel anomalies, such as tampering, rendez-vous between vessels and unusual routing [23]. For effective recognition, maritime event patterns should be defined in close collaboration with domain experts, and be amenable to continuous refinement by means of machine learning techniques [26].

In this chapter, we motivate and present a set of composite maritime event patterns in the language of the Run-Time Event Calculus, a composite event recognition system with a formal, declarative semantics [6]. The patterns have been evaluated in terms of predictive accuracy and efficiency, by domains experts and with the use of real, large AIS datasets [33]. The composite maritime events that were recognised by applying these patterns on the dataset of Brest, France [35, 36], are publicly available [32].

The remainder of this chapter proceeds as follows. In Sect. 8.2 we present the Run-Time Event Calculus, i.e. the composite event recognition engine used for illustration purposes throughout the chapter. Then, in Sect. 8.3 we present the maritime patterns in the language of the Run-Time Event Calculus. Section 8.4 presents the evaluation of the maritime patterns, while Sects. 8.5 and 8.6 summarise the chapter and point to the relevant literature.

8.2 Run-Time Event Calculus

The Event Calculus is a logic-based formalism for representing and reasoning about events and their effects [22]. The Run-Time Event Calculus (RTEC) is an open-source[1] logic programming (Prolog) implementation of the Event Calculus, that has been developed for composite event recognition [6]. RTEC has a formal, declarative semantics—composite event patterns are (locally) stratified logic programs [34]. Moreover, RTEC includes optimisation techniques for efficient pattern matching, such as 'windowing', whereby all input events that took place prior to the current window are discarded/'forgotten'. Details about the reasoning algorithms of RTEC, complexity and empirical analyses, may be found in [6, 33].

The time model in RTEC is linear and includes integer time-points. Following the logic programming convention, variables start with an upper-case letter, while predicates and constants start with a lower-case letter. Where F is a *fluent*—a property that is allowed to have different values at different points in time—the term $F = V$ denotes that fluent F has value V. Boolean fluents are a special case in which the possible values are true and false. For example, $trawling(Vessel) = $ true may be used to represent the trawling activity of some *Vessel*. holdsAt($F = V, T$) represents that fluent F has value V at a particular time-point T. holdsFor($F = V, I$) represents

---

[1]https://github.com/aartikis/RTEC.

Table 8.1 Main predicates of RTEC. '$F = V$' denotes that fluent F has value V

Predicate	Meaning
happensAt(E, T)	Event E occurs at time T
holdsAt($F = V, T$)	The value of fluent F is V at time T
holdsFor($F = V, I$)	I is the list of the maximal intervals for which $F = V$ holds continuously
initiatedAt($F = V, T$)	At time T a period of time for which $F = V$ is initiated
terminatedAt($F = V, T$)	At time T a period of time for which $F = V$ is terminated
union_all(L, I)	I is the list of maximal intervals produced by the union of the lists of maximal intervals of list L
intersect_all(L, I)	I is the list of maximal intervals produced by the intersection of the lists of maximal intervals of list L
relative_complement_all(I', L, I)	I is the list of maximal intervals produced by the relative complement of the list of maximal intervals I' with respect to every list of maximal intervals of list L

that I is the list of the maximal intervals for which $F = V$ holds continuously. holdsAt and holdsFor are defined in such a way that, for any fluent F, holdsAt($F = V, T$) if and only if T belongs to one of the maximal intervals of I for which holdsFor($F = V, I$).

Data stream elements are represented by means of the happensAt predicate, which expresses an instance of an event type. happensAt(*entersArea*(*Vesseal, AreaID*), 5), for example, denotes that a *Vessel* entered an area with *AreaID* at time-point 5. A knowledge base of maritime patterns in RTEC includes rules that define the effects of events with the use of the initiatedAt and terminatedAt predicates, and the values of the fluents with the use of the holdsFor predicate. Examples will be presented shortly. Table 8.1 summarises the main predicates of RTEC.

Fluents may be 'simple' or 'statically determined'. For a simple fluent F, $F = V$ holds at a time-point T if $F = V$ has been initiated by an event that has occurred at some time-point earlier than T, and has not been terminated at some other time-point in the meantime. This is an implementation of *the law of inertia*. To compute the intervals I for which $F = V$, that is, holdsFor($F = V, I$), we find all time-points T_s at which $F = V$ is initiated, and then, for each T_s, we compute the first time-point T_f after T_s at which $F = V$ is terminated.

Statically determined fluents are defined by means of application-dependent holdsFor rules, along with the interval manipulation constructs of RTEC: union_all, intersect_all and relative_complement_all. See Table 8.1 for a brief explanation of these constructs and Fig. 8.2 for an example visualisation. The interval manipulation constructs of RTEC support the following type of pattern: for all time-points T, $F = V$ holds at T if and only if some Boolean combination of fluent-value pairs holds at T. For some fluents, this is a much more concise pattern than the traditional style of Event Calculus representation of 'simple' fluents, i.e. identifying the various conditions under which the fluent is initiated and terminated, so that maximal intervals can then be computed using the domain-independent holdsFor, as explained in the

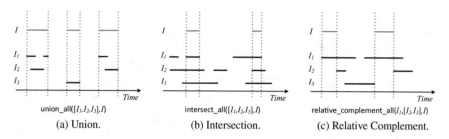

Fig. 8.2 A visual illustration of the interval manipulation constructs of RTEC. In these examples, there are three input streams, I_1, I_2 and I_3, coloured black. The output of each interval manipulation construct I is coloured light blue

paragraph above. Below, we discuss the representation of maritime patterns in RTEC as simple and statically determined fluents.

8.3 Maritime Patterns

We have been collaborating with domain experts in order to develop a set of patterns of maritime activity that allow for maritime situational awareness [33]. We begin by presenting a set of building blocks that will be later used for the construction of more involved patterns.

8.3.1 Building Blocks

8.3.1.1 Vessel Within Area of Interest

Calculating the time periods during which a vessel is in an area of some type, such as a Natura 2000, fishing or anchorage area, is particularly useful in maritime (e.g. fishing) patterns. Consider the formalisation below in the language of RTEC:

Table 8.2 Composite event/activity recognition. The input events are detected either at the spatial preprocessing step or by the trajectory synopsis generator (critical events). With the exception of *proximity*, all items of the input stream are instantaneous, while all output activities are durative

		Event/Activity	Description
Input	Spatial	*entersArea(Vessel, AreaID)*	A *Vessel* enters area with *AreaID*
		leavesArea(Vessel, AreaID)	A *Vessel* leaves area with *AreaID*
		proximity(Vessel₁, Vessel₂)	*Vessel₁* and *Vessel₂* are close
	Critical	*gap_start(Vessel)*	A *Vessel* stopped sending position signals
		gap_end(Vessel)	A *Vessel* resumed sending position signals
		slow_motion_start(Vessel)	A *Vessel* started moving at a low speed
		slow_motion_end(Vessel)	A *Vessel* stopped moving at a low speed
		stop_start(Vessel)	A *Vessel* started being idle
		stop_end(Vessel)	A *Vessel* stopped being idle
		change_in_speed_start(Vessel)	A *Vessel* started changing its speed
		change_in_speed_end(Vessel)	A *Vessel* stopped changing its speed
		change_in_heading(Vessel)	A *Vessel* changed its heading
Output	Composite	*highSpeedNC(Vessel)*	A *Vessel* has high speed near coast
		anchoredOrMoored(Vessel)	A *Vessel* is anchored or moored
		drifting(Vessel)	A *Vessel* is drifting
		trawling(Vessel)	A *Vessel* is trawling
		tugging(Vessel₁, Vessel₂)	*Vessel₁* and *Vessel₂* are engaged in tugging
		pilotBoarding(Vessel₁, Vessel₂)	*Vessel₁* and *Vessel₂* are engaged in pilot boarding
		rendezVous(Vessel₁, Vessel₂)	*Vessel₁* and *Vessel₂* are having a rendez-vous
		loitering(Vessel)	A *Vessel* is loitering
		sar(Vessel)	A *Vessel* is engaged in a search and rescue (SAR) operation

$$\text{initiatedAt}(\textit{withinArea}(\textit{Vessel}, \textit{AreaType}) = \text{true}, \ T) \leftarrow$$
$$\text{happensAt}(\textit{entersArea}(\textit{Vessel}, \textit{AreaID}), \ T),$$
$$\textit{areaType}(\textit{AreaID}, \textit{AreaType}).$$

$$\text{terminatedAt}(\textit{withinArea}(\textit{Vessel}, \textit{AreaType}) = \text{true}, \ T) \leftarrow \qquad (8.1)$$
$$\text{happensAt}(\textit{leavesArea}(\textit{Vessel}, \textit{AreaID}), \ T),$$
$$\textit{areaType}(\textit{AreaID}, \textit{AreaType}).$$

$$\text{terminatedAt}(\textit{withinArea}(\textit{Vessel}, \textit{AreaType}) = \text{true}, \ T) \leftarrow$$
$$\text{happensAt}(\textit{gap\_start}(\textit{Vessel}), \ T).$$

Recall that variables start with an upper-case letter, while predicates and constants start with a lower-case letter. *withinArea(Vessel, AreaType)* is a simple fluent indicating that a *Vessel* is within an area of some type. Simple fluents are defined by means of initiatedAt and terminatedAt predicates, expressing the conditions in which a fluent is, respectively, initiated and terminated. As mentioned in Sect. 8.2, according to the built-in representation of the law of inertia in RTEC, for a simple fluent F, $F = V$ holds at a time-point T if $F = V$ has been initiated by an event that has occurred at some time-point earlier than T, and has not been terminated at some other time-point in the meantime. The first body predicate of initiatedAt($F = V$, T) and terminatedAt($F = V$, T) rules is happensAt, expressing the event occurrence initiating/terminating $F = V$. The happensAt predicate may be followed by other happensAt predicates, expressing the occurrences of other (typically concurrent) events, holdsAt predicates and atemporal predicates. Details about the language of RTEC may be found in its manual.[2]

We chose to define *withinArea(Vessel, AreaType)* as a Boolean fluent, as opposed to a multi-valued one (i.e. *withinArea(Vessel)* = *AreaType*), since areas of different types may overlap, and thus a vessel may be within several areas at the same time. *entersArea(Vessel, AreaID)* and *leavesArea(Vessel, AreaID)* are input events computed at the spatial preprocessing step—see Fig. 8.1 and the top part of Table 8.2—indicating that a *Vessel* entered (respectively left) an area with *AreaID*. *areaType(AreaID, AreaType)* is an atemporal predicate storing the areas of interest of a given dataset. *withinArea(Vessel, AreaType)* = true is initiated when a *Vessel* enters an area of *AreaType*, and terminated when the *Vessel* leaves the area of *AreaType*. *withinArea(Vessel, AreaType)* = true is also terminated when the trajectory synopsis generator produces a *gap_start* event (see Fig. 8.1 and the middle part of Table 8.2), indicating the beginning of a communication gap (in the subsection that follows we discuss further communication gaps). In this case we chose to make no assumptions about the location of the vessel. With the use of rule-set (8.1), RTEC computes the *maximal intervals* during which a vessel is said to be within an area of some type.

8.3.1.2 Communication Gap

According to the trajectory synopsis generator, a communication gap takes place when no message has been received from a vessel for at least 30 minutes All such numerical thresholds, however, may be tuned, by machine learning algorithms [17], for example, to meet the requirements of the application under consideration. A communication gap may occur when a vessel sails in an area with no AIS receiving station nearby, or when the transceiver of the vessel is deliberately turned off. The rules below present a formalisation of communication gap:

---

[2]https://github.com/aartikis/RTEC/blob/master/RTEC_manual.pdf.

Table 8.3 Speed-related building blocks

Fluent	Min speed (knots)	Max speed (knots)
stopped(*Vessel*)	0	0.5
lowSpeed(*Vessel*)	0.5	5
movingSpeed(*Vessel*) = *below*	0.5	Min service speed of vessel type
movingSpeed(*Vessel*) = *normal*	Min service speed of vessel type	Max service speed of vessel type
movingSpeed(*Vessel*) = *above*	Max service speed of vessel type	–
trawlingSpeed(*Vessel*)	1.0	9.0
tuggingSpeed(*Vessel*)	1.2	15

$$
\begin{aligned}
&\mathsf{initiatedAt}(gap(Vessel) = nearPorts,\ T) \leftarrow \\
&\quad \mathsf{happensAt}(gap\_start(Vessel),\ T), \\
&\quad \mathsf{holdsAt}(withinArea(Vessel, nearPorts) = \mathsf{true},\ T). \\
&\mathsf{initiatedAt}(gap(Vessel) = farFromPorts,\ T) \leftarrow \\
&\quad \mathsf{happensAt}(gap\_start(Vessel),\ T), \\
&\quad \mathsf{not\ holdsAt}(withinArea(Vessel, nearPorts) = \mathsf{true},\ T). \\
&\mathsf{terminatedAt}(gap(Vessel) = \_Value,\ T) \leftarrow \\
&\quad \mathsf{happensAt}(gap\_end(Vessel),\ T).
\end{aligned}
\tag{8.2}
$$

gap is a simple, multi-valued fluent, *gap_start* and *gap_end* are input critical events (see the middle part of Table 8.2), 'not' expresses Prolog's negation-by-failure [12], while variables starting with '_', such as _*Value*, are 'free'. *withinArea* was defined in the previous section. We chose to distinguish between communication gaps occurring near ports from those occurring in the open sea, as the former usually do not have a significant role in maritime monitoring. According to rule-set (8.2), a communication gap is said to be initiated when the synopsis generator emits a 'gap start' event, and terminated when a 'gap end' is detected. Given this rule-set, RTEC computes the maximal intervals during which a vessel is not sending position signals.

8.3.1.3 Speed-Related Building Blocks

We have defined a number of speed-related building blocks that are useful in the specification of the more involved maritime patterns, which will be presented in the following section. Table 8.3 presents these building blocks. *stopped*(*Vessel*), for example, indicates that a *Vessel* has speed between 0 and 0.5 knots. All numerical thresholds have been set in collaboration with domain experts. Moreover, as mentioned earlier, these thresholds may be optimised for different monitoring applications.

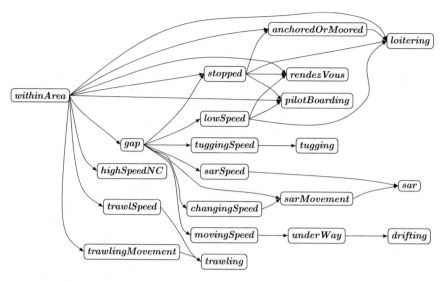

Fig. 8.3 Maritime pattern hierarchy

stopped (*Vessel*) is specified by means of the *stop_start* and *stop_end* input critical events (see Table 8.2). Similarly, *lowSpeed* (*Vessel*) indicates that a *Vessel* sails at a speed between 0.5 and 5 knots, and is defined by means of the *slow_motion_start* and *slow_motion_end* critical events. *stopped* (*Vessel*) and *lowSpeed* (*Vessel*) are independent of the vessel type. In contrast, the multi-valued *movingSpeed* (*Vessel*) fluent takes into consideration the minimum and maximum service speed—the speed maintained by a ship under normal load and weather conditions—of the type of *Vessel*. For example, *movingSpeed* (*Vessel*) = *normal* when the type of *Vessel* is cargo and its speed is between 9 and 15 knots. *trawlingSpeed* (*Vessel*) is restricted to fishing vessels, while *tuggingSpeed* (*Vessel*) is used for the specification of tugging, i.e. the activity of pulling a ship into a port.

8.3.2 Maritime Situational Indicators

We present a set of Maritime Situational Indicators expressing activities of special significance. Figure 8.3 displays the hierarchy of our formalisation, i.e. the relations between the indicators' specifications. In this figure, an arrow from fluent *A* to fluent *B* denotes that *A* is used in the specification of *B*. To avoid clutter, we omit the presentation of the input stream elements—items at the top part of Table 8.2—from Fig. 8.3. In what follows, we present a fragment of our formalisation.

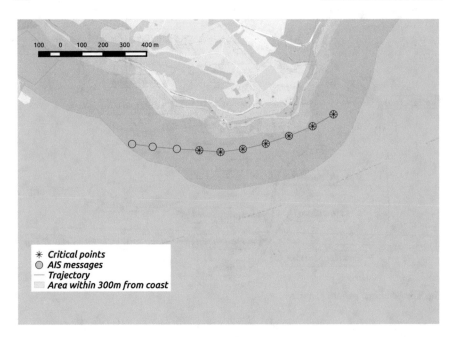

Fig. 8.4 A vessel near the port of Brest, France, with speed above the 5 knots limit. The marked circles denote the AIS position signals that are labelled as 'critical' by the synopsis generator

8.3.2.1 Vessel with High Speed Near Coast

Several countries have regulated maritime zones. In French territorial waters, for example, there is a speed limit of 5 knots for vessels or water-crafts within 300 meters from the coast. One of the causes of marine accidents near the coast is vessels sailing with high speed; thus the early detection of violators supports safety by improving the efficiency of law enforcement. Figure 8.4 displays an example of speed limit violation. A formalisation, in the language of RTEC, of this type of violation may be found below:

initiatedAt($highSpeedNC(Vessel)$ = true, T) ←

    happensAt($velocity(Vessel, Speed, \_CoG, \_TrueHeading)$, T),

    holdsAt($withinArea(Vessel, nearCoast)$ = true, T),

    $threshold(v_{hs}, V_{hs})$, $Speed > V_{hs}$.

terminatedAt($highSpeedNC(Vessel)$ = true, T) ←   (8.3)

    happensAt($velocity(Vessel, Speed, \_CoG, \_TrueHeading)$, T),

    $threshold(v_{hs}, V_{hs})$, $Speed \leq V_{hs}$.

terminatedAt($highSpeedNC(Vessel)$ = true, T) ←

    happensAt(end($withinArea(Vessel, nearCoast)$ = true), T).

$highSpeedNC(Vessel)$ is a Boolean simple fluent indicating that a *Vessel* is exceeding the speed limit imposed near the coast. *velocity* is input contextual information expressing the speed, course over ground (CoG) and true heading of a vessel (the use of the last two parameters will be illustrated in the following sections). This information is attached to each incoming AIS message. Recall that variables starting with '_' are free. $withinArea(Vessel, nearCoast)$ = true expresses the time periods during which a *Vessel* is within 300 meters from the French coastline (see rule-set (8.1) for the specification of *withinArea*). *threshold* is an auxiliary atemporal predicate recording the numerical thresholds of the maritime patterns. The use of this predicate supports code transferability, since the use of different thresholds for different applications requires only the modification of the *threshold* predicate, and not the modification of the patterns. end($F = V$) (respectively start($F = V$)) is an RTEC built-in event indicating the ending (resp. starting) points of the intervals during which $F = V$ holds continuously. According to rule-set (8.3), therefore, $highSpeedNC(Vessel)$ = true is initiated when a *Vessel* sails within 300 meters from the French coastline with speed above 5 knots. Moreover, $highSpeedNC(Vessel)$ = true is terminated when the speed of the *Vessel* goes below 5 knots, the *Vessel* sails away (further than 300 meters) from the coastline, or stops sending position signals (recall that *withinArea* is terminated/ended by *gap_start*).

8.3.2.2 Anchored or Moored Vessel

A vessel lowers its anchor in specific areas—for example, waiting to enter into a port, or taking on cargo or passengers where insufficient port facilities exist. See Fig. 8.5 for an example of a vessel stopped in an anchorage area, and the chapter of Andrienko and Andrienko for a discussion on archoring events [3]. Furthermore, vessels may be moored, i.e. when a vessel is secured with ropes in any kind of permanent fixture, such as a quay or a dock. Consider the specification below:

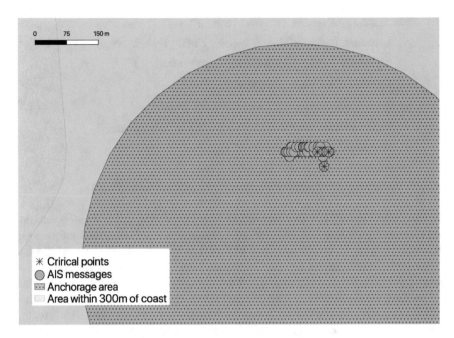

Fig. 8.5 Anchored vessel

$$
\begin{aligned}
&\text{holdsFor}(anchoredOrMoored(Vessel) = \text{true}, \ I) \leftarrow \\
&\quad \text{holdsFor}(stopped(Vessel) = farFromPorts, \ I_{sffp}), \\
&\quad \text{holdsFor}(withinArea(Vessel, anchorage) = true, \ I_{wa}), \\
&\quad \text{intersect\_all}([I_{sffp}, I_{wa}], I_{sa}), \quad\quad\quad\quad\quad\quad (8.4) \\
&\quad \text{holdsFor}(stopped(Vessel) = nearPorts, \ I_{sn}), \\
&\quad \text{union\_all}([I_{sa}, I_{sn}], I_i), \\
&\quad threshold(v_{aorm}, V_{aorm}), \quad intDurGreater(I_i, V_{aorm}, I).
\end{aligned}
$$

anchoredOrMoored(Vessel) is a statically determined fluent, i.e. it is specified by means of a domain-dependent holdsFor predicate and interval manipulation constructs—intersect_all and union_all in this case, that compute, respectively, the intersection and union of lists of maximal intervals (see Table 8.1 and Fig. 8.2). Recall that *stopped* is a fluent recording the intervals in which a vessel is stopped (see Table 8.3)—this may be far from all ports or near some port(s). *intDurGreater(I', V_t, I)* is an auxiliary predicate keeping only the intervals I of list I' with length greater than V_t. *anchoredOrMoored(Vessel)* = true, therefore, holds when the *Vessel* is stopped in an anchorage area or near some port, for a time period greater than some threshold (see V_{aorm} in rule (8.4)). The default value for this threshold was set to 30 minutes as suggested by domain experts. Alternatively, one may employ the techniques presented in [3] to set this threshold value.

Fig. 8.6 Interval computation example of *anchoredOrMoored*

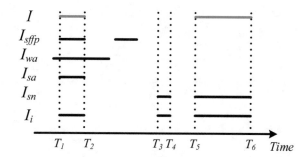

Figure 8.6 illustrates, with the use of a simple example, the computation of the *anchoredOrMoored* intervals. The displayed intervals I, I_{sffp}, etc., correspond to the intervals of rule (8.4); for instance, I_{sffp} expresses the list of maximal intervals during which a *Vessel* has stopped far from ports. In the example of Fig. 8.6, the second interval of I_i, i.e. $[T_3, T_4]$, is discarded since it is not long enough according to the V_{aorm} threshold.

8.3.2.3   Drifting Vessel

A vessel is drifting when its course over ground, i.e. the direction calculated by the GPS signal, is heavily affected by sea currents or harsh weather conditions. Typically, as illustrated in Fig. 8.7, when the course over ground deviates from the true heading of a sailing vessel, i.e. the direction of the ship's bow, then the vessel is considered drifting. Consider the formalisation below:

$$
\begin{aligned}
&\mathsf{initiatedAt}(drifting(Vessel) = \mathsf{true},\ T) \leftarrow \\
&\quad \mathsf{happensAt}(velocity(Vessel, \_Speed, CoG, TrueHeading),\ T), \\
&\quad angleDiff(CoG, TrueHeading, Ad), \\
&\quad threshold(v_{ad}, V_{ad}),\quad Ad > V_{ad}, \\
&\quad \mathsf{holdsAt}(underWay(Vessel) = \mathsf{true},\ T).
\end{aligned}
$$

$$
\begin{aligned}
&\mathsf{terminatedAt}(drifting(Vessel) = \mathsf{true},\ T) \leftarrow \qquad\qquad (8.5) \\
&\quad \mathsf{happensAt}(velocity(Vessel, \_Speed, CoG, TrueHeading),\ T), \\
&\quad angleDiff(CoG, TrueHeading, Ad), \\
&\quad threshold(v_{ad}, V_{ad}),\quad Ad \le V_{ad}. \\
&\mathsf{terminatedAt}(drifting(Vessel) = \mathsf{true},\ T) \leftarrow \\
&\quad \mathsf{happensAt}(end(underWay(Vessel) = \mathsf{true}),\ T).
\end{aligned}
$$

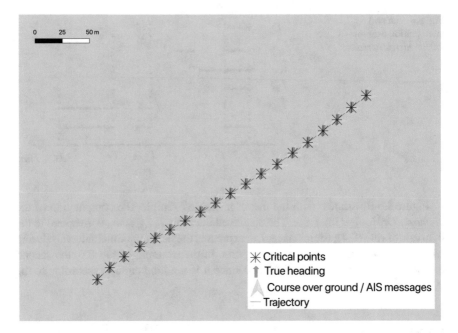

Fig. 8.7 A drifting vessel. In this example, all AIS position signals have been labelled as 'critical' (*change_in_heading*)

drifting is a Boolean simple fluent, while, as mentioned earlier, *velocity* is input contextual information, attached to each AIS message, expressing the speed, course over ground (CoG) and true heading of a vessel. *angleDiff*(*A*, *B*, *C*) is an auxiliary predicate calculating the absolute minimum difference *C* between two angles *A* and *B*. The intervals during which *underWay*(*V*) = true are computed by the union of the intervals during which the vessel is moving, i.e. *movingSpeed*(*V*) = *below*, *movingSpeed*(*V*) = *normal* and *movingSpeed*(*V*) = *above* (see Table 8.3). The use of *underWay* in the initiation and termination conditions of *drifting* (see rule-set (8.5)) expresses the constraint that stationary vessels cannot be drifting.

8.3.2.4 Trawling

Fishing exploits natural resources and thus needs to be regulated to safeguard fair access and sustainability. Maritime monitoring enables better regulation and monitoring of fishing activities. A common fishing method is trawling, involving a boat—trawler—pulling a fishing net through the water behind it. The trawler has steady (trawling) speed and a wide heading angle distribution. See Fig. 8.8 for an illustration. A formalisation of trawling movement may be found below:

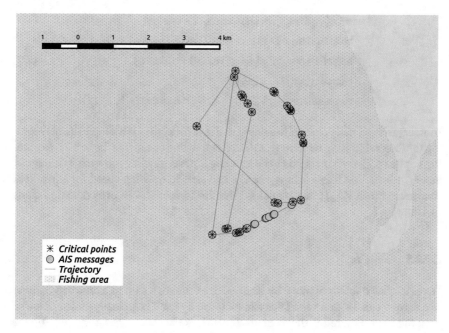

Fig. 8.8 A fishing vessel trawling

$$\text{initiatedAt}(trawlingMovement(Vessel) = \text{true}, \ T) \leftarrow$$
$$\text{happensAt}(change\_in\_heading(Vessel), \ T),$$
$$vesselType(Vessel, fishing),$$
$$\text{holdsAt}(withinArea(Vessel, fishing) = \text{true}), \ T). \tag{8.6}$$
$$\text{terminatedAt}(trawlingMovement(Vessel) = \text{true}, \ T) \leftarrow$$
$$\text{happensAt}(\text{end}(withinArea(Vessel, fishing) = \text{true}), \ T).$$

trawlingMovement is defined as a simple fluent, *change_in_heading* is an input critical event detected at the trajectory synopsis step, and *vesselType* is an auxiliary atemporal predicate recording the vessel types of a given dataset. *trawlingMovement* is subject to the 'deadlines' mechanism of RTEC, i.e. this fluent is automatically terminated after a designated period of time—10 minutes, in this pattern—has elapsed since its last initiation. We omit the corresponding RTEC declarations to simplify the presentation. As shown in rule-set (8.6), *trawlingMovement* is also terminated when the vessel in question leaves the fishing area. (In other applications, it may be desirable to relax the constraint of restricting attention to designated fishing areas.) Consequently, *trawlingMovement*(*Vessel*) is true as long as a fishing *Vessel* performs a sequence of heading changes, each taking place at the latest 10 minutes after the previous one, while sailing within a fishing area. Trawling may then be specified as follows:

$$\begin{aligned}
&\mathsf{holdsFor}(trawling(Vessel) = \mathsf{true},\ I) \leftarrow \\
&\quad \mathsf{holdsFor}(trawlingMovement(Vessel) = \mathsf{true},\ I_{tc}), \\
&\quad \mathsf{holdsFor}(trawlingSpeed(Vessel) = \mathsf{true},\ I_t), \hspace{3cm} (8.7) \\
&\quad \mathsf{intersect\_all}([I_{tc}, I_t], I_i), \\
&\quad threshold(v_{trawl}, V_{trawl}),\ \ intDurGreater(I_i, V_{trawl}, I).
\end{aligned}$$

trawling is a statically determined fluent and *trawlingSpeed* is a simple fluent record-ing the intervals during which a vessel sails at trawling speed (see Table 8.3). Accord-ing to rule (8.7), a vessel is said to be trawling if it has trawling movement (implying, among others, that it is a fishing vessel), and sails in trawling speed for a period of time greater than V_{trawl} (the default value for this threshold was set by domain experts to 1 hour).

8.3.2.5 Tugging

A vessel that should not move by itself—for example, a ship in a crowded harbour or a narrow canal—or a vessel that cannot move by itself is typically pulled or towed by a tug boat. Figure 8.9 shows an example. It is expected that during tugging the two vessels are close to each other and their speed is lower than normal, for safety

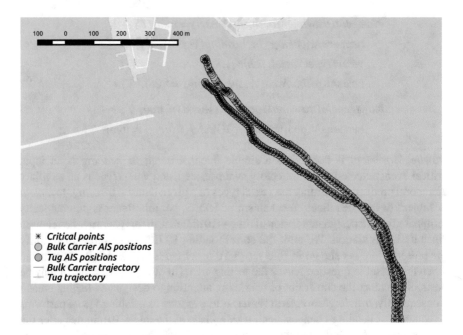

Fig. 8.9 Example of bulk carrier tugging

and manoeuvrability reasons. We have formalised tugging as follows:

$$
\begin{aligned}
&\text{holdsFor}(tugging(Vessel_1, Vessel_2) = \text{true},\ I) \leftarrow \\
&\quad oneIsTug(Vessel_1, Vessel_2),\quad \text{not } oneIsPilot(Vessel_1, Vessel_2), \\
&\quad \text{holdsFor}(proximity(Vessel_1, Vessel_2) = \text{true},\ I_p), \\
&\quad \text{holdsFor}(tuggingSpeed(Vessel_1) = \text{true},\ I_{ts1}), \\
&\quad \text{holdsFor}(tuggingSpeed(Vessel_2) = \text{true},\ I_{ts2}), \\
&\quad \text{intersect\_all}([I_p, I_{ts1}, I_{ts2}],\ I_i), \\
&\quad threshold(v_{tug}, V_{Tug}),\quad intDurGreater(I_i, V_{Tug}, I).
\end{aligned}
\tag{8.8}
$$

tugging is a relational fluent referring to a pair of vessels, as opposed to the fluents presented so far that concern a single vessel. *oneIsTug*($Vessel_1$, $Vessel_2$) is an auxiliary predicate that becomes true when one of $Vessel_1$, $Vessel_2$ is a tug boat. Similarly, *oneIsPilot*($Vessel_1$, $Vessel_2$) becomes true when one of $Vessel_1$, $Vessel_2$ is a pilot boat. (Piloting will be discussed shortly.) *proximity* is a durative input fluent computed at the spatial preprocessing step (see Table 8.2), expressing the time periods during which two vessels are 'close', i.e. their distance is less than 100 meters. *tuggingSpeed* is a simple fluent expressing the intervals during which a vessel is said to be sailing at tugging speed (see Table 8.3). According to rule (8.8), two vessels are said to be engaged in tugging if one of them is a tug boat, neither of them is a pilot boat, and, for at least V_{Tug} time-points, they are close to each other and sail at tugging speed.

8.3.2.6 Piloting

During piloting, a highly experienced sailor in navigation in specific areas—a maritime pilot—approaches with a pilot boat, boards and manoeuvres another vessel through dangerous or congested areas. Maritime pilots are navigational experts with knowledge of a particular area, such as its depth, currents and hazards. Piloting, therefore, is of major importance for maritime safety. A formalisation of pilot boarding may be found below:

$$
\begin{aligned}
&\mathsf{holdsFor}(pilotBoarding(Vessel_1, Vessel_2) = \mathsf{true},\ I) \leftarrow \\
&\quad oneIsPilot(Vessel_1, Vessel_2),\quad \mathsf{not}\ oneIsTug(Vessel_1, Vessel_2), \\
&\quad \mathsf{holdsFor}(proximity(Vessel_1, Vessel_2) = \mathsf{true},\ I_p), \\
&\quad \mathsf{holdsFor}(lowSpeed(Vessel_1) = \mathsf{true},\ I_{l1}), \\
&\quad \mathsf{holdsFor}(stopped(Vessel_1) = farFromPorts,\ I_{s1}), \\
&\quad \mathsf{union\_all}([I_{l1}, I_{s1}],\ I_1), \\
&\quad \mathsf{holdsFor}(lowSpeed(Vessel_2) = \mathsf{true},\ I_{l2}), \\
&\quad \mathsf{holdsFor}(stopped(Vessel_2) = farFromPorts,\ I_{s2}), \\
&\quad \mathsf{union\_all}([I_{l2}, I_{s2}],\ I_2), \\
&\quad \mathsf{intersect\_all}([I_1, I_2, I_p],\ I_f), \\
&\quad \mathsf{holdsFor}(withinArea(Vessel_1, nearCoast) = \mathsf{true},\ I_{nc1}), \\
&\quad \mathsf{holdsFor}(withinArea(Vessel_2, nearCoast) = \mathsf{true},\ I_{nc2}), \\
&\quad \mathsf{relative\_complement\_all}(I_f,\ [I_{nc1}, I_{nc2}],\ I_i), \\
&\quad threshold(v_{pil}, V_{pil}),\quad intDurGreater(I_i, V_{pil}, I).
\end{aligned}
\tag{8.9}
$$

pilotBoarding is a relational fluent referring to a pair of vessels, while *lowSpeed* is a fluent recording the intervals during which a vessel sails at low speed (see Table 8.3). relative_complement_all is an interval manipulation construct of RTEC (see Table 8.1 and Fig. 8.2). According to rule (8.9), *pilotBoarding(Vessel_1, Vessel_2)* holds when one of *Vessel_1, Vessel_2* is a pilot vessel, neither of them is a tug boat, *Vessel_1* and *Vessel_2* are close to each other and they are stopped or sail at low speed far from the coast. According to domain experts, the boarding procedure in pilot operations takes place far from the coast for safety reasons.

8.3.2.7 Vessel Rendez-Vous

A scenario that may indicate illegal activities, such as illegal cargo transfer, is when two vessels are nearby in the open sea, stopped or sailing at a low speed. See Fig. 8.10 for an illustration. A specification of 'rendez-vous', or 'ship-to-ship transfer', may be found below:

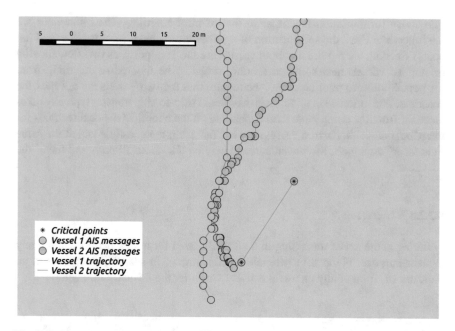

Fig. 8.10 Vessels in close proximity. In this example, the vessels started sailing at a low speed before they came close to each other. Hence, these critical events (*slow_motion_start*) are not displayed in the figure

$$
\begin{aligned}
&\mathsf{holdsFor}(rendezVous(Vessel_1, Vessel_2) = \mathsf{true},\ I) \leftarrow \\
&\quad \mathsf{not}\ oneIsTug(Vessel_1, Vessel_2),\quad \mathsf{not}\ oneIsPilot(Vessel_1, Vessel_2), \\
&\quad \mathsf{holdsFor}(proximity(Vessel_1, Vessel_2) = \mathsf{true},\ I_p), \\
&\quad \mathsf{holdsFor}(lowSpeed(Vessel_1) = \mathsf{true},\ I_{l1}), \\
&\quad \mathsf{holdsFor}(stopped(Vessel_1) = farFromPorts,\ I_{s1}), \\
&\quad \mathsf{union\_all}([I_{l1}, I_{s1}], I_1), \\
&\quad \mathsf{holdsFor}(lowSpeed(Vessel_2) = \mathsf{true},\ I_{l2}), \\
&\quad \mathsf{holdsFor}(stopped(Vessel_2) = farFromPorts,\ I_{s2}), \\
&\quad \mathsf{union\_all}([I_{l2}, I_{s2}], I_2), \\
&\quad \mathsf{intersect\_all}([I_1, I_2, I_p], I_f), \\
&\quad \mathsf{holdsFor}(withinArea(Vessel_1, nearPorts) = \mathsf{true},\ I_{np1}), \\
&\quad \mathsf{holdsFor}(withinArea(Vessel_2, nearPorts) = \mathsf{true},\ I_{np2}), \\
&\quad \mathsf{holdsFor}(withinArea(Vessel_2, nearCoast) = \mathsf{true},\ I_{nc1}), \\
&\quad \mathsf{holdsFor}(withinArea(Vessel_2, nearCoast) = \mathsf{true},\ I_{nc2}), \\
&\quad \mathsf{relative\_complement\_all}(I_f, [I_{np1}, I_{np2}, I_{nc1}, I_{nc2}], I_i), \\
&\quad threshold(v_{rv}, V_{rv}),\quad intDurGreater(I_i, V_{rv}, I).
\end{aligned}
\tag{8.10}
$$

The above formalisation is similar to that of pilot boarding. The differences are the following. First, the specification of *rendezVous* excludes pilot vessels (and tug boats). Second, we require that both vessels are far from ports, as two slow moving or stopped vessels near some port would probably be moored or departing from the port. Similar to pilot boarding, we require that the two vessels are not near the coastline. The rationale in this case is that illegal ship-to-ship transfer typically takes place far from the coast. Note that, depending on the chosen distance thresholds for *nearCoast* and *nearPorts*, a vessel may be 'far' from the coastline and at the same time 'near' some port. Moreover, a vessel may be 'far' from all ports and 'near' the coastline.

8.3.2.8 Loitering

Loitering is the act of remaining in a particular area for a long period without any evident purpose. Figure 8.11 presents an illustration. In sea, this behaviour is an indicator of a potentially unlawful activity. Consider the formalisation below:

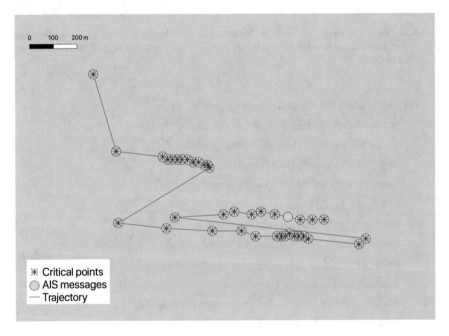

Fig. 8.11 A vessel loitering

$$\text{holdsFor}(loitering(Vessel) = \text{true},\ I) \leftarrow$$
$$\text{holdsFor}(stopped(Vessel) = farFromPorts,\ I_s),$$
$$\text{holdsFor}(lowSpeed(Vessel) = \text{true},\ I_l),$$
$$\text{union\_all}([I_s, I_l], I_{ls}),$$
$$\text{holdsFor}(withinArea(Vessel, nearCoast) = \text{true},\ I_{wa}),$$
$$\text{holdsFor}(anchoredOrMoored(Vessel) = \text{true},\ I_{am}),$$
$$\text{relative\_complement\_all}(I_{ls}, [I_{am}, I_{wa}], I_i),$$
$$\text{threshold}(v_{ltr}, V_{ltr}),\quad intDurGreater(I_i, V_{ltr}, I).$$

$$(8.11)$$

According to rule (8.11), a vessel is said to loiter when it is stopped or sails at a low speed far from ports and the coastline, and it is not anchored, for a period greater than V_{ltr} time-points. The default value for the V_{ltr} parameter was set to 30 minutes.

8.4 Empirical Analysis

For illustration purposes, we present an evaluation of the maritime patterns in terms of accuracy and efficiency using the real-world dataset of Brest, France [35, 36]. Figure 8.12 displays the geographical coverage of the dataset, while Table 8.4 outlines its characteristics. Approximately, the dataset consists of 18M AIS position signals from 5K vessels sailing in the Atlantic Ocean around the port of Brest, between October 1st, 2015, and March 31st, 2016. The spatial preprocessing module produced 374K spatio-temporal events linking vessels with, among others, 263 fishing areas and 1.2K Natura 2000 areas, as well as vessels with other vessels. The trajectory synopsis generator labelled 4.6M position signals as critical. See Fig. 8.1 for the steps required prior to composite maritime event recognition.

Table 8.4 Dataset characteristics

Attribute	Value
Period (months)	6
Vessels	5 K
Position signals	18 M
Spatio-temporal events	374 K
Critical position signals	4.6 M
Fishing areas	263
Natura 2000 areas	1.2 K
Anchorage areas	9
Ports	222

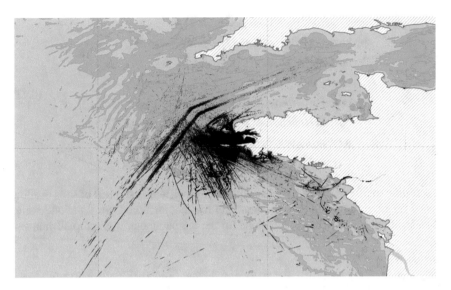

Fig. 8.12 Vessel positions from AIS streams in Brest, France (after the chapter of Etienne et al. [15])

The experiments were conducted using the open-source RTEC composite event recognition engine, under YAP-6 Prolog, in a machine running Ubuntu 16.04.3 LTS, with Intel Core i7-i7700 CPU @ 3.60GHz x 8 and 16 GB 2133 MHz RAM. The input of RTEC consisted of streams of the input events shown in Table 8.2, while instances of all composite events shown in this table were recognised [32].

8.4.1 Accuracy

Similar to most AIS datasets [45], the Brest dataset does not have an annotation of composite events. For this reason, domain experts were asked to provide feedback on the recognised instances of a representative subset of the formalised events: 'anchored or moored', 'trawling', 'tugging', 'pilot boarding' and 'rendez-vous' [25]. Since this evaluation process is highly time-consuming, as there are several thousand instances of recognised events, the experts provided feedback only for the first month of the dataset. Table 8.5 presents the number of True Positives (TP), False Positives (FP) and the Precision score for the selected events. False Negatives and hence Recall could not be computed due to the absence of complete ground truth. The results show nearly perfect scores. The four False Positives of *anchoredOrMoored* were caused by four vessels that continued transmitting position signals while on land. Concerning *rendezVous*, the experts stated that for two recognised instances of this pattern, there were too few position signals to classify the events as rendez-vous. Hence the two False Positives shown in Table 8.5.

Table 8.5 Precision based on expert feedback

Composite event	TP	FP	Precision
anchoredOrMoored(*Vessel*)	3067	4	0.999
trawling(*Vessel*)	29	0	1.000
tugging(*Vessel*)	117	0	1.000
pilotBoarding(*Vessel₁*, *Vessel₂*)	80	0	1.000
rendezVous(*Vessel₁*, *Vessel₂*)	52	2	0.963

8.4.2 Efficiency

Figure 8.13 presents the experimental results in terms of efficiency. The results concern the 'critical point stream', where the non-critical AIS messages have been discarded, as well as the 'enriched AIS stream', where all AIS messages have been retained (see Fig. 8.1). RTEC performs composite event recognition over a sliding window [6]. In these experiments, we varied the window size ω from 2 hours to 16 hours, while the slide step was set to 2 hours. Overlapping windows, as in the cases of $\omega = 4$, 8 and 16 hours in our experiments, are common in maritime monitoring, since AIS messages may arrive with (substantial) delay at the composite event recognition engine. This is especially the case for AIS messages arriving via satellites, i.e. when vessels sail out of the range of terrestrial antennas.

Figure 8.13a displays the average number of input events per window size, while Fig. 8.13b shows the average recognition times for all patterns. As expected, the performance is better in the smaller critical point stream rather than the enriched AIS stream. (The effects of trajectory compression on the predictive accuracy of composite event recognition are discussed in [33].) In both streams, composite event recognition is performed in real-time. For example, a window of 16 hours, including more than 50 K events in the critical point stream, is processed in less than a second in a single core of a standard desktop computer.

We may trivially parallelise composite event recognition by allocating different patterns to different processing units. Figure 8.13c shows the average recognition times per pattern. As discussed in Sect. 8.3, the maritime patterns form a hierarchy, in the sense that the specification of a pattern depends on the specification of other, lower-level patterns. For example, the specification of *tugging* depends on the specifications of *tuggingSpeed*, *gap* and *withinArea* (see Fig. 8.3). Thus, in Fig. 8.13c, the displayed recognition times of a pattern (for example, *tugging*) include the recognition times of all other patterns that contribute to its specification.

Figure 8.13c shows that the most demanding patterns are those of *trawling* and *sar*, i.e. search and rescue operations (see [33] for the specification of this pattern). This is due to the use of the 'deadlines' mechanism of RTEC in these patterns, whereby a fluent is automatically terminated some designated time after its last initiation.

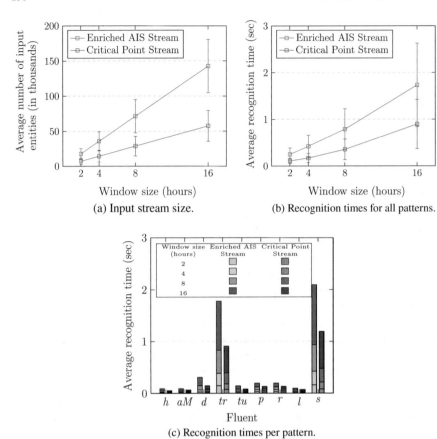

Fig. 8.13 Composite event recognition efficiency. In the bottom diagram, h, aM, d, tr, tu, p, r, l, and s stand for *highSpeedNC*, *anchoredOrMoored*, *drifting*, *trawling*, *tugging*, *pilotBoarding*, *rendezVous*, *loitering* and *sar* (search and rescue operations)

8.5 Summary

Composite event recognition systems support maritime situational awareness as they allow for the detection of dangerous, suspicious and illegal vessel activities. To illustrate the use of such systems, we presented a series of composite maritime event patterns in a formal language. For effective recognition, the maritime patterns have been developed in close collaboration with domain experts. Furthermore, the patterns have been evaluated with the use of real-world AIS datasets. The empirical analysis verified the quality of the patterns and showed that their use does not prohibit real-time performance.

8.6 Biliographical Notes

The models for composite event recognition are typically based on automata, tree structures and logic-based rules [5]. A survey of composite event recognition systems may be found in [13], while a tutorial on logic-based event recognition is documented in [7]. In this tutorial, the presented approaches are related to Chronicle Recognition, the Event Calculus and Markov Logic. Alevizos et al. presented a survey of techniques handling uncertainty in event recognition [2]. More recently, Giatrakos et al. performed a survey of techniques that handle effectively the volume, velocity and geographic distribution of Big Data in composite event recognition, discussing at the same time the expressivity of recognition languages [18]. Grez et al. have also been studying the expressivity of composite event recognition languages, with a focus on automata-based approaches, however [20, 21].

Composite event recognition systems have been employed in several application domains. Consider, for example, the recognition of attacks on computer network nodes given the TCP/IP messages [14], the recognition of dangerous driving and opportunities for refuelling in fleet management [44], human activity recognition from multimedia content [9, 40], traffic incident detection based on data from mobile sensors (for example, mounted on public transport vehicles and private cars) and stationary sensors (installed at intersections) [8], as well as fraud detection in credit card transactions [4, 39].

In the maritime domain, van Laere et al. presented various vessel anomalies as composite events, such as tampering, rendez-vous between vessels and unusual routing [23]. Snidaro et al. employed Markov Logic Networks to formalise maritime events and anomalies [42]. A system detecting, among others, when two vessels are in danger of colliding, was presented in [43]. The SUMO system integrates AIS data with images from synthetic aperture radars in order to detect illegal oil dumping, piracy and unsustainable fishing [19]. The use of RTEC for composite maritime event recognition is documented in [31, 33, 38]. Wayeb is a system that, in addition to composite event recognition, supports composite event *forecasting*, i.e. it calculates the intervals in the future within which a composite event may take place with some probability. The use of Wayeb in the maritime domain is presented in [1].

Since the original formulation of the Event Calculus [22], various dialects have been proposed in the literature, some of which exhibit features required by composite event recognition. The 'Macro-Event Calculus' includes axioms for computing the maximal intervals of fluents and composite (macro) event operators: sequence, disjunction, parallelism and iteration [10]. The 'Interval-based Event Calculus' includes event operators for sequence, disjunction, mutual exclusivity, conjunction, concurrency, negation, quantification and aperiodicity [29]. The 'Cached Event Calculus' [11] and the more recent 'Reactive Event Calculus' [28] compute and store the consequences of the data stream items as soon as they arrive. Query processing, therefore, amounts to retrieving the appropriate composite event intervals from memory. 'Prob-EC' [40] and 'MLN-EC' [41] are Event Calculus dialects based on ProbLog and Markov Logic Networks respectively; they have been developed for handing the uncertainty of event recognition.

Acknowledgements This chapter is based on [33]. We would like to thank Charilaos Akasiadis and Alexandros Troupiotis-Kapeliaris for their useful comments on this chapter.

References

1. Alevizos, E., Artikis, A., Paliouras, G.: Wayeb: a tool for complex event forecasting. In: Barthe, G., Sutcliffe, G., Veanes, M. (eds.) 22nd International Conference on Logic for Programming, Artificial Intelligence and Reasoning, EPiC Series in Computing, LPAR-22,vol. 57, pp. 26–35. EasyChair (2018)
2. Alevizos, E., Skarlatidis, A., Artikis, A., Paliouras, G.: Probabilistic complex event recognition: a survey. ACM Comput. Surv. **50**(5), 71:1–71:31 (2017)
3. Andrienko, N., Andrienko, G.: Visual analytics of vessel movement. In: Artikis, A., Zissis, D. (eds.) Guide to Maritime Informatics, chap. 5. Springer, Berlin (2021)
4. Artikis, A., Katzouris, N., Correia, I., Baber, C., Morar, N., Skarbovsky, I., Fournier, F., Paliouras, G.: A prototype for credit card fraud management: industry paper. In: Proceedings of the 11th ACM International Conference on Distributed and Event-based Systems, DEBS, pp. 249–260. ACM (2017)
5. Artikis, A., Margara, A., Ugarte, M., Vansummeren, S., Weidlich, M.: Complex event recognition languages: tutorial. In: Proceedings of the 11th ACM International Conference on Distributed and Event-based Systems, DEBS, pp. 7–10. ACM (2017)
6. Artikis, A., Sergot, M.J., Paliouras, G.: An event calculus for event recognition. IEEE Trans. Knowl. Data Eng. **27**(4), 895–908 (2015)
7. Artikis, A., Skarlatidis, A., Portet, F., Paliouras, G.: Logic-based event recognition. Knowl. Eng. Rev. **27**(4), 469–506 (2012)
8. Artikis, A., Weidlich, M., Gal, A., Kalogeraki, V., Gunopulos, D.: Self-adaptive event recognition for intelligent transport management. In: Hu, X., Lin, T.Y., Raghavan, V.V., Wah, B.W., Baeza-Yates, R., Fox, G.C., Shahabi, C., Smith, M., Yang, Q., Ghani, R., Fan, W., Lempel, R., Nambiar, R. (eds.) Proceedings of the 2013 IEEE International Conference on Big Data, pp. 319–325. IEEE Computer Society (2013)
9. Brendel, W., Fern, A., Todorovic, S.: Probabilistic event logic for interval-based event recognition. In: The 24th IEEE Conference on Computer Vision and Pattern Recognition, CVPR, pp. 3329–3336. IEEE Computer Society (2011)
10. Cervesato, I., Montanari, A.: A calculus of macro-events: progress report. In: 7th International Workshop on Temporal Representation and Reasoning, TIME, pp. 47–58. IEEE Computer Society (2000)
11. Chittaro, L., Montanari, A.: Efficient temporal reasoning in the cached event calculus. Comput. Intell. **12**, 359–382 (1996)
12. Clark, K.L.: Negation as failure. In: Gallaire, H., Minker, J. (eds.) Logic and Databases, pp. 293–322. Plenum Press (1978)
13. Cugola, G., Margara, A.: Processing flows of information: from data stream to complex event processing. ACM Comput. Surv. **44**(3), 15 (2012)
14. Dousson, C., Maigat, P.L.: Chronicle recognition improvement using temporal focusing and hierarchization. In: Veloso, M.M. (ed.) Proceedings of the 20th International Joint Conference on Artificial Intelligence, IJCAI 2007, pp. 324–329 (2007)
15. Etienne, L., Ray, C., Camossi, E., Iphar, C.: Maritime data processing in relational databases. In: Artikis, A., Zissis, D. (eds.) Guide to Maritime Informatics, chap. 3. Springer, Berlin (2021)
16. Etzion, O., Niblett, P.: Event Processing in Action. Manning Publications Company, Shelter Island (2010)

17. Fikioris, G., Patroumpas, K., Artikis, A.: Optimizing vessel trajectory compression. In: Maritime Big Data Workshop (MBDW) (2020)
18. Giatrakos, N., Alevizos, E., Artikis, A., Deligiannakis, A., Garofalakis, M.N.: Complex event recognition in the big data era: a survey. VLDB J. **29**(1), 313–352 (2020)
19. Greidanus, H., Alvarez, M., Santamaria, C., Thoorens, F.X., Kourti, N., Argentieri, P.: The sumo ship detector algorithm for satellite radar images. Remote. Sens. **9**(3) (2017). http://www.mdpi.com/2072-4292/9/3/246
20. Grez, A., Riveros, C., Ugarte, M.: A formal framework for complex event processing. In: Barceló, P., Calautti, M. (eds.) 22nd International Conference on Database Theory, ICDT, LIPIcs, vol. 127, pp. 5:1–5:18. Schloss Dagstuhl - Leibniz-Zentrum für Informatik (2019)
21. Grez, A., Riveros, C., Ugarte, M., Vansummeren, S.: On the expressiveness of languages for complex event recognition. In: Lutz, C., Jung, J.C. (eds.) 23rd International Conference on Database Theory, ICDT, LIPIcs, vol. 155, pp. 15:1–15:17. Schloss Dagstuhl - Leibniz-Zentrum für Informatik (2020)
22. Kowalski, R.A., Sergot, M.J.: A logic-based calculus of events. New Gener. Comput. **4**(1), 67–95 (1986)
23. van Laere, J., Nilsson, M.: Evaluation of a workshop to capture knowledge from subject matter experts in maritime surveillance. In: Proceedings of Fusion, pp. 171–178 (2009)
24. Luckham, D.C.: The Power of Events: An Introduction to Complex Event Processing in Distributed Enterprise Systems. Addison-Wesley Longman Publishing Co., Inc, USA (2001)
25. Masson, L.L., Barre, J.: Complex event recognition support for maritime monitoring in the context of the European datacron project. Final year project, French naval academy (2018)
26. Michelioudakis, E., Artikis, A., Paliouras, G.: Semi-supervised online structure learning for composite event recognition. Mach. Learn. **108**(7), 1085–1110 (2019)
27. Mills, C.M., Townsend, S.E., Jennings, S., Eastwood, P.D., Houghton, C.A.: Estimating high resolution trawl fishing effort from satellite-based vessel monitoring system data. ICES J. Mar. Sci. **64**(2), 248–255 (2007)
28. Montali, M., Maggi, F.M., Chesani, F., Mello, P., van der Aalst, W.M.P.: Monitoring business constraints with the event calculus. ACM Trans. Intell. Syst. Technol. **5**(1), 17:1–17:30 (2013)
29. Paschke, A., Bichler, M.: Knowledge representation concepts for automated SLA management. Decis. Support Syst. **46**(1), 187–205 (2008)
30. Patroumpas, K.: Online mobility tracking against evolving maritime trajectories. In: Artikis, A., Zissis, D. (eds.) Guide to Maritime Informatics, chap. 6. Springer, Berlin (2021)
31. Patroumpas, K., Alevizos, E., Artikis, A., Vodas, M., Pelekis, N., Theodoridis, Y.: Online event recognition from moving vessel trajectories. GeoInformatica **21**(2), 389–427 (2017)
32. Pitsikalis, M., Artikis, A.: Composite maritime events (version 0.1) [data set] (2019). https://zenodo.org/record/2557290
33. Pitsikalis, M., Artikis, A., Dreo, R., Ray, C., Camossi, E., Jousselme, A.: Composite event recognition for maritime monitoring. In: Proceedings of the 13th ACM International Conference on Distributed and Event-based Systems, DEBS, pp. 163–174 (2019)
34. Przymusinski, T.: On the declarative semantics of stratified deductive databases and logic programs. In: Foundations of Deductive Databases and Logic Programming. Morgan (1987)
35. Ray, C., Dréo, R., Camossi, E., Jousselme, A.L.: Heterogeneous integrated dataset for maritime intelligence, surveillance, and reconnaissance (version 0.1) (2018). https://zenodo.org/record/1167595
36. Ray, C., Dréo, R., Camossi, E., Jousselme, A.L., Iphar, C.: Heterogeneous integrated dataset for maritime intelligence, surveillance, and reconnaissance. Data Brief **25** (2019)
37. Santipantakis, G.M., Doulkeridis, C., Vouros, G.A.: Link discovery for maritime monitoring. In: Artikis, A., Zissis, D. (eds.) Guide to Maritime Informatics, chap. 7. Springer, Berlin (2021)
38. Santipantakis, G.M., Vlachou, A., Doulkeridis, C., Artikis, A., Kontopoulos, I., Vouros, G.A.: A stream reasoning system for maritime monitoring. In: Alechina, N., Nørvåg, K., Penczek, W. (eds.) 25th International Symposium on Temporal Representation and Reasoning, TIME, LIPIcs, vol. 120, pp. 20:1–20:17. Schloss Dagstuhl - Leibniz-Zentrum für Informatik (2018)

39. Schultz-Møller, N.P., Migliavacca, M., Pietzuch, P.R.: Distributed complex event processing with query rewriting. In: Gokhale, A.S., Schmidt, D.C. (eds.) Proceedings of the Third ACM International Conference on Distributed Event-Based Systems, DEBS. ACM (2009)
40. Skarlatidis, A., Artikis, A., Filipou, J., Paliouras, G.: A probabilistic logic programming event calculus. Theory Pract. Log. Program. **15**(2), 213–245 (2015)
41. Skarlatidis, A., Paliouras, G., Artikis, A., Vouros, G.A.: Probabilistic event calculus for event recognition. ACM Trans. Comput. Log. **16**(2), 11:1–11:37 (2015)
42. Snidaro, L., Visentini, I., Bryan, K., Foresti, G.L.: Markov logic networks for context integration and situation assessment in maritime domain. In: 15th International Conference on Information Fusion, FUSION, pp. 1534–1539. IEEE (2012)
43. Terroso-Saenz, F., Valdes-Vela, M., Skarmeta-Gomez, A.F.: A complex event processing approach to detect abnormal behaviours in the marine environment. Inf. Syst. Front. **18**(4), 765–780 (2016)
44. Tsilionis, E., Koutroumanis, N., Nikitopoulos, P., Doulkeridis, C., Artikis, A.: Online event recognition from moving vehicles: application paper. Theory Pract. Log. Program. **19**(5–6), 841–856 (2019)
45. Tzouramanis, T.: Navigating the ocean of publicly available maritime data. In: Artikis, A., Zissis, D. (eds.) Guide to Maritime Informatics, chap. 2. Springer, Berlin (2021)

Part IV
Applications

Chapter 9
Uncertainty Handling for Maritime Route Deviation

Anne-Laure Jousselme, Clément Iphar, and Giuliana Pallotta

Abstract Detecting and classifying anomalies generally contributes to Maritime Situation Awareness and highly benefits from the combination of multiple sources, as correlating their output allows detecting inconsistencies in vessels' behaviour. In particular, detecting the route followed by vessels, and identifying off-route vessels is a challenging problem which requires a first characterisation of maritime routes (the normalcy) and secondly the association of a vessel track to an existing route. In this context, adequate uncertainty representation and processing is crucial for this higher-level task where the operator analyses information in conjunction with background knowledge. The maritime anomaly detection solution is framed into a mathematical uncertainty theory, encoding thus semantics in both uncertainty representation and reasoning. The choice of the theory together with the associated calculus defines then directly the output of the algorithm and the result presented to the user. In this chapter, we dissect six classical Uncertainty Representation and Reasoning Techniques (URRTs), each solving the problem of track to route association. In their basic form, the URRTs are framed into the three uncertainty theories of probabilities, belief functions and fuzzy sets, which capture different feature of information deficiencies, of uncertainty, imprecision, graduality. The different URRTs are qualitatively evaluated according to their expressiveness along these deficiencies and quantitatively evaluated according their trueness, precision and certainty when processing real AIS data with route labels.

A.-L. Jousselme (✉) · C. Iphar
NATO STO Centre for Maritime Research and Experimentation, La Spezia, Italy
e-mail: anne-laure.jousselme@cmre.nato.int

C. Iphar
e-mail: clement.iphar@cmre.nato.int

G. Pallotta
Lawrence Livermore National Laboratory, Livermore, CA, USA
e-mail: pallotta2@llnl.gov

© Springer Nature Switzerland AG 2021
A. Artikis and D. Zissis (eds.), *Guide to Maritime Informatics*,
https://doi.org/10.1007/978-3-030-61852-0_9

9.1 Introduction

In the field of Maritime Situation Awareness (MSA) [5], detecting and classifying vessels' abnormal behaviour is a challenging and crucial task at the core of the compilation of the maritime picture [26, 27]. It requires not only the extraction of relevant contextual patterns-of-life information shaped for instance as maritime routes or loitering areas [36], but also the real time monitoring of the maritime traffic by a set of sensors mixing cooperative self-identification systems (such as the Automatic Identification System (AIS)) [37, 38, 45] and non-cooperative systems such as coastal radars or satellite imagery, to overcome the possible spoofing of the AIS signal [41]. In many cases, intelligence information is of great help to refine and guide the search in the huge amount of data to be processed, filtered and analysed.

In order to take informed decisions, the operator needs to get good quality information. Furthermore, the operator needs to understand additional characteristics of the provided information, including for instance, how that information has been obtained, processed, or what the context of its creation was. In particular, understanding how an anomaly detector came up with an alert is of great importance to the Vessel Traffic System operator. More specifically, the operator would benefit from knowing which reference data were used, which sources were processed, if the information and associated uncertainty were obtained in an objective (e.g. derived from statistics) or subjective (e.g. provided by a human) manner, whether the decision process considered the sources' quality and how, if the contextual information was considered in the decision, what the meaning of numerical output values expressing uncertainty was, or what the underlying logical reasoning providing the answer was. Second-order information quality (e.g. uncertainty about uncertainty assessment such as probability intervals) may also be highly valuable. For example, probability maps about possible threats could be supplemented by uncertainty assessments about the validity of the probability values, represented as intervals or error estimations on algorithm performance. The benefit of including these different information quality dimensions is twofold: on the one hand, they increase the operator's situation awareness and, on the other hand, they improve trust in the use of the system.

To characterise the outputs provided to operators by some information system, the standard performance criteria of algorithms such as precision, accuracy, false alarm rate, area under the receiver operating characteristic curve, timeliness or computational cost [2, 28, 56] may not be sufficient to cover the spectrum of performance criteria of systems and should be complemented by others to cover the interaction of humans and systems. For instance, some criteria such as explainability, adaptability, simplicity, or expressiveness could be considered as well. These criteria and others are defined and articulated for instance in the Uncertainty Representation and Reasoning Framework (URREF), for the specific purpose of uncertainty handling in fusion algorithms [10].

As a step toward a formal analysis of uncertainty representation and reasoning techniques, the work presented in this chapter aims at comparing some classical uncertainty models. We compare six different approaches, hereafter called Uncertainty Representation and Reasoning Techniques (URRTs), to combine pieces of information from a set of heterogeneous sources, as the core of a maritime anomaly detector for route deviation. In complement to comparative analyses (e.g. [4, 23]), this chapter identifies additional comparison elements which may have an impact on the behaviour (and performances) of the fusion algorithm.

The maritime anomaly detection problem is first introduced in Sect. 9.2 together with some associated uncertainty-related challenges, while an overview of information deficiencies and some uncertainty theories is provided in Sect. 9.3. The six URRTs are introduced in Sect. 9.4 as alternative basic fusion schemes to solve the problem of detection of off-route vessels, with an emphasis on the uncertainty representation. The six URRTs are compared in Sect. 9.5 through classical but complementary quality criteria (trueness, precision, certainty), processing a real AIS dataset. We conclude in Sect. 9.6 on further challenges and readings.

9.2 Maritime Anomaly Detection and Route Deviation

We herein use the term "maritime anomaly" to indicate a *deviating* behaviour from traffic normalcy. More specifically, the analysis of spatio-temporal data streams from the Automatic Identification System (AIS) once processed, provides some normalcy about the maritime traffic (i.e. summarised as the maritime routes), which can be ultimately transformed into usable and actionable knowledge when detecting and characterising inconsistencies or ambiguities against the current traffic [17, 34, 55]. We briefly introduce in this section the problem of route extraction, which builds the normalcy models, in our case, traffic normalcy. We then introduce the problem of associating a vessel to a pre-defined maritime route selected from the extracted system of *synthetic maritime routes* as the core ingredient of the anomaly detection. We conclude this section with some uncertainty challenges related to the way we represent the maritime routes which affects the maritime anomaly detection.

9.2.1 Synthetic Maritime Route Extraction

We introduce below some vocabulary further used in this chapter: *maritime route, waypoints, synthetic maritime route, synthetic waypoints, route prototype, vessel (AIS) contact*. We also describe the way synthetic maritime routes are extracted from AIS data.

A *maritime route* may be defined as the prescribed course to be travelled by a vessel from a specific point of origin to a specific destination. It includes an origin and a destination, and possibly a series of *waypoints* through which or close to which, the vessel plans to pass.

The analysis and synthesis of the activity at sea, as patterns of life, is referred to as *synthetic maritime routes* and summarises the normal maritime traffic over a given period of time, a given area and a specified set of employed sensors (or sources). To some extent, the synthetic route summarises the maritime route vessels actually used in the past. The Traffic Route Extraction and Anomaly Detection (TREAD) tool presented in [36] implements an unsupervised classification approach which we here use to derive a set of the maritime traffic routes by processing spatio-temporal data streams from terrestrial and/or satellite AIS receivers. A *TREAD synthetic route* is then defined by a starting point and an ending point, together with a subset of intermediate *synthetic waypoints*, describing a physical path on a portion of the sea. If the area under surveillance is captured by a big enough bounding box, the route starting and ending points [1, 17] are the centroids of stationary areas, which can be coastal areas such as ports, offshore areas such as islands, offshore platforms, or open-sea areas such as fishing areas. The TREAD algorithm first reconstructs the single-vessel trajectories by linking the vessels' *positional contacts* (each AIS message is a vessel contact and contains positional information) and then clusters the trajectories followed by vessels into groups having the same starting and ending points. Each of these clusters represents a synthetic maritime route.

Several representations of a synthetic route exist, including the subset of raw contacts (the cluster), the set of raw trajectories, and the average or median trajectory, computed as a unique abstract trajectory from the spatial AIS contacts. In the following, we will call *route prototype*, the average or median trajectory representing a synthetic maritime route. The basic uncertainty around this route prototype is computed using the trajectories of all the vessels which transited along that route in the given time window, as it will be detailed in the following and illustrated in Table 9.3.

While only temporal streams of positional information are processed to extract the set of maritime routes, they can be further characterised by additional features representing the traffic of vessels composing them, such as speed, type or heading. The associated uncertainty characterisation of the route along these features can be more or less complex, ranging from simple average values, to added variance parameters, histograms, estimated complete probability distributions, and sets of distributions. The maritime traffic, and thus the set of routes, may be influenced by meteorological conditions (some areas may be avoided), season, economical context (ships may decide their destination based on the current stock market linked to their cargo) or areas of conflict. Also, in order to derive the average path (i.e. route prototype) from the route cluster, an extent parameter is included in the algorithm which allows adjusting the search range radius dynamically, thus enhancing the

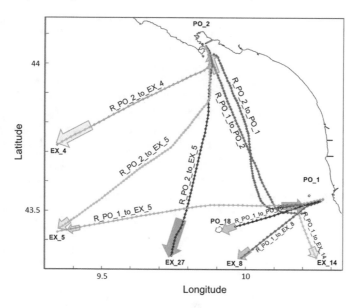

Fig. 9.1 Historical route prototypes extracted via the TREAD algorithm [36] in the area between La Spezia and Livorno, Italy, from AIS data (Jan 1–Feb 20 2013). Synthetic routes are named by combining the two labels corresponding to the origin and destination ports or areas, as displayed on each route

computation of intermediate synthetic waypoints while still avoiding issues such as land crossing.

We display below the synthetic maritime routes extracted from two AIS datasets: (1) AIS data collected in the area of La Spezia (Italy) from January to February 2013 (Fig. 9.1), and (2) AIS data collected in the area of Brest (France) from October to November 2015 (Fig. 9.2). The used AIS data are part of two reference datasets published respectively at CMRE [35] (dataset (1)), and on Zenodo [39], described in [40]. Dataset (2) is the dataset used throughout this book.

> **Remark ! Important**
>
> The set of synthetic routes thus summarises some kinematic patterns of life of vessels over a given period of time and region, possibly layered by specific vessel types (e.g. fishing vessels, tankers, passenger vessels). This synthetic information and associated uncertainty characterises part of the context or background knowledge for the problem of route association and detection of anomalies at sea. It provides a *reference* or *normalcy* against which the current vessel contacts will be compared, and the anomalies detected.

9.2.2   The Vessel to Route Association Problem

A route deviation detector is designed to help the Vessel Traffic System operator to (1) associate vessels to existing routes (and possibly predict their destination), and

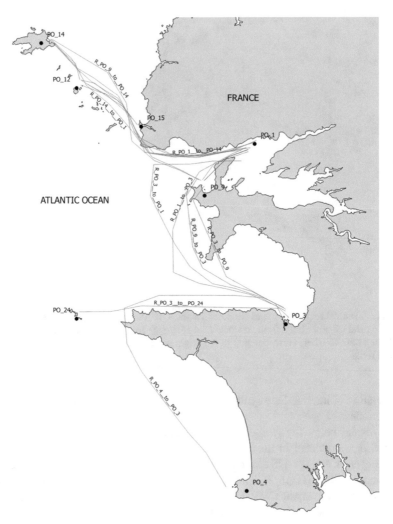

Fig. 9.2 A selection of synthetic routes, extracted via the TREAD algorithm [36] off the coast of Brittany, from AIS data (October 1st–November 29th 2015) collected from an antenna located in Brest, France

(2) detect abnormal behaviours to be further investigated. We consider a vessel V possibly observed by a series of heterogeneous sources such as a coastal radar and its associated tracker, a synthetic aperture radar image associated with an automatic target recognition algorithm or a human analyst, a visible camera operated by a human analyst, the AIS information sent by the vessel itself or some intelligence source. We formalise the vessel to route association problem as a *fusion problem* where several sources provide pieces of information about the association (or not)

of the vessel to a pre-computed synthetic route. Table 9.1 lists the notation used in this chapter.

Let $\mathcal{A}$ be the set of vessel features of interest, about which information is either provided by some sources, or to be inferred. For solving the problem of route association, we consider the following *five features*: the position (latitude, longitude), the Course Over Ground (COG), the Speed Over Ground (SOG), the Type of vessel, the Length of the vessel, that we will match to the equivalent features characterising the synthetic maritime route possibly followed by the vessel. Let X denote the set of uncertain variables corresponding to features of $\mathcal{A}$, X_i the variable of X corresponding to feature $i \in \mathcal{A}$. We denote by Ω_i the domain of definition of X_i containing the set of its possible values, by $x_i \in \Omega_i$ a singleton of Ω_i and by $A_i \subseteq \Omega_i$ a subset of Ω_i. Let Ω be the corresponding space, defined as the Cartesian product of the Ω_i corresponding to vessel features of interest at a given timestamp t. Also, $\mathbf{x}$ denotes a set of information items about some features in $\mathcal{A}$ at a specific instant in time t.[1] This notation of information item encompasses the general case where sources may provide some uncertainty about their statement and thus an element of $\mathbf{x}$ denotes a source statement as a single measurement (precise and certain), as a probability vector (expressing some uncertainty interpreted as provided by the source itself), or as a natural language expression (possibly vague), etc. Table 9.2 lists some examples. In the specific case of precise and certain measurements defined over a scale of real numbers, $\mathbf{x}$ would simply be a vector of real values.

Let $\Omega_R = \{R_0, R_1, \ldots, R_K\}$ be the finite set of labels for the possible routes followed by the vessel V, where R_k for $k = 1, \cdots, K$ is the label for a pre-computed synthetic route and R_0 represents "none of the K routes". The R_0 class gathers the events of "The vessel is physically off-route", "The vessel is on the route, in the reverse traffic", "The speed is not compatible with the route followed", "The type of the vessel is not compatible with the route followed", representing some maritime situational indicators of possible interest to the Vessel Traffic System operator.

> The fusion algorithm Ψ to be designed aims at establishing a mapping between the observation space Ω_o and the decision space Ω_R, $\Psi : \Omega_o \longrightarrow \Omega_R$ such that $\hat{R} = \Psi(\mathbf{x})$ is the route label assigned to V represented by $\mathbf{x}$. When the five feature values describing V, gathered within $\mathbf{x}$, are used as input for the function Ψ, the class $\hat{R}$ is estimated as the route to which V belongs.
> R_k, for $k = 1, \ldots, K$, is the label to be assigned by the fusion algorithm and corresponds to the event "The vessel V follows route R_k", while R_0 is a rejection class corresponding to "The vessel V follows no specific route".

---

[1] In this chapter, we consider that each feature value estimation is provided by a single source (while in general several sources may provide information about the same feature), and we focus on the fusion of all (singular) observations obtained at the same instant in time t. This thus lightens the notations by omitting subscripts t for the time and s for the source.

Table 9.1 Notation used in this chapter

V	A vessel of interest
$\mathcal{A}$	The set of vessel features, either observed or to be inferred
i	A feature of $\mathcal{A}$
X	The set of uncertain variables
X_i	The uncertain variable of X corresponding to feature $i \in \mathcal{A}$
Ω_i	The domain of definition of X_i (set of all possible values)
x_i	A singleton of Ω_i
A_i	A subset of Ω_i
$\overline{A_i}$	The complement set of A_i
Ω	The space generated by the Cartesian product of the Ω_i
t	A timestamp
$\mathbf{x}$	A set of information items gathering observations about V
Ω_R	The set of labels for the synthetic routes
R_k	The label of the synthetic maritime route K
$\hat{R}$	The estimated route, $\hat{R} \in \Omega_R$
Ω_o	The observation (feature) space
Ψ	A fusion scheme $\Psi : \Omega_o \longrightarrow \Omega_R$
$\mathbf{r}^{(k)}$	A prototype route representing the route k
$w^{(k)}$	The width of $\mathbf{r}^{(k)}$
$\bar{s}$	The mean of the random variable X_s
σ	The standard deviation of the random variable X_s
$\mathcal{N}(\bar{s}, \sigma)$	A normal distribution of mean $\bar{s}$ and standard deviation σ
σ_{ij}	A covariance between features i and j
$p(A)$	The probability of A
$m(A)$	The mass (Basic Probability Assignment) of A in evidence theory
$\text{Bel}(A)$	The belief of A in evidence theory
$\text{Pl}(A)$	The plausibility of A in evidence theory
$\text{BetP}(A)$	The pignistic transformation of m
μ_A	A fuzzy set, in fuzzy set theory
$\mu_i^{(k)}(x)$	The degree of membership of x to R_k
$d^{(E)}(\mathbf{x}, R_k)$	The Euclidean distance between $\mathbf{x}$ and R_k
$d^{(M)}(\mathbf{x}, R_k)$	The Mahalanobis distance between $\mathbf{x}$ and R_k
$E(\cdot)$	The expectation operator
ε_j	A decision threshold
α_i	A weight for feature i
$\text{Tru}(\Psi)$	The measure of trueness
$\text{Imp}(\Psi)$	The measure of imprecision
$\text{Unc}(\Psi)$	The measure of uncertainty

Table 9.2 Examples of source statements expressed about different vessel features

Feature i	x	Type of statement
Speed Over Ground (knots)	10.3	Single measurement (precise and certain)
Type (Cargo Tanker Passenger Other)	[0.2 0.1 0.7 0]	Probability vector
Length	"Big vessel"	Natural language

Remark ! Important

The underlying reasoning is that any observed feature combined with background knowledge contributes to a global belief (or disbelief) that V is following a pre-established route from Ω_R. Indeed, if all the observed (measured) features match the corresponding feature values of a specific existing route, then the corresponding route label is assigned to the vessel. If some "inconsistency" exists between the set of observed features and the route features, then V is assigned to no route and an anomaly is reported (label R_0). "Inconsistency" can arise for instance if the distance between $\mathbf{x}$ and each of the R_k is too high, or if the set of compatible routes according to one feature (e.g. the speed) does not match the set of compatible routes according to another feature (e.g. the type of the vessel). The same set of pieces of information is then be used for two purposes:

(1) associating a vessel to a route, under the assumption that all the sources are reliable and
(2) detecting anomalies, under the assumption that an inconsistency among the set of estimated features would reveal a possible anomalous behaviour of interest.

However, information is inherently imperfect (incomplete or imprecise, uncertain, gradual, granular [16]) and inconsistencies may arise either from source limitations (e.g. gaps in or weak coverage of sensors, limited reasoning abilities, storage limitations, false detections or identifications, and lack of reliability in general) or from malevolent behaviour of the vessel such as deception. The appropriate detection and identification of anomalies highly relies on the technique for fusing the different pieces of information and detecting inconsistencies, which include the handling of uncertainty.

9.2.3 Uncertainty in Maritime Anomaly Detection

We discuss in this section the uncertainty linked to the different representations of the synthetic routes, on two AIS datasets previously introduced and displayed in Figs. 9.1 and 9.2.

As already mentioned, a synthetic maritime route identifies the portion of the sea where vessels have been observed travelling between a pre-defined origin point and destination point and can for instance be represented by a cluster of vessel contacts (i.e. positions from AIS messages). From this set of contacts (i.e. the cluster) several derived representations can be further extracted, more or less complex, more or less rich, more or less precise. As an example, each route can be represented by a route prototype, as a series of intermediate synthetic waypoints with associated average heading, width, etc. Additional features characterising the traffic can be further extracted such as the distribution of speed, length and type of vessels traveling on this route. Routes are by nature, uncertain objects and the characterisation and representation of their uncertainty is of primary importance for a proper use of this information for the anomaly detection task, as well as for communicating with the end-user. The end-user is for instance the Vessel Traffic System operator who is monitoring a specific area. Figure 9.3 provides some statistics characterising the synthetic routes from dataset (1) in the La Spezia area, displayed in Fig. 9.1, along with the different route features. The geographical displacements of vessel positions with respect to the average route (the prototype) shows the dispersion of trajectories along the average, suggesting a variable width for the route. The width could be defined as an interval within which the vessels usually traveled. The distribution of Speed Over Ground (SOG) exhibits a kind of bell shape, showing that vessels traveling that route have a mean speed value of 13 knots. The distribution of ship length itself, is rather uniform with a gap between circa 25m to 75m. The vessel type distribution exhibits a high proportion of Cargo vessels, with some Passenger and Tanker vessels, among the ships which transited along that route in the given time window. The histograms displayed in Fig. 9.3 highlight the need for different models to capture uncertainty along the different features. The diversity of features and their distributions emphasise that a unique uncertainty representation for all features would not be appropriate and suggest rather individual uncertainty representations suited to the specific features.

Table 9.3 lists several examples of simple uncertain representations for the different routes along with the different features that can be encoded in a compact way, building a kind of "dictionary of routes". Each synthetic route R_k may be represented by a route prototype $\mathbf{r}^{(k)}$, itself characterised by a series of feature values corresponding to either the mean or the most frequent trajectory of the cluster. Those features are *precise* and *certain* values to which some *imprecision* or *uncertainty* can be added for a richer representation, based on the statistical information extracted from the raw AIS messages. The route width $w^{(k)}$ may be defined as the maximum of the distances of each route point (i.e. vessel positions associated to the route) to the closest synthetic waypoint on the synthetic route. It defines an area where the vessels have been observed transiting in the past.

The uncertainty representations may vary with respect to the data field of interest to reflect the continuous, interval or categorical nature of the feature. For instance, it seems that the distribution of the speed variable X_S for route R_6 can be defined by the tuple $(\bar{s}_6; \sigma_1^{(s)})$ representing the mean and standard deviation of speed values estimated on the training dataset used to build R_6. With the additional assumption of a Gaussian (normal) model, these two parameters would completely define one estimation of a probability distribution for X_S. This model may not fit for instance the length distribution, which could be better captured by a Mixture-of-Gaussian (MoG), with two distinct modes. This is the case for routes followed by two types of vessels of two categories of length. Simpler models could be simple intervals for

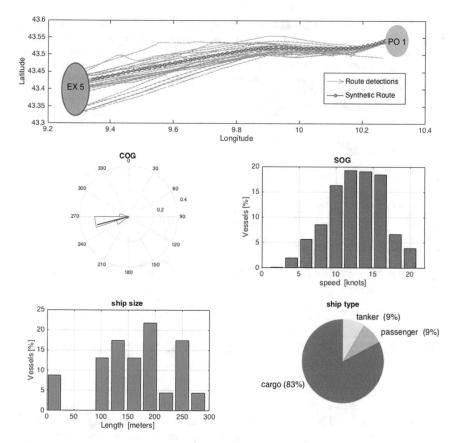

Fig. 9.3 An example of multi-dimensional uncertainty representation for Route R_6 going from port PO_1 to exit point EX_5 in the area of La Spezia (see Fig. 9.1). Top panel—geographical displacements of vessel positions with respect to the average route; middle left panel—distribution of Course Over Ground (COG); middle right panel—distribution of Speed Over Ground (SOG); bottom left panel—distribution of ship length; bottom right panel—frequency of types of ships which transited along that route in the given time window. The corresponding uncertainty models are reported in Table 9.3

lengths of vessels, or pairs of intervals. More sophisticated techniques of joint density estimation, or models of dynamics of vessels, considering as well the interaction between speed and position (e.g. [33, 42]) would be even better suited. The type of vessel is itself a categorical feature (expressed on a non-numerical and non-ordered scale), which could be best represented by a probability mass function on the different types.

In some cases, the amount of data (e.g. number of trajectories) building the cluster may not be large enough to estimate reliable distributions and thus considering second-order uncertainty could be appropriate. Also some AIS fields, especially the ones entered manually, are often missing or misspelled. For instance, the destination

Table 9.3 Dictionary of routes with examples of simple associated uncertainty representations, for the La Spezia dataset: List of synthetic waypoints $\{WP\}$ (as a set of positions) and associated width, average Course Over Ground $(\overline{COG})$ and standard deviation, distribution of Speed Over Ground (SOG) as a normal distribution $\mathcal{N}$ with parameters $\bar{s}$ (mean) and σ (standard deviation), vessel length intervals, vessel type probability distribution over $\{Cargo; Tanker, Fishing, Passenger; Other\}$

Route	Synthetic route $\mathbf{r}^{(k)}$	Traffic information statistics			
	POSITION ± WIDTH (KM)	$\overline{COG}$±STD	SOG (KNOTS)	LENGTH (M)	TYPE (FREQUENCY)
R_1	$\{WP\}^{(1)} \pm 2.77$	297 ±85°	$\mathcal{N}(10, 4)$	[80 : 100]; [260 : 300]	[0.8 0.1 0 0 0.1]
R_2	$\{WP\}^{(2)} \pm 3.01$	140 ±49°	$\mathcal{N}(11, 4)$	[0 : 130]; [250 : 300]	[0.37 0.13 0 0 0.5]
R_3	$\{WP\}^{(3)} \pm 1.19$	185 ±11°	$\mathcal{N}(12, 2)$	[120 : 250]	[0.75 0 0 0 0.25]
R_4	$\{WP\}^{(4)} \pm 2.86$	221 ±31°	$\mathcal{N}(15, 4)$	[100 : 200]; [260 : 350]	[0.97 0 0 0 0.03]
R_5	$\{WP\}^{(5)} \pm 5.08$	209 ±19°	$\mathcal{N}(11, 1)$	[100 : 300]; [200 : 210]	[1 0 0 0 0]
R_6	$\{WP\}^{(6)} \pm 1.91$	255 ±18°	$\mathcal{N}(13, 4)$	[0 : 25]; [110 : 300]	[0.82 0.09 0 0.09 0]
R_7	$\{WP\}^{(7)} \pm 1.25$	210 ±90°	$(\mathcal{N}(10, 2); \mathcal{N}(18, 2))$	[50 : 100]; [120 : 200]	[0.93 0 0 0.07 0]
R_8	$\{WP\}^{(8)} \pm 0.86$	225 ±14°	$(\mathcal{N}(11, 2); \mathcal{N}(19, 2))$	[100 : 150]; [190 : 240]	[0.38 0.24 0 0.38 0]
R_9	$\{WP\}^{(9)} \pm 0.98$	244 ±21°	$\mathcal{N}(11, 3)$	Not reported	[0 0 0 1 0]

may not be specified or may not be valid [3], the Estimated Time of Arrival (ETA) may not be updated. The positional and kinematic information being automatically sent is more reliable but can suffer from incompleteness due to a lack of coverage of the AIS receivers [44] resulting in missing reports for a certain period of time. The lack of reception of the AIS signal [5] may also arise from an intentional manipulation, to conceal some activity [19] (e.g. hiding fishing area, keep hidden from pirates, smuggling). Finally, the AIS signal can be spoofed for instance shifting the positional information to another area, or by modifying the MMSI or IMO identifier of the vessel [41]. Previous studies have demonstrated that roughly 5% of AIS data is generally inconsistent (see e.g. [30]).

The statistics extracted from the raw AIS dataset may serve two purposes: on the one hand, they can be used as the basic ingredient for the generic uncertainty representation captured by the route objects and, on the other hand, they are possibly transferred to express some uncertainty about new singular measurements. For instance, uncertainty extracted from the histograms of the different features (SOG, COG, Length, Type) as illustrated in Fig. 9.3, can be further interpreted as likelihood functions $p(X_i = x | R_k)$, where X_i is the uncertain variable associated to feature i and R_k is a route, and approximated by different models. The exploitation of such information will be further discussed in Sect. 9.4.

> **Remark ! Important**
>
> The consideration of different imperfections of information (or information deficiencies) is crucial in the design of maritime anomaly detection solutions. Note that it requires a prior understanding of the origins of uncertainty, of the kinds of imperfection, of the type of information (be it relevant to a population of situations such as the maritime routes or to a single one such as the AIS contact of a specific vessel at a specific instant in time) to properly interpret the estimates produced by the algorithms available to the user. We refer the reader to [16] for more explanations.

9.3 Uncertainty Handling

9.3.1 Information Deficiencies

Throughout this chapter, we will refer to the four information deficiencies or defects of *incompleteness* (or yet imprecision), *uncertainty*, *graduality* and *granularity*, following Dubois and Prade's typology [16], to which we add *trueness* as another dimension.

- *Imprecision* refers to a set of possible values: the bigger the size of the set, the higher the imprecision. It represents the inability of the source to provide a single value or to discriminate between several values. Imprecision is interpreted as a type of incompleteness as it arises from a lack of information. For instance, the statement "The vessel is following either route R_1 or R_2" is imprecise and provides only incomplete information not allowing to answer the question "What is the route followed by the vessel?".
- *Uncertainty* arises when an agent does not know (or partially knows) if a proposition is true or false. It can be expressed by a degree (or a set of degrees) of belief assigned to a specific (or set of) value(s) to be "true". It may origin from a lack of knowledge (epistemic uncertainty) or from the variability of an underlying process (aleatory uncertainty). For instance, the probability distribution over the set of possible types of vessels T_i as output by some classifier can be interpreted as an uncertainty expression, i.e. expressing a set of (normalised) degrees of belief in the truth of the propositions "The vessel is of type T_i".
- *Graduality* arises usually from linguistic expressions and induces propositions with some possible degrees of truth (i.e. non Boolean). That kind of imperfection allows a proposition to be more or less true or more or less false. For instance, "The vessel is fast" is a gradual information item, using the gradual predicate "fast", and is typically represented by fuzzy sets. As we will illustrate later, "on a maritime route" can be considered as a gradual predicate making maritime routes ill-defined objects.
- *Granularity* refers to the support over which the proposition is defined, i.e. to the set of pre-established possible values. Granularity refers to the partition granules used in the definition of a set. For instance, the set Ω_1 defined as

{$FishingVessel$; $NotFishingVessel$} describing exhaustively the types of vessels has a rougher granularity than the set Ω_2 defined as {$FishingVessel$, $Cargo$, $Tanker$, $Other$} covering also exhaustively the possible types of vessels.

- *Trueness* is considered here as the criterion relating a piece of information (either input or output) to truth or to some reference value. The notion of trueness covers two different aspects: (1) *how close* the results are to "truth", especially in case of measurements on continuous scales and (2) *how frequently* the results correspond to "truth", especially in case of nominal scales. The first aspect captures some graduality of truth (fuzzy event), while the second aspect captures some uncertainty about the truth (crisp event). For example, the F_β measure (see Eq. (9.16)) quantifies some notion of trueness aggregating true positive and false negative rates. In the case of a classifier producing two types of outputs, On-route and Off-route vessels, a True Positive is a vessel "truly" On-route and correctly classified as On-route by the classifier, while a False Negative is a vessel "truly" On-route but wrongly classified as Off-route by the classifier.

All these information deficiencies are somehow related while their exact relationship is not clear. Imprecision (or precision) and uncertainty (or certainty) are opposed [50]: "I'm certain that the speed of the vessel is between 3 and 6 knots" (Certain but imprecise statement) versus "I'm not certain that the speed of the vessel is 5 knots" (Precise but uncertain statement). Precision and trueness are often associated in the performance assessment of systems, and are gathered under the term *accuracy* in ISO 5725 [20], referring to a series of independent tests. Granularity constrains imprecision as it sets the basis for imprecision expression: a rough support (low granularity) would offer less levels for imprecision expression than a thin support (high granularity). Graduality expresses some notion of imprecision, usually through some natural language expression. Uncertainty expresses a degree of belief about the trueness of the event. Graduality expresses different levels of truth about the (fuzzy) event.

These five deficiencies (or imperfections) will be used in the following to characterise both input and output information of the fusion technique.

9.3.2 Some Uncertainty Theories: A Brief Overview

Mathematical theories for representing and reasoning with uncertainty, or *uncertainty theories*, typically include, but are not limited to, probability theory, fuzzy set theory, possibility theory, belief function theory, rough set theory and imprecise probability theory (see [16] for a survey). In the following, we will consider the three mathematical frameworks of probabilities, belief functions and fuzzy sets. We provide in this section some brief theoretical background while details on application to solve the practical problem of vessel to route association will be given in Sect. 9.4.

Let us denote by $\Omega = \{x_1, \ldots, x_n\}$ a set of exhaustive[2] and exclusive[3] classes (or hypotheses). In the problem of route association, the exhaustivity property of Ω_R guarantees that the vessel will be assigned a class, either one of the synthetic maritime routes R_1 to R_K or an Off-route label denoted as R_0. The exclusivity property guarantees that the vessel is following a single route (or no route). Of course, these properties can be relaxed, leading to extensions of the mathematical frameworks considered. An uncertain variable[4] X represents some unknown state of the world and takes values in Ω.

The properties of all functions introduced herein correspond to some possibly desirable behaviours of the uncertainty handling models within the fusion method to be designed. One of the tasks of the designer is to identify and select the uncertainty representation together with the associated mathematical framework in order to meet the requirements of the expected underlying logic of the fusion method.

Referring to the information deficiencies introduced previously, probabilities convey the notion uncertainty only, belief functions and plausibility functions of evidence theory convey both uncertainty and imprecision, while fuzzy sets convey the notion of graduality which can be assessed using, for instance, distance measures.

9.3.2.1 Probability Theory

Probability theory is the most standard mathematical framework for handling uncertainty, with two main interpretations of probability measures, either *frequencist* (objective estimation) or *subjective* (belief-based assessment from humans). A probability measure P assigns to subsets of Ω named *events*, a real value in the interval $[0; 1]$ such that:

$$P\left(\bigcup_{i=1}^{\infty} A_i\right) = \sum_{i=1}^{\infty} P(A_i), \tag{9.1}$$

where $\{A_i\}$ is a set of disjoint subsets of Ω, with $P(\emptyset) = 0$ and $P(\Omega) = 1$.

If $\overline{A}$ denotes the negation (or complement) of A, i.e. $\overline{A} = \Omega \backslash A$, the additivity property expressed by Eq. (9.1) implies in particular that $P(A) + P(\overline{A}) = 1$. A state

---

[2]The set of classes is collectively exhaustive if at least one of the classes is true: $\Omega = \bigcup_{x_i \in \Omega} x_i$.

[3]The set of classes is mutually exclusive if only one of them is true: $x_i \cap x_j = \emptyset, \forall i \neq j$.

[4]The term "uncertain variable" is used here to emphasise the origin of uncertainty which may be due to randomness (i.e. random variable) or to a lack of knowledge (i.e. epistemic variable).

of complete ignorance about the value of x is usually represented by a uniform distribution over Ω such that $P(\{x\}) = \frac{1}{|\Omega|}$, for all $x \in \Omega$, where $|.|$ denotes the set cardinality. The additivity property is what distinguishes probability measures from other non-additive measures such as belief and plausibility functions from evidence theory or possibility and necessity functions from possibility theory.

Bayes' rule is the classical way of updating probabilities in light of new evidence:

$$P(A|E_1, E_2) = \frac{P(E_1, E_2|A)P(A)}{P(E_1, E_2)} \qquad (9.2)$$

where E_1 and E_2 are two pieces of evidence, provided for instance by some observations. We also have $P(E_1, E_2) = \sum_i P(E_1, E_2|A_i)P(A_i)$ if $\{A_i\}$ is a collection of events forming a partition of Ω and $P(E_1, E_2|A)$ is the likelihood of observing both E_1 and E_2 under the condition A. In our case, A would be the event $X_R = R_k$ ("the vessel follows R_k") and E_1 and E_2 could be $X_S = 13$ and $X_\theta = 240$ respectively standing for "the vessel speed is 13 knots" and "the vessel COG is 240 degrees".

9.3.2.2 Evidence Theory

In evidence theory (Dempster-Shafer theory, or belief functions theory) [11, 46], the uncertainty is expressed by a series of non-additive measures. In particular, the belief of an event A is super-additive i.e. $\mathrm{Bel}(A \cup B) > \mathrm{Bel}(A) + \mathrm{Bel}(B)$ and the plausibility of an event A is sub-additive i.e. $\mathrm{Pl}(A \cup B) < \mathrm{Pl}(A) + \mathrm{Pl}(B)$. Relaxing the additivity axiom of probabilities allows to express some second-order uncertainty, meaning some uncertainty about the probability measure itself. Under this interpretation of evidence theory as an imprecise probability theory, $\mathrm{Bel}(A)$ and $\mathrm{Pl}(A)$ are respectively lower and upper bounds of an unknown probability $P(A)$:

$$\mathrm{Bel}(A) \leq P(A) \leq \mathrm{Pl}(A), \forall A \subseteq \Omega.$$

The uncertainty functions Bel and Pl are dual i.e. $\mathrm{Bel}(A) = 1 - \mathrm{Pl}(\overline{A})$, and the length of the interval $[\mathrm{Bel}(A); \mathrm{Pl}(A)]$ expresses some level of ignorance. A state of complete ignorance is represented by the vacuous Basic Belief Assignment $m(\Omega) = 1$ or equivalently by $[\mathrm{Bel}(A); \mathrm{Pl}(A)] = [0; 1]$ for all $A \subset \Omega$, $A \neq \emptyset$ and $A \neq \Omega$, which is distinct from the uniform distribution. The underlying distribution of a belief function is no longer defined over the singletons of Ω but rather over its powerset 2^Ω. The Basic Belief (or Probability) Assignment (BBA or BPA) m assigns to each event A of Ω a real value in $[0; 1]$ with the only constraint that:

$$\sum_{A \subseteq \Omega} m(A) = 1$$

$m(A)$ is the degree of belief assigned exactly to A. For instance, if 0.7 is the belief degree assigned to the type *Cargo* for a vessel by a classifier, the probability theory would require 0.3 to be assigned the complement of this singleton (i.e. all the other classes but $Cargo$):

$$P(Cargo) = 0.7 \text{ and } P(\overline{Cargo}) = 0.3 \quad (9.3)$$

Instead, in evidence theory, without any evidence *against* the class $Cargo$, the remaining mass would be assigned to Ω:

$$m(Cargo) = 0.7 \text{ and } m(\Omega) = 0.3.$$

Such an expression captures *only* the evidence available and does not make any further assumption on its contrary (in this case "not a Cargo"). $m(\Omega)$ represents the weight of ignorance. Due to the additivity axiom, probability theory imposes the remaining mass to be assigned to "not a Cargo". Evidence theory offers more flexibility to represent uncertainty and ignorance. m induces a family of probability distributions compatible with the corresponding lower and upper bounds (i.e. the belief and plausibility), including P from Eq. (9.3).

The open-world assumption [53] relaxes the exhaustivity of the original Dempster-Shafer model, allowing the empty set to have a non-null mass. That means that other classes than the ones initially considered in Ω can actually be true. The interpretation of a non-null mass on the empty set allows to revise the original set of classes considered as possible for the type of vessel, for instance. We may first consider classifying only Cargos and Tankers, while another type is popping up. It is interesting in our practical case of route association as this empty set would then act as a rejection class for "off-route" vessels as it will be illustrated later in Sect. 9.4.6.

The belief and plausibility functions can then respectively defined, for all $A \subseteq \Omega$, as:

$$\text{Bel}(A) = \sum_{B \subseteq A} m(B) \quad \text{and} \quad \text{Pl}(A) = \sum_{B \cap A \neq \emptyset} m(B) \quad (9.4)$$

While $m(A)$ expresses the degree of belief committed exactly to A, $\text{Bel}(A)$ is the degree of belief committed to all subsets of A and $\text{Pl}(A)$ is the degree of belief committed to all subsets which are consistent with A. The *pignistic probability* [48] transforms a Basic Belief Assignment m into a probability distribution compatible with the interval $[\text{Bel}(A); \text{Pl}(A)]$, and is defined for any singleton of Ω as:

$$\text{BetP}(x) = \sum_{x \in A} \frac{m(A)}{|A|} \quad (9.5)$$

The pignistic transformation distributes in an additive and uniform way the mass of sets A to each singleton x composing it. Such a transformation is necessary when one wants to decide among the singletons.

Two functions m_1 and m_2 provided by two independent sources can be conjunctively combined as:

$$m_{12}(A) = \sum_{B \cap C = A} m_1(B)m_2(C) \qquad (9.6)$$

A non-null mass on the empty set after conjunctive combination denotes a conflict between the two belief functions combined and $m(\varnothing)$ can be interpreted as the weight of conflict between two independent but partially reliable sources, acting then an indicator to a possible anomaly. For instance, if Source 1 says the vessel is either a Cargo or a Tanker, and Source 2 says it is a Passenger, this conflict could be interpreted as one of the sources being wrong. Alternatively as discussed above, the mass of the empty set could be interpreted under the open-world assumption as another type of vessel not originally considered. This interpretation will be used later to detect off-route vessels, such as those which observed features inducing conflicting information, meaning that no route among the selected ones matches all the features of the vessel.

9.3.2.3 Fuzzy Sets Theory

Fuzzy set theory was developed by Zadeh [58] to model vague information such as human language descriptions ("*small, large, quick, young*"), and concepts which cannot be defined by an interval with strict limits, such as "*the vessel is small*" or "*the vessel has a quite low speed*".

A fuzzy set is a set with imprecise (not well defined) boundaries, i.e. *fuzzy* and extends the concept of classical (crisp) set. In classical set theory, a subset $A \subseteq \Omega$ is referred to as a crisp set and is represented by a characteristic function μ_A such that $\mu_A : \Omega \to \{0; 1\}$ with $\mu_A(x) = 1$ if $x \in A$ and $\mu_A(x) = 0$ if $x \notin A$. A fuzzy set is defined by a membership function $\mu_A(x)$ which is a generalisation of the characteristic function and can take its values in the interval [0, 1], when normalised. A fuzzy set μ in Ω is a set of ordered pairs:

$$\mu = \{(x, \mu(x)|x \in \Omega)\}$$

$\mu(x)$ is the degree of membership of x to the fuzzy set μ.

Compared to probabilities and belief functions which define degrees of belief regarding the occurrence (or truth) of an event, being itself either true or false, fuzzy sets define degrees of truth for events which are thus allowed to be more or less true. Fuzzy sets are classically combined using *min* and *max* operators, acting as conjunctive and disjunctive operators respectively.

9.4 Uncertainty Representation and Reasoning Techniques

Six different Uncertainty Representation and Reasoning Techniques (URRTs) are presented below and applied to maritime anomaly detection. The URRTs described herein are very basic and simple schemes, far less complete than the ones reported in the literature addressing the problems of maritime anomaly detection or route association. However, this deliberate simple exposure is aimed at *"dissecting"* the underlying uncertainty representation and reasoning, as a first step toward comparison and better understanding. The operators for aggregating the different estimations along the different features are also discussed. These will be named hereafter the *fusion operators.*

We also provide a comparative description of the expressiveness of the six URRTs, assessed along the types of information deficiency introduced in Sect. 9.3.1 (*graduality*, *uncertainty* and *imprecision*). *Granularity* is captured by the set of possible values for all variables from Ω (the set of all hypotheses) which is kept constant for all the methods, thus will not be discussed. *Trueness* will be illustrated in reference to some ground-truth in the experiment presented in Sect. 9.5.

9.4.1 URRT#1: Pattern matching—Euclidean

Intuitively, the closer an observed vessel is to the centroid of the route, the more likely it is to belong to the route. A standard pattern matching approach computes the Euclidean distance between $\mathbf{x}$, i.e. the vector of the observed feature values for a given vessel, and each of the routes of Ω_R as:

$$d^{(E)}(\mathbf{x}, R_k) = \sqrt{\left(\mathbf{x} - \mathbf{r}^{(k)}\right)' \left(\mathbf{x} - \mathbf{r}^{(k)}\right)} \tag{9.7}$$

$$= \sqrt{\sum_{i \in \mathcal{A}} \left(x_i - r_i^{(k)}\right)^2}$$

where $\mathcal{A}$ is the set of features, $\mathbf{r}^{(k)}$ is the prototype corresponding to route R_k (see Table 9.3), defined in the feature space Ω_o and $\mathbf{x}'$ is the transpose vector of $\mathbf{x}$. The ith components of $\mathbf{x}$ and $\mathbf{r}^{(k)}$ are denoted by x_i and $r_i^{(k)}$ respectively. The quantity $\left(d_i^{(k)}\right)^2 = \left(x_i - r_i^{(k)}\right)^2$ can be interpreted as an inverse *degree of match* of the observation x_i to the equivalent prototypical element of R_k, that we denote as $r_i^{(k)}$: the lower the square distance, the higher the degree of match of the vessel to that route. Thus, if a vessel is traveling close to a route R_k, its local distance $d_1^{(k)}$ along the positional feature will be very small. If, for instance, it is also traveling in the same direction than Route R_k, its local distance along the course over ground feature will also be small, hence a global small distance to that route.

This distance can be used to define a degree of membership of the vessel to that route. Adopting a similarity view of fuzzy sets [6, 15], the degree of membership $\mu_i^{(k)}$ of the vessel to the route R_k, according to feature i, can be defined through $d_i^{(k)}$ as, for instance:

$$\mu_i^{(k)} = \exp\left(-\left(d_i^{(k)}\right)^2\right)$$

which tends toward 0 whenever the distance tends toward infinity and equals to 1 if the distance is null. That means we assume the vessel is belonging "more or less" to Route R_1, up to the degree $\mu_i^{(1)}$. Equation (9.7) can then be written as:

$$d^{(E)}(\mathbf{x}, R_k) = \sqrt{-\sum_{i \in \mathcal{A}} \log\left(\mu_i^{(k)}\right)} \qquad (9.8)$$

where $\mu_i^{(k)} \in]0; 1]$ is a normalised degree of membership. Equation (9.8) is a bisymmetrical continuous strictly monotonous mean [8]. The fusion operator in Eq. (9.7) and equivalently in Eq. (9.8) is a sum (disjunction) which averages local dissimilarities with R_k along the different features.

We then consider the following decision rule:

$$\hat{R} = \begin{cases} \arg\min_k d^{(E)}(\mathbf{x}, R_k) & \text{if } d^{(E)}(\mathbf{x}, R_k) < \varepsilon_1 \\ R_0 & \text{otherwise} \end{cases} \qquad (9.9)$$

where ε_1 is a threshold to be set according to the operator's needs or expectations, representing some tolerance over the global distance (or degree of membership) over the $|\mathcal{A}|$ features. In practice, ε_1 can be deduced from some aggregation of the individual thresholds ε_1^i for each feature. According to the decision rule in Eq. (9.9), all synthetic routes with distance to the observed vessel less then the threshold ε_1 will be selected as possible candidates for the vessel. This decision rule allows some imprecision in the decision space as it can lead to a set of possible routes, without identifying a single one. An anomaly is detected (i.e. class R_0 selected) if the vessel does not match any route, meaning that the distances to all routes are higher than the threshold ε_1. Many anomaly detection approaches are based on distance computation as an implementation of the notion of "closeness to normalcy" (e.g. [56]). Semantic distances can also be used to assess the different meanings between attributes (e.g. [7]).

URRT#1 Information deficiency handling. In URRT#1, the route representation is considered as precise and certain since the route prototypes are defined by single values (either the mean, or the mode for the vessel type); the dependency between the uncertain variables corresponding to each observed feature is not considered, nor are the possible links between routes, such as the link between "sibling routes" connecting the same ports but in reverse direction. The sources' uncertainty about their singular declaration at t is not considered; the sources' reliability is not represented, nor is any second-order uncertainty. URRT#1 captures a single imperfection type as a notion of graduality through a distance measure, the route prototype being considered as a reference: the distance to route can be interpreted as an inverse degree of membership of $\mathbf{x}$ to R_k, such that the smaller the distance to the route, the higher the degree of membership to that route. From this generic information, a singular imperfection is further derived combining with the observation of the vessel. That means that the uncertainty we may have about the observed vessel following a specific route at the current instant in time (singular information) originates from uncertainty from historical data, in combination with the precise and certain observation about that vessel at the current instant in time. The fusion of the estimations along the different features is performed through the distance definition by a sum operator acting as an average of inverse of similarities along the different features of $\mathcal{A}$: The higher the local similarities, the lower the global distance and the higher the membership of $\mathbf{x}$ to R_k. However, this aggregation is valid for numerical features only and needs to be adapted for categorical features, such as the type of vessel.

9.4.2 URRT#2: Pattern Matching—Mahalanobis

A modified version of the Euclidean pattern matching scheme is obtained by using the Mahalanobis distance:

$$d^{(M)}(\mathbf{x}, R_k) = \sqrt{(\mathbf{x} - \mathbf{r}^{(k)})' \Sigma^{-1} (\mathbf{x} - \mathbf{r}^{(k)})} \qquad (9.10)$$

where Σ is the covariance matrix of the random vector $\mathbf{X}$ associated to $\mathbf{x}$, whose coordinates are random variables X_is. The superscript $^{-1}$ denotes the inverse matrix. The element $\sigma_{i,j}$ of Σ is the covariance of X_i and X_j defined as $E(X_i, X_j) - E(X_i)E(X_j)$ where E is the expectation operator such that $E(X) = \sum xp(X = x)$ for a discrete random variable X. The same decision rule Eq. (9.9) as for the Euclidean pattern matching is used. However, another threshold ε_2 must be used instead of ε_1, based on the covariance matrix.

As in Eq. (9.7), the fusion operator in Eq. (9.10) is a disjunction but including weights which would discount the local individual dissimilarities relatively to the variance of their corresponding feature, and pairs of errors relatively to their covariance.

The Euclidean and Mahalanobis distances in Eq. (9.7) and Eq. (9.10) are well suited to features defined over numerical and continuous scales while other measures are better suited to nominal variables such as the type of vessel. More suitable distance measures are usually based on the aggregation of individuals for each feature, possibly using other definitions than the square difference (e.g. [18]). Other distances such as the log-normal probability density (e.g. [2]) would account for the routes statistics as well. The Mahalanobis distance is used in [31] to associate vessel tracks to maritime routes.

URRT#2 Information deficiency handling. The extension of URRT#1 using the Mahalanobis distance as described by URRT#2, accounts for both the spread of the routes along the different features (through the individual standard deviations σ_i) and the dependency between variables (through the covariances $\sigma_{i,j}$). The variance can be interpreted as a measure of imprecision regarding X_i. The covariance describes how the variables vary with each other, measures the dependency between them, and expresses then some statistical uncertainty on the link between X_i and X_j. Compared to URRT#1, URRT#2 considers some imperfection about the reference objects (the routes). Still, there is no consideration of singular uncertainty about the observations at t, except for the graduality measured by the distance to the prototype route.

9.4.3 URRT#3: Probability-based—Bayesian

In the standard Bayesian approach to fusion, the function $p(\mathbf{X} = \mathbf{x}|R_k)$ represents the likelihood of observing a specific set of feature values $\mathbf{x}$ on a given route R_k, and is usually derived from past observations used to compute the routes. For instance, referring to Table 9.3, $p(X_5 = Cargo|R_1) = 0.8$ is the likelihood of observing a Cargo ship on Route R_1. The different observations are combined following Bayes' rule (as introduced by Eq. (9.2)):

$$P(R_k|\mathbf{x}) \propto p(R_k) \prod_{i \in \mathcal{A}} p(x_i|R_k), \forall R_k \in \Omega_R \qquad (9.11)$$

under the assumption of *independent and identically distributed observations*. $\mathcal{A}$ is the set of features. $p(R_k)$ is some prior probability that the vessel follows a specific route, under the assumption of *independent features*. The resulting posterior probability $P(R_k|\mathbf{x})$ represents some belief that the route followed by the vessel of interest is R_k given that we currently observe $\mathbf{x}$. A normalisation factor ensures that a probability distribution is obtained. Equation (9.11) is known as Naïve Bayes[5] model in classification. The decision rule is the Maximum A Posteriori (MAP) probability:

$$\hat{R} = \begin{cases} \arg\max_k p(R_k|\mathbf{x}) \text{ if } p(R_k|\mathbf{x}) > \varepsilon_3 \\ R_0 \text{ otherwise} \end{cases} \qquad (9.12)$$

where ε_3 is a threshold: if the posterior probability is too uniformly distributed among the routes, then no clear matching is detected and an anomaly is returned. The Bayesian reasoning scheme is at the basis of the Bayesian network approach proposed for instance in [21].

The fusion operator is a conjunctive operator, i.e. the product of individual likelihoods. It has the property of decreasing very fast to 0 as the number of features to be combined increases. Also, the result is exactly 0 if only one likelihood is null.

URRT#3 Information deficiency handling. As in URRT#1, the independence assumption between variables applies to the Bayesian approach of URRT#3. No consideration for the source's reliability or self-confidence ; the measurement itself is assumed both certain and precise by the source. Uncertainty is considered over the mapping between Ω and Ω_R where the likelihoods $p(x_i|R_k)$ describe how likely it is to obtain some specific measurement given that the vessel follows route R_k. Prior

---

[5]The model is named "naïve" because it is based on the assumption of independent features.

uncertainty about routes is explicitly considered by $p(R_k)$ which could be based on other contextual information such as meteorological or seasonal information. The fusion is done through a product operator which has the drawback of decreasing very rapidly to 0 once one of the likelihoods is very low. This rule is named "severe" for that reason [24], since it is very sensitive to one source's negative opinion. The product is a conjunctive operator (corresponding to a logical AND) making the underlying assumption either that all the measurements are correct, or that all the sources are reliable. Including the source's reliability about measurements is a direct extension of URRT#3 (see for instance [29]), as well as considering the dependencies between variables. The final assessment expresses some *uncertainty* degree that the vessel is actually following route R_k.

9.4.4 URRT#4: Probability-based - Non-Bayesian

In a still probabilistic but non-Bayesian approach, each measured feature is considered providing some evidence about the membership of $\mathbf{x}$ to a given route R_k. For instance, $p_s(R_k) = p(R_k|x_s)$ is the contribution of the speed observation to the membership of the vessel V to R_k and is interpreted as the probability that V belongs to R_k given (or according to) the estimated speed. Then, the observations are aggregated by a weighted sum as:

$$p(R_k|\mathbf{x}) = \sum_{i \in \mathcal{A}} \alpha_i p(R_k|x_i), \forall R_k \in \Omega_R \qquad (9.13)$$

where $\alpha_i \in [0, 1]$ is a weight reflecting the confidence in the decision computed by the individual sources, and possibly be deduced from $p(x_i)$, or the relevance of the features to the fusion problem (for instance, the position and heading may be given a higher weight than the vessel type). Each individual posterior can be estimated by Bayes' rule applied to single features, i.e. $p(R_k|x_i) \propto p(x_i|R_k)p(R_k)$. Essentially, the distinction between URRT#3 and URRT#4 is the disjunctive operator (sum) used in URRT#4 rather than the conjunctive operator (product) used in URRT#3. The decision rule is then Eq. (9.12).

URRT#4 Information deficiency handling. In the probabilistic non-Bayesian approach of URRT#4, the individual probabilities $p(x_i|R_k)$ are assumed to provide local belief degrees toward each route. They are summed up to give a global belief so that the higher the belief degree according to each feature, the higher the global belief. URRT#4 does not consider the dependency between features. However, some notion of source's *reliability* can be captured by the weights α_i that can be derived from some likelihood measures extracted from a confusion matrix. This expresses some *second-order uncertainty* about the source's declaration at t. The combination rule is a disjunction (logical OR) and is known to be less sensitive to estimation errors (unreliable sources), and to single source's opinion [24] making the approach more robust.

9.4.5 URRT#5: Transferable Belief Model (TBM) Model-Based

The reasoning scheme considered here is the one proposed in [12, 43] within the Transferable Belief Model (TBM) framework and making use of the Generalised Bayes Theorem (GBT) [49] as the combination rule, given by the following plausibility measure for a subset of routes A:

$$Pl(A|\mathbf{x}) = 1 - \prod_{R_k \in A} (1 - Pl(\mathbf{x}|R_k)), \forall A \subseteq \Omega_R \qquad (9.14)$$

where $Pl(A) = \sum_{A \cap B \neq \varnothing} m(B)$ is the plausibility of $A \subseteq \Omega_R$, with m being a Basic Belief Assignment (see Sect. 9.3.2). $Pl(A|\mathbf{x})$ is the conditional plausibility of A and is interpreted as the maximum confidence that can be assigned to A (i.e. that the route followed is one of the subset A) given that $\mathbf{x}$ has been observed. More details are given in [43]. As introduced in Sect. 9.3.2, pairs of plausibility and belief values can be interpreted as intervals over the probability of a subset of routes $A \subseteq \Omega_R$. As illustrated in Sect. 9.2.3, the lack of data may render the probability distributions difficult to estimate precisely and accounting for some second-order uncertainty may be justified in this case. The decision rule requires two steps: (a) the transformation of the Pl measure into a probability distribution over Ω_R (e.g. the pignistic probability defined in Eq. (9.5)) such that (b) the MAP rule of Eq. (9.12) can be applied (with the appropriate threshold). The fusion operator is again a conjunctive operator with similar properties than the ones described in Sect. 9.4.3.

URRT#5 Information deficiency handling. URRT#5 may be seen as an extension of URRT#3 where upper probabilities (i.e. plausibility functions) are used rather than probabilities. The plausibility function $Pl(\mathbf{x}|R_k)$ models some imprecision about the (assumed precise but unknown) likelihood function $p(\mathbf{x}|R_k)$ (itself capturing some uncertainty) used in URRT#3. Equation (9.14) is obtained under the assumption of a vacuous prior on Ω_R, meaning that no prior uncertainty on routes is considered. The output of the Generalised Bayesian Theorem being also a plausibility function, assigns plausibility values to *subsets* of routes and captures thus some imprecision over Ω_R. It defines then *second-order uncertainty* by means of a couple belief-plausibility measures expressing some uncertainty about the posterior event $(R_k|\mathbf{x})$. This second-order uncertainty is not considered in the traditional Bayesian approach where the probability estimations are considered precise and certain. Other equivalent approaches exist framed into imprecise probability or robust Bayesian frameworks.

9.4.6 URRT#6: Belief Functions—Database Query

Similarly to the probabilistic non-Bayesian URRT#4, each observed feature x_i of $\mathbf{x}$ is assumed to provide some evidence about route R_k being followed by vessel V. The uncertainty about the route followed by the vessel is modeled by belief functions rather than probabilities. Each observation x_i is regarded as a query to Ω_R such that only the items (i.e. routes) satisfying the observations are retrieved, to form a set of possible routes A_i according to x_i. A_i is the subset of routes satisfying the query x_i:

$$A_i = \{R \in \Omega_R | x_i \in \Omega_i\}$$

For instance, A_1 is the set of routes compliant with an average measured speed of their points of 5 knots, meaning that the likelihood (as estimated in Table 9.3) of a vessel traveling at 5 knots for routes from A_1 is not null, and may be higher than a certain threshold. The mapping between the observation space Ω and decision space Ω_R assigns to any singleton of Ω a *subset* of Ω_R, and thus named *multi-valued*. Let us consider that some singular information (at the current instant in time) about the vessel type is provided under the form of a probability mass function, for instance by a classifier: $\mathbf{p}_T = [0.4\ 0.3\ 0\ 0.3\ 0]$, where T denotes the vessel type and $p^{(T)}(X_T = \text{Cargo}) = 0.4$ is the probability that the observed vessel is a Cargo type, as estimated based on current observations. This uncertainty is transferred to the corresponding subsets of Ω_R previously defined by the multi-valued mapping, and defines a Basic Belief Assignment m_T over Ω_R, where the numerical weight $m_T(A_i) = p_T(x_i)$ is interpreted as the degree of belief that can be assigned to A_i and to none other subset of A_i. Then, as a result of computation (for the sake of the example), $A_{\text{Cargo}} = (R_1, R_2, R_3, R_4, R_5)$ is the set of routes possibly followed by cargo vessels and is assigned a weight of 0.4. Equivalently, $A_{\text{Tanker}} = (R_2, R_3, R_5)$ and $m(A_{\text{Tanker}}) = 0.3$ and $A_{\text{Passenger}} = (R_2, R_5)$ and $m(A_{\text{Passenger}}) = 0.3$. This multi-valued mapping does not induce a probability distribution over Ω_R but a Basic Belief Assignment.

The resulting Basic Belief Assignment m over Ω_R is obtained by combining the individual contributions of each feature by the conjunctive rule of Eq. (9.6), which is based on the intersection between sets, where weights are assigned to conjunctions of sets of routes A_i and A_j. For instance, if the set of compatible routes according the vessel position is (R_1, R_2) and the set of compatible routes according to the vessel type is (R_1, R_4), then the conjunctive rule will assign a high belief degree to R_1, the common element to both subsets.

The decision rule is similar to Eq. (9.12) but considers the weight of conflict as a criterion for anomaly detection:

$$\hat{R} = \begin{cases} \arg\max_k \text{BetP}(R_k) \text{ if } m(\emptyset) < \varepsilon_4 \\ R_0 \text{ otherwise} \end{cases} \tag{9.15}$$

where BetP is the pignistic transformation [48] of m and ε_4 a threshold value. The quantity $m(\varnothing)$ is the mass of the empty set after combination of the different Basic Belief Assignments built upon the different observations along the different features and represents the global weight of conflict between all the sources (or features). For instance, if the set of compatible routes according the vessel position is (R_1, R_2) and the set of compatible routes according to the vessel type is (R_3, R_4), some conflict will be detected, because $(R_1, R_2) \cap (R_3, R_4) = \varnothing$. This could highlight a case where a passenger vessel is following a route classically followed by cargo vessels.

URRT#6 Information deficiency handling. In URRT#6, the uncertainty output by the sources about the measurement provided at t is considered (singular information). Rather than a single (precise and certain) measure, each source outputs a probability distribution over the set of values of their respective feature, inducing a plurality of multi-valued mappings over Ω_R when querying the "dictionary" of routes. The multi-valued mappings define some imprecision over the set of routes, since to a single value in Ω_i corresponds a subset A of Ω_R. The main characteristic of this scheme is to deal with subsets of routes, in a qualitative way, with an additional quantification. The explicit notion of *conflict* is a way to detect inconsistencies between the subsets of routes compatible with each feature. The fusion is performed through a conjunctive rule, assuming independent and totally reliable sources. The uncertainty captured is only some "self-uncertainty" of the source itself which could be interpreted as "I'm not certain that the type of the vessel is a Cargo", for instance. Note that the reliability of the source is rather a meta-information on the *general* ability of the source to provide correct information. A source can thus be reliable and uncertain.

The type of imperfection handled by URRT#1 and URRT#2 is graduality, meaning that the route is considered as an object with fuzzy boundaries, to which vessels belong more or less. With this interpretation, the distance measure provides an aggregated inverse degree of membership of the vessel to a given route: If the distance is low then the vessel belongs to the route with a high degree of membership. Instead, the other methods (URRT#3 to URRT#6) express a "degree of belief" that the vessel is following the route. This is a difference between a binary event (URRT#3 to URRT#6) and a fuzzy event (URRT#1 and URRT#2). This semantic aspect highlights the need for a clear semantics for the concept of *maritime route*, whether it means either "following a specific path and heading to a specific destination" (binary event) or "being positioned on a portion of the sea with ill-defined boundaries" (fuzzy event).

9.5 Quantitative Comparison

The qualitative analysis presented in the previous section is now complemented by a quantitative analysis based on the AIS dataset used throughout the book. We first identify and define a series of criteria and measures for quantitative assessment of the six URRTs discussed in this chapter, in line with the information deficiencies discussed previously.

9.5.1 Evaluation Criteria

We consider the quality criteria of TRUENESS, PRECISION and CERTAINTY. To measure TRUENESS we use the standard F_β-score:

$$\text{Tru}(\Psi) = F_\beta(\Psi) = \frac{(1+\beta^2)TP}{(1+\beta^2)TP + \beta^2 FN + FP} \tag{9.16}$$

where TP, FN and FP are the number of true positives, false negatives and false positives respectively. The parameter $\beta \in \mathbb{R}$ allows weighting the two types of errors, i.e. the false negatives and false positives.

IMPRECISION and UNCERTAINTY quantify how much a URRT is non-specific and uncertain respectively. They are assessed through the (normalised) Hartley measure and Shannon entropy:

$$\text{Imp}(\Psi) = \frac{1}{\log_2(|\Omega_R|)} \log_2(|A|) \tag{9.17}$$

$$\text{Unc}(\Psi) = -\frac{1}{\log_2(|\Omega_R|)} \sum_{R \in \Omega_R} p(R) \log_2(p(R)) \tag{9.18}$$

where $|.|$ denotes the cardinality of sets and p is the probability distribution over the set of routes before decision is taken. The equations above are normalised versions of the measures, such that 1 means maximally imprecise or uncertain respectively. In Eq. (9.17), A is the set of compatible routes according to the corresponding decision criterion.

9.5.2 Dataset of Tracklets

The six URRTs are tested on the AIS dataset used throughout the book [40], the routes of which are illustrated in Fig. 9.2. From this dataset of AIS contacts, a tracklet dataset has been extracted for the 60-day period from October 1st to November 29th 2015. A *tracklet T* is a set of five consecutive AIS contacts originating from the same vessel.

Tracklets have been labeled with their route previously computed by the TREAD software [36], providing some ground truth about the maritime route followed. The tracklet dataset includes 800 tracklets of 5 consecutive AIS data contacts, 400 of them being extracted from a selection of 21 routes, and 400 extracted from the other contacts, considered off-route with respect to the 21 selected routes. The 4000 points corresponding to the 800 tracklets have then been removed from the original data set and the TREAD software was run again on the pruned dataset. The fact to run the software a second time (over the modified set of points) enables the generation of the same 21 routes (those 21 routes are shown in Fig. 9.2) without the points that were extracted. Thus, it enables the labelled dataset to be completely independent from the set of routes. It was assumed, as the difference of 4000 points is negligible with respect to the whole dataset, that the number of contacts removed would not have changed the labeling of the routes. The synthetic maritime routes computed with the dataset were then isolated and used in the 6 URRTs computations.

9.5.3 Results and Discussion

We present here results of route deviation detection, thus considering two classes only, R_0 the class of off-route tracklets, not belonging to one of the 21 selected routes, and R_1, the class of on-route tracklets, belonging to one of the 21 routes. The representation of the synthetic routes used for these experiments is a cluster of contacts, as introduced in Sect. 9.2.1. The distance from the tracklet T to a route R in URRT#1 and URRT#2 is computed with Hausdorff distance along each feature, defined as:

$$d_i^{(H)}(T, R) = \max \left\{ \sup_{x_i \in T} \inf_{r_i \in R} d_i(x_i, r_i), \sup_{r_i \in R} \inf_{x_i \in T} d_i(x_i, r_i) \right\}$$

where d_i is the simple Euclidean distance, x_i and r_i are the values along feature i for a contact of the tracklet T and of the route R respectively. The aggregation is then done following either Eq. (9.7) for URRT#1 or Eq. (9.10) for URRT#2 defined in Sect. 9.4.

Figure 9.4 displays the output quality results on a spider (radar) graph, with the three criteria of TRUENESS as in Eq. (9.16), CERTAINTY in Eq. (9.18) and PRECISION in Eq. (9.17). The best method is the one displaying the triangle with highest area in the graph.

The TRUENESS criterion as measured by F_1 aggregates the TPR (True Positive Rate) and FNR (False Negative Rate) and hides thus the contribution of each corresponding type of errors. The true positive (resp. negative) rate is the number of true positive TP (resp. negative TN) over the number of real positive cases P (resp. negative cases N). Table 9.4 expands the criterion of TRUENESS by displaying additional measures to the TPR, as the TNR, the F_1, F_2 and $F_{0.5}$ measures. While F_1 assigns

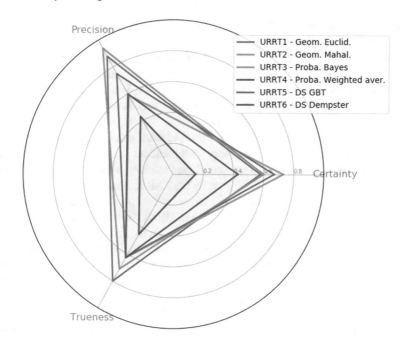

Fig. 9.4 Spider graph of three output quality criteria for the six URRTs

Table 9.4 Trueness measures for the six URRTs

	TPR	TNR	F_1	F_2	$F_{0.5}$
URRT#1	1.00	0.50	0.80	0.91	0.71
URRT#2	0.81	0.53	0.71	0.77	0.66
URRT#3	0.56	0.77	0.63	0.58	0.67
URRT#4	0.49	0.31	0.45	0.47	0.43
URRT#5	0.56	0.77	0.63	0.58	0.67
URRT#6	0.56	0.77	0.63	0.58	0.67

equal weights to false negatives and true positives, F_2 gives more emphasis on false negatives and $F_{0.5}$ attenuates the influence of false negatives.

Through these criteria and associated performance measures, we observe that URRT#1 provides very good results in terms of PRECISION and good results in terms of TRUENESS, but the CERTAINTY value is rather low. That means that, although before decision a single route was detected as compatible with the set of observed features and that it was generally the correct route, the score (once converted to probability) was not that high for that route. Analysing Table 9.4, it appears that the global TRUENESS is mainly affected by a very low TNR while the TPR is 1.

The extension of URRT#1 to the Mahalanobis distance (URRT#2) provides similar results in terms of CERTAINTY, PRECISION, and TRUENESS. Slight differences are however observed: The TPR is lower, meaning that on-route vessels are less frequently detected. Similarly, the TNR is lower, resulting in a slightly lower TRUENESS; while PRECISION is improved and CERTAINTY is the same. Indeed, it appears that considering the dependency between the features in the observation space, as opposed to the naive independence assumption under URRT#1, leads to a slight change in performances, while we cannot draw any relevant conclusions due to the marginal changes resulting in different dimensions.

The Bayesian approach URRT#3 and its evidential extension URRT#5 provide quite similar performance results, notably with identical TRUENESS and very similar CERTAINTY values. Compared to the prototype matching approaches URRT#1 and URRT#2, the TPR has dropped to 56% while the TNR increased to 77%. That means that these two approaches assign vessels to routes less correctly than the matching approaches. However the CERTAINTY is higher. URRT#3 and URRT#5 use the likelihoods extracted from the routes' statistics as a basis for uncertainty representation. No probability distribution estimation method was applied and the likelihoods were simply extracted from the histograms. The evidential approach based on the Generalised Bayes Theorem (URRT#5) uses plausibility functions instead of probabilistic likelihoods and allows by that to account for some imprecision on the probability distributions. This approach is particularly interesting when the amount of data available does not guarantee a reliable estimation of the probability distribution. Indeed, some routes are built upon only a few trajectories and their uncertainty may be better represented by lower and upper bounds of unknown probability distributions (as provided by belief and plausibility measures respectively) or simply by crisp intervals, rather than a single precise probability distribution.

The weighted average of probabilities (URRT#4) provides the worst results along the three criteria, and the TNR is the worst of all approaches. From Table 9.4, it appears that the bad performance of URRT#4 is mainly due to a very low TPR (the only under 50%). That means that on-route and off-route vessels are randomly detected as either on-route and off-route, making this approach very uninformative. The disjunctive operator $(+)$ averages the posterior probabilities and a very low probability along one feature (denoting an anomaly) would be diluted among other higher probabilities. This dilution would make it more difficult to detect the directional and kinematic anomalies. As mentioned previously, the disjunctive operator is a rather cautious fusion operator, more suited to a consensus. We should not however conclude that URRT#4 is not a good approach, as its strength is to be robust to errors and unreliable sources, something that was not reflected in our dataset.

The evidential approach using the conjunctive rule (URRT#6) provides a mediocre TPR value with similar TRUENESS values to URRT#3 and URRT#5. The CERTAINTY and PRECISION are both quite low meaning that the decision was taken with still a high hesitation. URRT#6 is the only method whose rejection criterion is based on a measure of conflict. The conflict is represented by the empty set between subsets of routes compatible with different features. In case this intersection is empty, no route is actually detected as compatible and the tracklet is classified as off-route. From Table 9.4 it appears that this was correct in 77% of the cases, as displayed by the TNR value.

Finally, note that all the URRTs except URRT#6 rely on *generic imperfection* only, i.e. the routes and related information extracted from historical AIS data. URRT#6 is the only approach (again, as currently implemented) which accounts for the uncertainty expressed at the current instant in time t (*singular information*). All the other approaches rely on uncertainty derived from past observations (historical data). In case a source expresses some uncertainty about its declaration, this singular uncertainty could be considered in this approach. If we assume receiving highly truthful information (for instance, the AIS position), then the uncertainty at the current instant in time may not be very relevant and could be ignored. However, in case we have serious doubt about the veracity of the information provided (for instance, the type of the vessel which can be easily deceived) a proper account of singular uncertainty may be relevant. Whatever the solution, it remains important not confusing generic and singular uncertainty as their meaning is very different.

> **Remark ! Attention**
>
> The results presented here should be read as serving the purpose of understanding how the different approaches handle uncertainty, and their impact on different performance criteria. The application of such techniques (and their more elaborated counter-part) to maritime anomaly detection requires deeper work.

9.6 Conclusions

In this chapter, we compared six uncertainty and reasoning techniques (URRTs) in their basic implementation, in their ability to handle uncertainty in representation and fusion. We selected a variety of classical and simple schemes from (or adapted from) the literature which are all good candidates to solve the problem of maritime route association, and route deviation. A first qualitative analysis highlighted that the methods have different expressiveness, capturing different information deficiencies, i.e. uncertainty, imprecision, graduality, granularity.

The implementation of the URRTs to detect off-route vessels, processing real AIS data, allowed us to illustrate that the classical evaluation criteria should not be assessed in isolation. Rather, quantifying jointly trueness, precision and certainty highlights the respective strengths and weaknesses of the methods. Additionally, the different types of information deficiency constitutes another dimension to be considered in the assessment of a solution.

In summary, rather than identifying a "winner" approach, the comparison between the URRTs presented herein aimed at highlighting their *differences* and possible *complementarity* in uncertainty representation and reasoning. The approaches have been kept simple for a clearer understanding and further development of each approach in conjunction with a good characterisation of the uncertainty underlying the problem and data at hand could lead to the design of efficient algorithms with interpretable results for detecting the anomalies at sea.

9.7 Bibliographical Notes

In this chapter, we saw that rather than competitors, the different models for uncertainty representation and reasoning are dedicated to different types of information and uncertainty, and encode different reasoning for different problems.

The reader is encouraged to consider the following (classical) works providing some survey or comparison of the different mathematical frameworks, while relevant references are highlighted below. Evaluating or comparing uncertainty *calculi* in the absolute is not a trivial task because these make different fundamental assumptions about the nature and interpretation of uncertainty they aim at representing or processing (see for instance [16, 25]). Fundamental and global formal evaluation and analysis have been in particular presented in [14, 25, 47, 51, 54, 57] and more recently in [16]. It appears for instance that probability, possibility and fuzzy set theory are not comparable since they are appropriate to deal with different types of uncertainty. Rather than competitors, they appear to be "complementary theories of uncertainty that utilise distinct types of uncertainty for expressing deficient information" [25]. Belief functions [46, 53] are "[…] aimed directly at modeling incomplete evidence, but certainly not incomplete knowledge", and designed to handle singular uncertainty while they are inadequate for dealing with statistical knowledge [16, 46]. Fusion rules have their own meaning and application constraints as well. While being an update rule (i.e. to update probabilities in light of new evidence), Bayes' rule is also widely used for fusion purposes (e.g. [9, 29]). However, Bayes' rule is not applicable in case of probable knowledge, unanticipated knowledge and introspective knowledge [13]. In Shafer's view, Dempster's rule is specifically dedicated to combine uncertain and imprecise singular information, such as testimonies. Dempster's rule should also be applied only to independent and reliable sources [46, 52].

Each of the URRT described in this chapter could be enriched to solve the problem of vessel-to-route association. As examples only, the reliability of the sources is classically considered in URRT#6 by introducing discounting (or reinforcement) operations for belief functions such as those described in [32]. Also, the reasoning scheme of Equation (9.2) in URRT#3 can be enriched by considering the reliability of the sensors in providing accurate measurements, and introducing factors $p(Z_i|X_i)$ where Z_i is the measurement provided by the source while the true value was X_i, as proposed in [29] for instance. URRT#3 can be implemented as a Bayesian network (e.g. [21]) where the dependency between variables is considered. A Bayesian network has the advantage of a better *transparency* in the reasoning for the user, which is an interesting assessment criterion to be considered. Moreover, the *computational cost* is improved by local computations.

Acknowledgements The authors wish to thank the NATO Allied Command Transformation (NATO-ACT) for supporting the CMRE project on Data Knowledge and Operational Effectiveness (DKOE). Some material of this chapter is published in [22].

References

1. Andrienko, N., Andrienko, G.: Visual analytics of vessel movement. In: Artikis, A., Zissis, D. (eds.) Guide to Maritime Informatics, chap. 5. Springer (2021)
2. Auslander, B., Gupta, K.M., Aha, D.W.: A comparative evaluation of anomaly detection algorithms for maritime video surveillance. In: Proceedings of SPIE, Sensors, and Command, Control, Communications, and Intelligence (C3I) Technologies for Homeland Security and Homeland Defense X, vol. 8019. Orlando, Florida, USA (2011)
3. Ben Abdallah, N., Iphar, C., Arcieri, G., Jousselme, A.L.: Fixing errors in the ais destination field. In: Proceedings of the OCEANS 2019 Marseille Conference (2019)
4. Benavoli, A., Ristic, B.: Classification with imprecise likelihoods: A comparison of TBM, random set and imprecise probability approach. In: Proceedings of the 14th International Conference on Information Fusion (2011)
5. Bereta, K., Chatzikokolakis, K., Zissis, D.: Maritime reporting systems. In: Artikis, A., Zissis, D. (eds.) Guide to Maritime Informatics, chap. 1. Springer (2021)
6. Bilgiç, T., Türkşen, I.B.: Measurement of Membership Functions: Theoretical and Empirical Work, pp. 195–227. Springer US, Boston, MA (2000)
7. Blasch, E., Dorion, E., Valin, P., Bossé, E.: Ontology alignment using relative entropy for semantic uncertainty analysis. In: Proceedings of the IEEE National Aerospace and Electronics Conference (NAECON), pp. 140–148. Fairborn, OH (2010)
8. Bloch, I.: Information combination operators for data fusion: a comparative review with classification. IEEE Trans. Syst. Man Cybern. - Part A: Syst. Hum. **26**(1), 52–67 (1996)
9. Challa, S., Koks, D.: Bayesian and Dempster-Shafer fusion. Sādhanā **29**, 145–176 (2004)
10. Costa, P., Jousselme, A.L., Laskey, K., Blasch, E., Dragos, V., Ziegler, J., de Villiers, J.P., Pavlin, G.: URREF: uncertainty representation and reasoning evaluation framework for information fusion. J. Adv. Inf. Fus. **13**(2), 137–157 (2018)
11. Dempster, A.: Upper and lower probabilities induced by multivalued mapping. Ann. Math. Stat. **38**, 325–339 (1967)
12. Denoeux, T., Smets, P.: Classification using belief functions: Relationship between case-based and model-based approaches. IEEE Trans. Syst. Man Cybern. - Part B: Cybern. **36**(6), 1395–1406 (2006)
13. Diaconis, P., Zabell, S.: Some alternative to Bayes's rule. In: Grofman, B., Owen, G. (eds.) Proceedings of the Second University of California, Irvine, Conference on Political Economy, pp. 25–38 (1986)

14. Dubois, D., Prade, H.: Fuzzy sets and probability: Misunderstandings, bridges and gaps. In: Proceedings of the 2nd IEEE International Conference on Fuzzy Systems, pp. 1059–1068 (1993)
15. Dubois, D., Prade, H.: The three semantics of fuzzy sets. Fuzzy Sets Syst. **90**, 141–150 (1997)
16. Dubois, D., Prade, H.: Formal representations of uncertainty, vol. Decision-making - Concepts and Methods, chap. 3, pp. 85–156. ISTE, London, UK & Wiley, Hoboken (2009). Invited paper
17. Etienne, L., Ray, C., Camossi, E., Iphar, C.: Maritime data processing in relational databases. In: Artikis, A., Zissis, D. (eds.) Guide to Maritime Informatics, chap. 3. Springer (2021)
18. Gupta, K.M., Aha, D.W., Moore, P.: Case-based collective inference for maritime object classification. In: Proceedings of the Eighth International Conference on Case-Based Reasoning, pp. 443–449. Springer, Seattle, WA (2009)
19. Iphar, C., Ray, C., Napoli, A.: Uses and misuses of the automatic identification system. In: Proceedings of the OCEANS 2019 Marseille Conference (2019)
20. ISO 5725: Accuracy (trueness and precision) of measurement methods and results - part 1: Introduction and basic principles. Technical report, ISO International Standardization (2011). standardsproposals.bsigroup.com/Home/getPDF/830
21. Johansson, F., Falkman, G.: Detection of vessel anomalies - a Bayesian network approach. In: Proceedings of the 3rd International Conference on Intelligent Sensors, Sensor Networks and Information, pp. 395–400. Melbourne, Qld (2007)
22. Jousselme, A.L., Pallotta, G.: Dissecting uncertainty handling techniques: Illustration on maritime anomaly detection. J. Adv. Inf. Fus. **13**(2), 158–197 (2018)
23. Karlsson, A., Johansson, R., Andler, S.F.: Characterization and empirical evaluation of Bayesian and credal combination operators. J. Adv. Inf. Fus. (2011)
24. Kittler, J.: Combining classifiers: a theoretical framework. Pattern Anal. Appl. **1**(18–27), (1998)
25. Klir, G.J.: Probability-possibility transformations: a comparison. Int. J. Gen. Syst. **21**, 291–310 (1992)
26. Lane, R.O., Nevell, D.A., Hayward, S.D., Beaney, T.W.: Maritime anomaly detection and threat assessment. In: Proceedings of the 13th International Conference on Information Fusion. Edinburgh, UK (2010)
27. Laxhammar, R.: Anomaly detection for sea surveillance. In: Proceedings of the International Conference on Information Fusion. Firenze, Italy (2008)
28. Laxhammar, R.: Anomaly detection in trajectory data for surveillance applications. Ph.D. thesis, School of Science and Technology at Örebro University (2011)
29. Leung, H., Wu, J.: Bayesian and Dempster-Shafer target identification for radar surveillance. IEEE Trans. Aerosp. Electron. Syst. **36**(2), 432–447 (2000)
30. Liu, B., de Souza, E.N., Matwin, S., Sydow, M.: Knowledge-based clustering of ship trajectories using density-based approach. In: Proceedings of the IEEE International Conference on Big Data (Big Data), pp. 603–608 (2014). https://doi.org/10.1109/BigData.2014.7004281
31. Mazzarella, F., Vespe, M., Santamaria, C.: SAR ship detection and self-reporting data fusion based on traffic knowledge **12**, 1685–1689 (2015)
32. Mercier, D., Quost, B., Denœux, T.: Refined modeling of sensor reliability in the belief function framework using contextual discounting. Inf. Fus. **9**, 246–258 (2008)
33. Millefiori, L., Braca, P., Bryan, K., Willett, P.: Modeling vessel kinematics using a stochastic mean-reverting process for long-term prediction. IEEE Trans. Aerosp. Electron. Syst. **52**(5), 2313–2330 (2016)
34. Pallotta, G., Jousselme, A.L.: Data-driven detection and context-based classification of maritime anomalies. In: Proceedings of the 18th International Conference on Information Fusion. Washington, D. C. (USA) (2015)
35. Pallotta, G., Vespe, M.: Vessel traffic dataset from ground-based AIS receiver Castellana: description and use. Technical Report CMRE-DA-2014-001, NATO STO CMRE (2014). NATO UNCLASSIFIED
36. Pallotta, G., Vespe, M., Bryan, K.: Vessel pattern knowledge discovery from AIS data - A framework for anomaly detection and route prediction. Entropy **5**(6), 2218–2245 (2013)

37. Patroumpas, K.: Online mobility tracking against evolving maritime trajectories. In: Artikis, A., Zissis, D. (eds.) Guide to Maritime Informatics, chap. 6. Springer (2021)
38. Pitsikalis, M., Artikis, A.: Composite maritime event recognition. In: Artikis, A., Zissis, D. (eds.) Guide to Maritime Informatics, chap. 8. Springer (2021)
39. Ray, C., Dréo, R., Camossi, E., Jousselme, A.L.: Heterogeneous integrated dataset for maritime intelligence, surveillance, and reconnaissance (version 0.1) [data set] (2018). https://doi.org/10.5281/zenodo.1167595
40. Ray, C., Dréo, R., Camossi, E., Jousselme, A.L., Iphar, C.: Heterogeneous integrated dataset for maritime intelligence, surveillance, and reconnaissance. Data in Brief **25**, (2019). https://doi.org/10.1016/j.dib.2019.104141
41. Ray, C., Gallen, R., Iphar, C., Napoli, A., Bouju, A.: DeAIS project: Detection of AIS spoofing and resulting risks. In: OCEANS 2015 - Genova, pp. 1–6 (2015). https://doi.org/10.1109/OCEANS-Genova.2015.7271729
42. Ristic, B., La Scala, B., Morelande, M., Gordon, N.: Statistical analysis of motion patterns in AIS Data: Anomaly detection and motion prediction. In: Proceedings of the 11th Conference on Information Fusion (FUSION). Cologne, Germany (2008)
43. Ristic, B., Smets, P.: Target classification approach based on the belief function theory. IEEE Trans. Aerosp. Electron. Syst. **41**(2), 574–583 (2005)
44. Salmon, L., Ray, C., Claramunt, C.: Continuous detection of black holes for moving objects at sea. In: Proceedings of the 7th ACM SIGSPATIAL International Workshop on GeoStreaming, IWGS '16, pp. 2:1–2:10. ACM, New York, NY, USA (2016). https://doi.org/10.1145/3003421.3003423
45. Santipantakis, G.M., Doulkeridis, C., Vouros, G.A.: Link discovery for maritime monitoring. In: Artikis, A., Zissis, D. (eds.) Guide to Maritime Informatics, chap. 7. Springer (2021)
46. Shafer, G.: A Mathematical Theory of Evidence. Princeton University Press (1976)
47. Shafer, G., Tversky, A.: Languages and designs for probability judgment. Cognit. Sci. **9**, 309–339 (1985)
48. Smets, P.: Constructing the pignistic probability function in a context of uncertainty. Uncertain. Artif. Intell. **5**, 29–39 (1990). Elsevier Science Publishers
49. Smets, P.: Belief functions: the disjunctive rule of combination and the generalized Bayesian theorem. Int. J. Approx. Reason. **9**, 1–35 (1993)
50. Smets, P.: Imperfect information: Imprecision - uncertainty. In: Motro, A., Smets, P. (eds.) Uncertainty Management in Information Systems. From Needs to Solutions, pp. 225–254. Kluwer Academic Publishers (1997)
51. Smets, P.: Probability, possibility, belief: Which and where? In: Smets, P. (ed.) Quantified Representation of Uncertainty & Imprecision, vol. 1, pp. 1–24. Kluwer, Doordrecht (1998)
52. Smets, P.: Analyzing the combination of conflicting belief functions. Inf. Fus. **8**, 387–412 (2007)
53. Smets, P., Kennes, R.: The transferable belief model. Artif. Intell. **66**, 191–234 (1994)
54. Sombé, L.: Reasoning under incomplete information in artificial intelligence. Int. J. Intell. Syst. (Special Issue) **5**. Wiley (1990)
55. Tampakis, P., Sideridis, S., Nikitopoulos, P., Pelekis, N., Theodoridis, Y.: Maritime data analytics. In: Artikis, A., Zissis, D. (eds.) Guide to Maritime Informatics, chap. 4. Springer (2021)
56. de Vries, G.K.D., van Someren, M.: Machine learning for vessel trajectories using compression, alignments and domain knowledge. Expert Syst. Appl. **39**(18), 13426–13439 (2012)
57. Walley, P.: Measures of uncertainty in expert systems. Artif. Intell. **83**, 1–58 (1996)
58. Zadeh, L.A.: Fuzzy sets. Inf. Control **8**, 338–353 (1965)

Chapter 10
Maritime Network Analysis: Connectivity and Spatial Distribution

César Ducruet, Justin Berli, Giannis Spiliopoulos, and Dimitris Zissis

Abstract In this chapter, we apply conventional graph-theory and complex network methods to a sample of port and inter-port shipping flows at and amongst the top 50 European ports in 2017. Such methods help to detect the main topological and geographic structures of this network in order to answer three main questions. First, why are certain port nodes better connected than others? Such a level of hierarchy is best approached by testing the scale-free and rich-club dimension of the network. For this we measure node connectivity in various ways, from local to global indices, all confirming inequality in traffic distribution. Second, what is the influence of cargo specialisation or diversity on the network structure? This relates to the concepts of multiplexity and assortativity, i.e. the ability of nodes to diversify their activity or to specialise. Two principal layers are analysed and compared, namely cargo and bulk, showing that larger ports and links are more diversified. Lastly, what are the substructures or geographic patterns underlying the distribution of maritime flows? To answer this, we examine the influence of physical distance on connectivity and on the emergence of subnetworks.

C. Ducruet (✉) · J. Berli
Centre National de la Recherche Scientifique (CNRS), Paris, France
e-mail: cdu@parisgeo.cnrs.fr

J. Berli
e-mail: justinberli@gmail.com

G. Spiliopoulos · D. Zissis
MarineTraffic, Oxford, UK
e-mail: giannis.spiliopoulos@marinetraffic.com

D. Zissis
e-mail: dzissis@aegean.gr

D. Zissis
University of the Aegean, Syros, Greece

© Springer Nature Switzerland AG 2021
A. Artikis and D. Zissis (eds.), *Guide to Maritime Informatics*,
https://doi.org/10.1007/978-3-030-61852-0_10

10.1 Introduction

The analysis of communication and transport systems in mathematics dates back to the birth of graph theory when Leonard Euler analysed the street network of Konigsberg in the 18th century. Much later, the basic tools of graph theory were introduced in geography and regional science to study mainly road and railway networks. The emphasis on physical infrastructure networks left aside other network types for decades, such as maritime networks [9]. Physical networks are typically planar, i.e. one systematically witnesses a node at the intersection between two links. Maritime networks, like airline networks, are non-planar, as their links may cross without creating a node (port, airport). This has many implications on the evolution and growth of maritime networks. In this sense such networks are more closely related to what physicists defined in the late 1990s as complex networks, namely scale-free and small world networks, where a handful of very large nodes (or hubs) dominate numerous secondary nodes [4].

Studying maritime flows as complex networks is fruitful in many ways. As ports often advertise their activity by the number of overseas connections, the measurement and international comparison of connectivity is one step beyond individual assumptions using home-made calculations, made by port authorities, for instance. Second, the network dimension can potentially shed new light on the factors underlying growth or decline, strengths and weaknesses of ports, which can be useful to shipping lines for their network design, but also to ports, in their search for a better attractiveness in a context of fierce competition. This relational perspective is important as it considers ports being connected and animated by other forces than economic and managerial, or a process known in complexity sciences as "self-organisation". This process postulates that local environments through their mutual connections give birth to a system that is more than the sum of its parts. Another advantage of complex networks is the possibility to measure such ideas using relatively simple methods of descriptive statistics. Other factors shaping port hierarchy and competition can thus be compared with such results, such as market location, hinterland accessibility, and efficient supply chain integration that is known from the press and official reports as well as specialised scholarly works.

The remainder of this chapter is organized as follows. Section 10.2 outlines the network construction procedure from Automatic Identification System (AIS) data and introduces key concepts of maritime network analysis. Section 10.3 is a network-analytical approach to maritime flows by measuring global and local connectivity indices. It compares our results with those obtained in earlier works and conveys new findings about Europe's network structure in terms of centralization, hierarchy and multiplexity. Finally, we map inter-port flows from the perspective of graph visualization on the one hand, and flow mapping on the other, to better understand the role of space and geography in shaping such flows.

Fig. 10.1 A small example
network with five vertices
and five edges

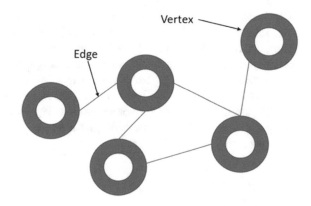

10.2   Data and Methodology

10.2.1   Background

In general terms, a graph is an ordered pair $G = (V, E)$ comprising:

- V a set of vertices (also called nodes or points);
- $E \subseteq \{\{x, y\}|(x, y) \in V^2 \wedge x \neq y\}$ a set of edges (also called links or lines), which are unordered pairs of vertices.

A network is simply a collection of sentences that describe relationships, in the following way: A ship X can travel from "Antwerp to Rotterdam". The simple phrase in Fig. 10.1 is a basic unit of network analysis called a dyad. Every dyad denotes a single relationship—an edge in traditional graph theory. The nouns in the phrase represent ports involved in the relationship—these are called vertices, while the connection represents the existence of a route between them. Maritime networks can be constructed on the principles below:

- Ports are considered vertices (or nodes).
- Trips between them or routes are edges (or relations).
- Relationships can be binary or valued by e.g. the frequency of trips.
- Each node has a degree such as the frequency of incoming or outcoming trips.

Network analysis can be regarded as "a set of techniques with a shared methodological perspective, which allow researchers to depict relations among actors and to analyze the social structures [considering a vessel trip as a social phenomenon] that emerge from the recurrence of these relations" [7].

Numerous methods and indicators have appeared in the bibliography in recent years. Applying a full set of measures to the maritime network is unrealistic; therefore, a careful selection of the most relevant indicators is necessary. Complex network measures are mainly descriptive statistics applied to a communication system made of nodes and links. Local measures are those at node (port) level, while global measures refer to the whole network (see Fig. 10.2 for an illustration). As for any other (transport) network, maritime flows are considered as a graph with ports as nodes (or vertices) connected by inter-port voyages as links (or edges). The weight of connections among ports is the number of vessel trips. The difference between network and graph is that the graph exists in a relatively abstract topological space regardless of the real pattern of infrastructure or circulation (Fig. 10.3). Such a dimension allows to apply a variety of measures and algorithms at the global and local levels (see [11] for a synthesis).

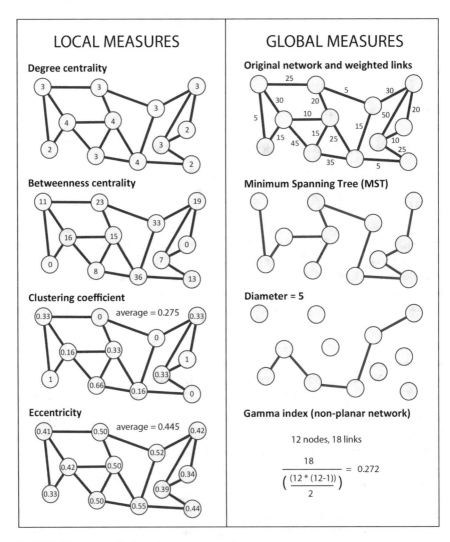

Fig. 10.2 Illustration of selected complex network measures

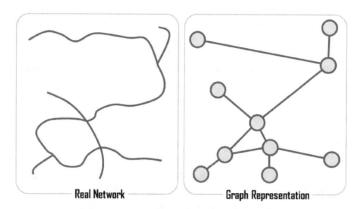

Fig. 10.3 Schematic representation of a network and a graph (after [17])

Below are the local measures that we adopt:

- Degree centrality: number of adjacent (or neighbor) nodes. The maximum value in Fig. 10.2 is 4.
- Betweenness centrality: number of occurrences of a node on all shortest paths. A shortest path is the topological length (number of links) between two nodes with the minimum number of stops in between. Betweenness centrality is a measure of global accessibility.
- Clustering coefficient: for each node, it is the proportion of connected neighbors in the total maximum possible number of connected neighbors. In Fig. 10.2, nodes with 0 value have unconnected neighbors and nodes with value 1 see their neighbors mutually connected.
- Eccentricity: also coined 'closeness centrality' in the literature, is the inverse of the average length of all shortest paths from one node to all other nodes. Values closer to 1 indicate nodes in close proximity to the topological center of the network.

The global measures adopted for our analysis are the following:

- Diameter: length (number of edges) of the longest shortest path, which equals to 5 in Fig. 10.2.
- Average eccentricity, average clustering coefficient and average shortest path length.
- Assortativity coefficient: correlation (ij) between the degree centrality of connected nodes (i) and (j). The network is said to be assortative for a positive correlation (nodes of similar degree centrality connect with each other) and disassortative in the case of negative correlation (nodes of dissimilar degree centrality connect with each other).
- Gamma index or density: proportion of existing links in the total maximum possible number of links. In Fig. 10.2, the 18 existing links account for 27.2% of the maximum possible number of links between the 12 nodes.

- Rich-club coefficient: gamma index (or density) among top ports/nodes divided by gamma index (or density) for the whole network. Top ports are usually defined as those with a degree centrality higher than average.
- Degree distribution: exponent of the power-law slope in a log-log plot of degree centrality (x-axis) and cumulated distribution of nodes (y-axis). A hierarchical network or "scale-free" network is characterized by a power-law distribution, with an exponent value between 1 and 3.

There are numerous algorithms which serve to reveal hidden information from a network. This chapter picks two of the most common ones with the goal to compare two distinct approaches based on the same dataset. The first algorithm extracts a so-called minimum spanning tree (MST), i.e. the most efficient "line" connecting all nodes according to a desired criterion. Based on the weight of links, as seen e.g. in the top-right of Fig. 10.2, we use the multiplicative inverse of flows per link (i.e. $1/N_{calls}^{i,j}$, where $N_{call}^{i,j}$ is the number of vessels that travelled from port i to port j), as a base for the application of this algorithm, to reveal which optimal route actually carries the largest traffic. The second method is a partitioning algorithm called single linkage analysis (SLA). This also results into a tree-like structure often split into connected components (or nodal regions), each of them being more or less centered upon one main node, as initially applied in geography by Nystuen and Dacey [16]. SLA only retains for each node its largest traffic link, a method that proved useful to detect hub-like structures and barrier effects. All in all, the underlying hypothesis of such analyses is that space matters, should it be Euclidean distance, coastal morphology, nautical distance, or some other geographic feature, in the distribution of maritime networks.

10.2.2   From AIS Data to Maritime Network

The first step in our analysis is transforming tracking data, such as AIS data, into a network-related data representation. From a computational perspective, we would like to convert AIS data into network edge lists, such as the following:

From	To	Value (Frequency of trips)
A	B	5
A	D	3
A	E	3
B	A	5

For this study, we make use of AIS data collected in EU waters in 2017, with a focus on cargo vessels (AIS Ship Type:70-79) and tankers (AIS Ship Type:80-89), travelling both to and from the top 50 European ports, with respect to the number of arrivals of cargo and tanker vessels combined. Converting our given data into the required form requires primarily counting the number of ships arriving to port B from port A, to port C from port A, and so on and so forth for all ports in our dataset.

Even though AIS messages include both departure and destination port information, it cannot be considered as a reliable source of information, as it is manually entered by the vessel's crew, without following a specific standard, making it thus prone to errors. Instead, we calculate port visits by making use of the World Port Index dataset,[1] which contains the location and physical characteristics of major ports and terminals worldwide. The aim is to assign to all positional data accurate port arrival information. As an additional step and so as to improve result accuracy, we exclude smaller vessels that only operate within the given port area, by only calculating arrivals of vessels that have a valid IMO number and gross tonnage greater than 5000 tons.

We transformed the network into an undirected graph by merging the directional links serving two of the same ports into one single link, and added their respective number of calls (e.g. Amsterdam-Rotterdam plus Rotterdam-Amsterdam). Thus we simplified the graph and decreased the 'statistical noise' by reducing the number of links. Given that AIS messages are received in ascending timestamp order, we identified changes from "in-port" areas to "open sea" and vice-versa, and characterized them as port-call start and port-call end with respect to the direction of the transition (i.e. moving from open-sea to in-port defines a port-call start and the opposite direction defines its end). See, e.g., the Chapter of Andrienko and Andrienko [1] for another way of identifying port-call starts and ends.

Next, given that we have identified all vessel's arrivals (port-call start) and departures (port-call end), we examined them in pairs with ascending order of occurrence per vessel, to extract information about the time interval of each port-call. We excluded from further analysis all cases (both start and end of a port call) where the duration of the port call is too small (less than 15 min), or the vessel is not immobilized during the port-call, as it's not feasible for large vessels to conduct any activity related to commodity transportation if those requirements are not satisfied. Subsequently, we repeated the analysis in pairs for the remaining sets of port-calls to extract inter-port voyage information and respectively the port-connection (pairs of departure and arrivals). We aggregated information for each port-connection per ship-type (cargo or bulk) with respect to the number of voyages conducted and the distinct number of vessels performing the trip for each connection. At the end of the above process, the network is represented as an edge list such as that presented in Table 10.1.

10.2.3 Network Visualization

We use a Geographic Information System (GIS) called Geoseastems to calculate nautical distances and visualize the network [6], as well as a graph visualization software called Tulip [3]. The GIS was specifically designed to map maritime flows without crossing land areas and calculate descriptive statistics based on actual inter-

---

[1] https://msi.nga.mil/Publications/WPI.

Table 10.1 Top port 5 connections based on the number of distinct vessels for the dry cargo market (MarineTraffic data as computed by the Gephi software [5])

From	To	Voyages	Number of distinct vessels
VALENCIA	BARCELONA	1084	280
HAMBURG	ROTTERDAM	664	258
LE HAVRE	ANTWERP	694	244
ANTWERP	HAMBURG	736	236
ANTWERP	ROTTERDAM	476	187

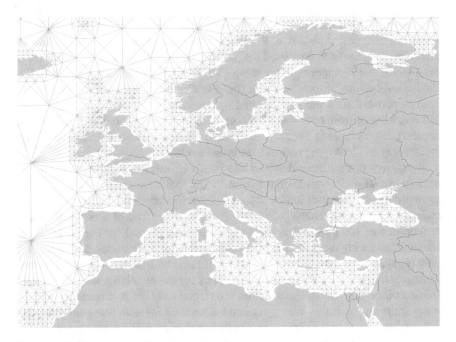

Fig. 10.4 The European maritime grid

port flows. The maritime grid thus provides an approximation of paths taken by ships while respecting the geographic constraints of coastlines. This allows defining how to segment and partition the continuous ocean space where ships circulate by using a virtual grid. This model rests on a regular worldwide meshing composed of eight squares, based on which the meshing was refined with an iterative process, subdividing each square intersecting a land mass into four same sized squares, resulting in a denser meshing near coastlines (see Fig. 10.4).

As ports needed to be connected to the grid, we decided to group them by clusters based on their geographic proximity. The centroid of every cluster of points is calculated and linked to the maritime grid; every port, except isolated ones, is linked to its respective cluster, consequently offering us a fully functional network upon which we can calculate shortest paths between pairs of ports using Dijkstra's algorithm. Finally, one last enhancement of the grid was to include main rivers to link river ports to the maritime network, and thus ensure the network's completeness. Figure 10.4 was thus used to project the AIS port-to-port matrix, including every potential path a vessel can use to sail across the sea and connect ports with each other. Clusters as well as links between ports and the main grid are not represented, as it would conceal the main network and impact overall readability. Last but not least, visualizing the network for a given date or period of time requires to calculate shortest paths between each pair of ports.

10.3 Network Topology and Connectivity

10.3.1 Global Structure and Hierarchical Effects

We first investigate the inequality of the network's structure by looking at the distribution of degree centrality among ports. Scale-free networks are known to contain a few large hubs and a majority of small nodes. The degree distribution (Fig. 10.5a) is obtained by plotting the degree centrality of nodes (x axis) against the cumulated number of nodes (y axis) in a log-log diagram and inserting a power-law function to obtain the slope exponent of the line. Interestingly, each network (whole, dry and liquid) exhibits scale-free properties as the exponent value ranges between 1 and 3 [4]. This means that despite the absence of small and medium-sized ports in the study sample, the European maritime network is highly hierarchical and organized around a few main hubs. In addition, we observe that although exponent values are close among the three networks, the dry cargo network is the most concentrated, with a value of 1.328 compared with 1.277 for the whole network, and only 1.14 for the liquid cargo network. This result is in accordance with the nature of the flows, as dry cargo includes containers, i.e. the most hierarchical configuration of maritime transport, since it is very much tied to transshipment hubs. Container shipping is highly selective and liner shipping networks have become more and more concentrated around major hubs, in the search for economies of scale and route rationalization [8, 14].

We also tested the relationship between traffic size and degree centrality (Fig. 10.5b) but it is not in accordance with already existing works, such as [13]. The whole European network exhibits zero correlation, while the correlation increases when considering each layer independently. Regardless of degree centrality, the statistical distribution of links and node weights based on vessel calls (Fig. 10.5c) exhibits a power-law distribution only at node level for the whole network and the liquid cargo layer. All other slope exponents fall under 1, so that their topological structure remains hierarchically moderate (for an empirical validation of statistical laws obeying maritime networks, see [12]).

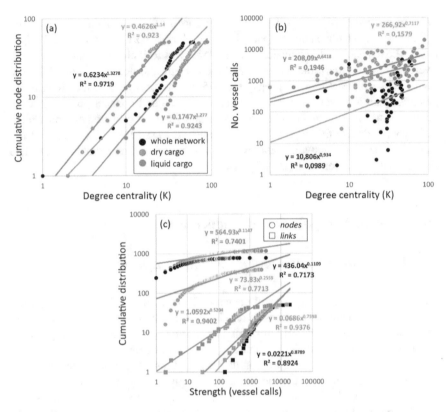

Fig. 10.5 Statistical distribution of the European maritime network (MarineTraffic data as computed by the TULIP software [3])

European maritime connectivity may also be illustrated by plotting node-level variables and checking their compatibility. The outcome is often a good manner to conclude to specific features of networks in general. As shown in Fig. 10.6a, the clustering coefficient is inversely proportional to the degree centrality of nodes, since large-degree nodes (i.e. nodes with many links) often act as hubs. While the inverse relationship is verified for the liquid cargo network (i.e. determination coefficient of 0.576 and Pearson correlation coefficient of −0.759), the whole network and the dry cargo network exhibit nearly zero correlation. This is rather surprising given the stronger hierarchical structure of the dry cargo network. The results of Fig. 10.6b based on nearest neighbors' degree do not exhibit specific trends. Hu and Zhu, in [13], demonstrated that this correlation should be meaningful as a proof of the disassortative behavior of the network under study. The power-law relationships between betweenness centrality and degree centrality (Fig. 10.6c) is by far the strongest, with slope exponents over 2 for all networks, compared with 1.66 for Hu and Zhu [13]. This means that ports having a wider set of local connections (degree centrality) are

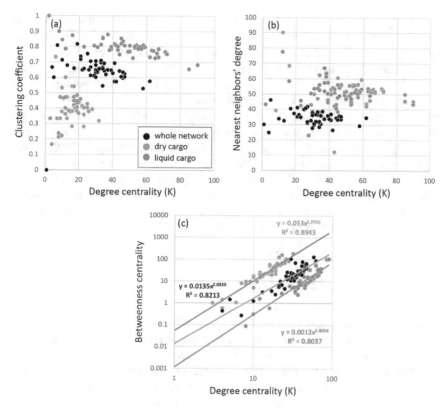

Fig. 10.6 Centrality correlations of the European maritime network (MarineTraffic data as computed by the TULIP software [3])

often located at the crossroads of maritime routes within Europe, and they have a strong global accessibility (betweenness centrality).

Lastly, the rich-club effect or "rich get richer" phenomenon is worthy of investigation. As indicated in Sect. 10.2, the rich-club coefficient corresponds to the density (Gamma index) of the "top network" divided by the density (Gamma index) of the whole network. The result should be larger than 1 to conclude to the existence of rich-club properties. Such properties are important as they indicate whether the network is denser at the top of the hierarchy. The threshold over which the "top nodes" are chosen to calculate the Gamma index of the "top network" is generally the average degree centrality. Our results demonstrate that larger ports (i.e. degree centrality over 46) are more tightly connected with each other than the bottom of the hierarchy, while the whole network of the top 50 ports has a rich-club coefficient around 1 (0.93). The rich-club coefficient of 2.283 for larger ports, i.e. larger ports are nearly three times more connected than the whole network, is much higher than for smaller ports, i.e. only 1.15, near the network's average.

Other global-level analyses help to understand the network's structure. One of them is assortativity, based on the average degree centrality of nodes' neighbors. The entire graph can be said disassortative as the correlation coefficient between node degree and average neighbor degree is negative. Our results show that the whole network is disassortative (-0.178) but the correlation coefficient has a relatively low statistical significance. It means that on average, larger nodes tend to connect smaller nodes and vice-versa, thereby suggesting a hierarchical structure. The same phenomenon applies to the two subnetworks of dry cargo (-0.154) and liquid cargo (-0.150). Nevertheless, the fact that the significance is slightly higher for the whole network means that hierarchical tendencies are stronger in the composite network, which is in accordance with theory. More precisely, [18] demonstrated that when the layers or subnetworks connect with each other through mainly the same nodes, the whole network becomes both more centralized but more vulnerable to disruptions of all kinds (e.g. natural disasters, labor strikes, wars).

10.3.2 Multiplexity and Assortativity

One simple way to detect the influence of cargo diversity or specialization on maritime traffic and network structure is the averaging of indices (i.e. betweenness centrality, clustering coefficient, eccentricity, and degree centrality) and traffic per port quantiles using the specialization rate. This rate corresponds to the maximum percentage taken by one of the two traffic categories, dry cargo and liquid cargo, in the total number of calls for each port. Percentiles may be more or less precise, from 5 to 2 classes (Table 10.2). Based on our results on two classes of traffic share (i.e. 56.0–87.5% and 87.8–100.0%), we observe a strong effect of specialization or diversity on connectivity: more diversified ports (i.e. having a maximum of 87.5% traffic in one of the two cargo types) are more accessible (betweenness), have stronger hub functions (clustering coefficient), stand closer to the network's topological center of gravity (eccentricity), are more connected on average (degree centrality), and handle more vessel calls than more specialized ports. The results are thus motivating going further. Yet, the same results obtained for 5 quantiles show that the most diversified ports (first quantile, 56–72.4%) are always less central and dynamic than the second quantile, so that there is no linear relationship between specialization and connectivity/traffic. Across European ports, the linear relationship between traffic diversity and traffic size is blurred by the fact that many of the largest ports are actually specialized in one specific cargo type (e.g. Amsterdam and dry bulks, Bremerhaven and vehicles, Gioia Tauro and containers) while many secondary ports keep being diversified, notably those located away from the contestable hinterland of the Le Havre–Hamburg North European Range (e.g. islands, British Isles, Scandinavia, Southern Europe).

Our next results (Fig. 10.7) exhibit a strong specialization of nodes more than links given their respective determination coefficient. Typical examples include Augusta (Italy), a Sardinian port specialized in liquid cargo, and on the other side container transshipment hubs such as Felixstowe, Piraeus, and Tilbury handling mainly dry

Table 10.2 Traffic specialization and network connectivity (MarineTraffic data as computed by the Wessa software [20])

		Average scores				
		Betweenness centrality	Clustering coefficient	Eccentricity	Degree centrality	No. of calls
Share of the largest cargo type in total traffic (% calls)	56.0–72.4	17.3	0.785	0.755	53.2	3269
	74.7–82.0	35.1	0.779	0.739	55.8	4988
	82.1–90.0	19.5	0.788	0.794	52.9	3621
	90.6–98.7	9.9	0.811	0.680	42.4	2289
	99.3–100	9.8	0.829	0.582	31.9	2999
	56.0–87.5	23.5	0.786	0.755	53.9	4034
	87.8–100	12.8	0.813	0.658	39.8	2817

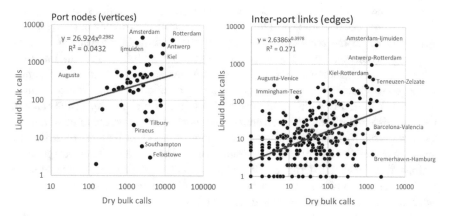

Fig. 10.7 Bilayered structure of the main European maritime network

cargo. The fact that node correlation remains lower than link correlation reveals a significant gap with usual spatial networks where largest nodes connect multiple layers. It means that each subnetwork has its own specificities in terms of geographic distribution and topological structure. Here it is more the link level that exhibits such a trend, with 0.271 compared with 0.043 of determination coefficient, meaning that on average, and despite a low probability, links (edges) handling dry bulk cargo handle a comparable proportion of liquid cargo. But this is not the case for the majority of linkages, as domestic links (e.g. Bremerhaven-Hamburg; Barcelona-Valencia) handle more cargo than international links such as Kiel-Rotterdam and Antwerp-Rotterdam. In the latter cases, much cargo is moved by inland transport such as road and river/canal, thereby lowering the maritime connection.

10.4 Maritime Network and Spatial Embedding

10.4.1 Substructures Within the European Maritime Network

Last but not least, the extent to which the main European maritime network also exhibits small-world properties has been tested in Table 10.3. Small-world networks have been defined by Watts and Strogatz [19] as graphs with a relatively higher average clustering coefficient and lower average shortest path length (ASPL) than a random network of identical size. It is important to check the applicability of this model to maritime networks, as small-world properties suggest the existence of tightly connected communities and crucial links. Degree distribution in random graphs is known to follow a normal (i.e. Gaussian) distribution, contrary to scale-free and small-world networks. As a matter of fact, these conditions are fulfilled in our case study, because inter-port links are not randomly distributed but correspond, as said above for the scale-free dimension, to the patterns of a particular industry (shipping) where spatial and economic features are most influential. The maritime network is thus easier to navigate (lower diameter and ASPL) and more tightly connected (higher density and average clustering coefficient). The average number of links per node (average degree centrality) is much higher than in the random network, also underlining the unevenness and hierarchical structure of the maritime network. Table 10.3 also confirms, for dry cargo, important features of scale-free networks as defined by [4] in their seminal paper. The fact that the dry cargo layer exhibits much lower clustering coefficient and density is also in accordance with our previous results showing the stronger scale-free dimension of this particular layer, especially in terms of containers and related hubs.

When it comes to the Minimum Spanning Tree (Fig. 10.8, left), we observe a core-periphery structure combined with a North-South division, whereby North European ports stand out by their higher betweenness centrality and optimal situation than their southern counterparts. This is in line with the fact that Rotterdam, Antwerp and Hamburg are Europe's largest ports in terms of total throughput, thanks to extended intermodal services connecting a vast continental hinterland. They are also the ports receiving the largest vessels for cargo redistribution through sea-sea and sea-land

Table 10.3 Topological properties of the European maritime network versus a random network (MarineTraffic data as computed by the Gephi software [5])

Network	Nodes	Links	Average clustering coefficient	Diameter	Average shortest path length	Average degree centrality	Density (Gamma index)
Whole network	50	1,165	0.800	3	1.366	46.60	0.951
Dry cargo network	50	387	0.401	4	1.781	15.48	0.316
Liquid cargo network	50	778	0.842	3	1.526	31.12	0.635
Random graph	50	62	0.039	11	4.029	1.24	0.050

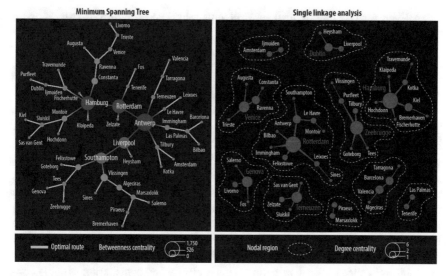

Fig. 10.8 Substructures within the main European maritime network, 2017 (MarineTraffic data computed by the TULIP software [3])

transshipment. This "Le Havre-Hamburg maritime range" or North European range is prolonged towards Southampton, which is another important gateway port along this corridor, the busiest sea lane of the world. Southern Europe is partitioned into subcomponents (e.g. Adriatic, West Mediterranean, Atlantic), each of them being relatively peripheral on the optimal route. This overall pattern confirms that European maritime traffic is much concentrated at large mainland ports, as main trunk lines tend to bypass Mediterranean and Atlantic ports. The inclusion of small and medium-sized ports in the dataset may, however, increase the position of certain Mediterranean hubs such as Algeciras, Marsaxlokk and Valencia, as these ports operate for a noticeable share of their activity through sea-sea transshipment to/from smaller ports.

The results obtained from single linkage analysis provide complementary evidence on the role of the geographic factors influencing network and node connectivity (Fig. 10.8, right). As such, there is an overwhelming importance of spatial proximity in the emergence of nodal regions. Except from the one dominated by Rotterdam, Europe's largest port, which extends to the Iberian Peninsula, all nodal regions are confined in the vicinity of the connected ports, a typical example being the nexus (or dyad) Las Palmas-Tenerife, but are not necessarily bound to the national level, as seen with Piraeus-Marsaxlokk, Genoa-Fos, Venice-Constanta, Dublin-Liverpool, Kiel-Klaipeda and Zeebrugge-Goteborg. Nevertheless, such results confirm that even in the case of supposedly footloose flows and networks, distance plays a very important role in the configuration of maritime networks. This also corresponds to an economic and operational reality because the two main factors that control the hub-and-spokes patterns are port concentration and the high sailing frequency between the hub and the spokes, and both factors are governed from each region's characteristics. Even for less hierarchical substructures, physical proximity is still the main explanatory factor.

10.4.2 Flowmap of the European Maritime Network

The generated map (Fig. 10.9) displays arcs width and node size by their proportionality to ship traffic, i.e. the number of vessel calls. Color and grayscale indicate the share of dry cargo versus total traffic. The importance of container traffic (69,402) in comparison with tanker traffic (15,620) which is approximatively 4.5 times higher, has been taken into consideration to construct classes. The cartography of the European maritime network shows a strong traffic concentration in Northern Europe, especially along the aforementioned "Le Havre–Hamburg" range, as the crucial link within a longer corridor from the English Channel to the Baltic, including the main Benelux and German ports. To corroborate previous findings, proximity between ports seems to greatly influence traffic distribution. This is clear when considering strongly connected port pairs such as Dublin-Liverpool, Valencia-Barcelona, Genoa-Livorno and Las Palmas-Tenerife. Dry cargo appears to be dominant between those local/regional transport routes, while liquid cargo gains importance on longer-distance transportation. This is revealed by the main north/south corridor along Europe's Atlantic façade,

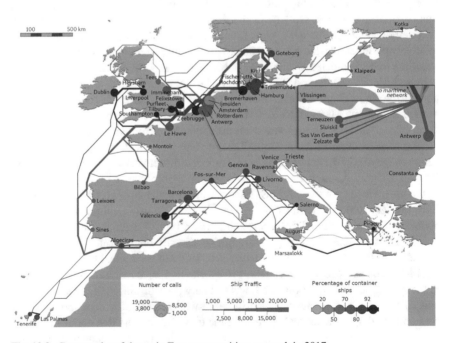

Fig. 10.9 Cartography of the main European maritime network in 2017

where those two types of traffic are more balanced. Only two Portuguese ports, Sines and Leixoes, stand apart as they focus more on liquid cargo, Sines in particular being traditionally a petrochemical complex. From the Gibraltar Straits and Algeciras in particular, traffic volume on the main route regularly decreases up to Marsaxlokk, Piraeus and Constanta. Yet, one can also observe the South European range between Valencia and Livorno (Leghorn), as another important port system made of "multi-port gateway regions" [15], especially for containers.

Another interesting finding from this cartography is the importance of smaller but yet top European port regions such as North Adriatic, Black Sea, East Baltic, and most of all, the British Isles. The latter, for instance, have become increasingly connected to international markets, for containers, by feeder vessels to/from the Rotterdam hub. Yet, recent policies such as the Liverpool mega-port project and the London Gateway initiative seek to promote direct calls from large shipping lines. Last but not least, we observe that port size and port connectivity are not necessarily in direct relationship with traffic specialization, as Northern Adriatic ports handle more liquid bulks, but also Amsterdam and Sines (a former petro-chemical complex), as well as Tees and Immingham, being the main ports of industrially declining economic bases. The strong emphasis on container transshipment is another example of specialization. This applies to ports serving wide and urbanized hinterlands, such as Valencia (Madrid), Tilbury and Felixstowe (London), but also "pure" transshipment ports like Marsaxlokk in Malta. Other cases include dry cargo ports specialized in automobiles (Bremerhaven), being the focus of global terminal handling operators like COSCO (c.f. Antwerp) or CMG-CGM (c.f. Marsaxlokk), or multi-purpose ports doing both dry and bulk cargoes such as Le Havre, Rotterdam, and Antwerp.

10.5   Summary

This chapter investigated the topological, statistical and spatial structure of the European maritime network. AIS data was transformed into an origin-destination matrix connecting the top 50 European ports, allowing us to apply a set of graph-theoretical and complex network methods. This analysis confirmed the scale-free and small-world topological structure of the network, whereby maritime flows concentrate at a handful of large hubs, to be redistributed within densely connected subregions. European hub ports exhibit a stronger correlation between global and local connectivity compared with previous works at the global scale [13]. As a consequence, the network is much disassortative and marked by rich-club effects. The extraction of the network's backbone and the flowmap suggest a strong core-periphery organization of the European maritime network, as a diversified corridor converges at northern mainland gateways, bypassing a number of Mediterranean and Atlantic ports.

10.6   Bibliographical Notes

A broad overview of concepts, methods and areas of application of Transportation Systems is available online at [17]. A practical exploration of maritime networks, port efficiency and hinterland connectivity, with a focus on the Mediterranean Sea, may be found at [2]. In [10] the shipping patterns, shipping flows and their multidisciplinary importance are discussed in both depth and scale under new evidence based on real data. Network analysis and graph theory topics can be found in [4, 7, 18, 19], while the tools used in this chapter for network analysis and visualisation are presented in [3, 5, 20]. Finally, some other some works that also focus on global maritime networks and flows are [9, 12, 13, 15], while in [8, 14] the authors address two economic and efficiency topics of maritime networks; the former focuses on the economies of scale for container ships, and the latter on the network economics for the same market.

References

1. Andrienko, N., Andrienko, G.: Visual analytics of vessel movement. In: Artikis, A., Zissis, D. (eds.) Guide to Maritime Informatics, Chap. 5. Springer, Berlin (2021)
2. Arvis, J.F., Vesin, V., Carruthers, R., Ducruet, C.: Maritime Networks, Port Efficiency, and Hinterland Connectivity in the Mediterranean. World Bank Publications, Washington, D.C. (2018)
3. Auber, D.: Tulip—a huge graph visualization framework. In: Graph Drawing Software, pp. 105–126. Springer, Berlin (2004)
4. Barabási, A.L., Albert, R.: Emergence of scaling in random networks. Science **286**(5439), 509–512 (1999)
5. Bastian, M., Heymann, S., Jacomy, M.: Gephi: an open source software for exploring and manipulating networks. In: 3rd International AAAI Conference on Weblogs and Social Media (2009)
6. Berli, J., Bunel, M., Ducruet, C.: Sea-land interdependence in the global maritime network: the case of Australian port cities. Netw. Spat. Econ. **18**(3), 447–471 (2018)
7. Chiesi, A.M.: Network analysis. In: Wright, J.D. (ed.) International Encyclopedia of the Social & Behavioral Sciences, 2nd edn, pp. 518–523. Elsevier, Oxford (2015). https://doi.org/10.1016/B978-0-08-097086-8.73055-8
8. Cullinane, K., Khanna, M.: Economies of scale in large containerships: optimal size and geographical implications. J. Transp. Geogr. **8**(3), 181–195 (2000)
9. Ducruet, C.: Maritime flows and networks in a multidisciplinary perspective. Maritime Networks: Spatial Structures and Time Dynamics, pp. 3–26 (2015)
10. Ducruet, C.: Advances in Shipping Data Analysis and Modeling: Tracking and Mapping Maritime Flows in the Age of Big Data. Routledge, Abingdon (2017)
11. Ducruet, C., Lugo, I.: Structure and Dynamics of Transportation Networks: Models, vol. 347. Sage Publications, Thousand Oaks (2013)
12. Gastner, M.T., Ducruet, C.: The distribution functions of vessel calls and port connectivity in the global cargo ship network. Maritime Networks: Spatial Structures and Time Dynamics. Routledge Studies in Transport Analysis, pp. 242–261 (2015)
13. Hu, Y., Zhu, D.: Empirical analysis of the worldwide maritime transportation network. Phys. A: Stat. Mech. Its Appl. **388**(10), 2061–2071 (2009)
14. Notteboom, T.E.: Container shipping and ports: an overview. Rev. Netw. Econ. **3**(2), (2004)

15. Notteboom, T.E.: Concentration and the formation of multi-port gateway regions in the European container port system: an update. J. Transp. Geogr. **18**(4), 567–583 (2010)
16. Nystuen, J.D., Dacey, M.F.: A graph theory interpretation of nodal regions. Pap. Reg. Sci. Assoc. **7**(1), 29–42 (1961). https://doi.org/10.1007/BF01969070
17. Rodrigue, J.P.: The Geography of Transport Systems. Taylor & Francis, Milton Park (2016). https://transportgeography.org/?page_id=5976. Accessed April 2020
18. Vespignani, A.: The fragility of interdependency. Nature **464**(7291), 984–985 (2010)
19. Watts, D.J., Strogatz, S.H.: Collective dynamics of 'small-world' networks. Nature **393**(6684), 440–442 (1998)
20. Wessa, P.: Free statistics software. Office for research development and education, version 1.21. **1**, 23–r7 (2018). https://www.wessa.net/. Accessed July 2020

Chapter 11
Shipping Economics and Analytics

Roar Adland

Abstract In theory, spatial shipping data can be used to track variables that influence the supply or demand for maritime transportation. In real applications, the limitations of using AIS data alone for economic analysis soon become apparent, though it remains a very useful tool. In this chapter we review how AIS data can be used for data-driven market analysis in shipping. In particular, we outline how AIS data can be used to track commodity flows and key variables for fleet efficiency, supply and demand. We also emphasize the limitations of ship tracking data for commercial analysis. The chapter deals solely with the economics and analytics of bulk shipping (i.e. bulk carriers or tankers) as it represents the purest case of a perfectly competitive market where observed changes in supply and demand should have some bearing on the economic outcomes.

11.1 Introduction

The price of seaborne transportation (the freight rate) is set in an equilibrium between supply and demand [18]: ships offering a transportation service and cargoes that need to be moved from A to B. The freight rate is set in a decentralized spot market, where shipowners and cargo owners (charterers) negotiate the pricing of individual voyages bilaterally through middlemen called shipbrokers. In the very short run, this spot freight market consists of local micro-auctions where cargoes requiring transportation are matched with ships that are able to make a certain time window at the loading port. While such auctions can be thought of as local in nature, ships are frequently contracted (fixed) for the next cargo while they are still several weeks' sailing away from the loading port. Because there is some flexibility in terms of when a vessel and a cargo need to be fixed, there is no strong link between the physical location of a vessel and her ability to participate in these auctions. This is an important point that we will revisit later. We also note that while it is common to merely count ships (as a proxy for supply) and cargoes (as a proxy for demand)

R. Adland (✉)
Norwegian School of Economics, Bergen, Norway
e-mail: roar.adland@nhh.no

© Springer Nature Switzerland AG 2021
A. Artikis and D. Zissis (eds.), *Guide to Maritime Informatics*,
https://doi.org/10.1007/978-3-030-61852-0_11

when considering the current market balance in a region, the proper measure of demand and supply is 'tonnemiles', that is, the product of the distance that the ship sails with cargo onboard (the *laden voyage*) and the amount of cargo onboard. The reason is that carrying a greater volume of cargo over longer distances requires more ships to satisfy the demand. Over a slightly longer time horizon (i.e. in the order of weeks), the fleet is able to reposition itself globally in an attempt to take advantage of regional imbalances in the demand and supply of transportation such that very short-term regional rate differences are reduced, making the spot freight market more geographically integrated.

What we observe as ship movements using AIS data is therefore a result of a simple economic mechanism: Firstly, ships need to satisfy the demand for seaborne transportation, resulting in some well-known global trading patterns, such as flows of iron ore from Australia to China or crude oil exported from the Arabian Gulf. Secondly, shipowners and operators must attempt to reposition their empty ships in space and time such that they maximize the vessels' earnings, subject to the constraints imposed by the cargo flows at the macro level. In its most basic form, this means moving the ships to regions where there is a perceived future undersupply of ships and therefore a potential for higher freight rates, relative to the global average. Advances in remote sensing, such as the tracking of ships using Automated Identification System (AIS) data, have opened up new and interesting possibilities to optimize the commercial management of ships in this manner.

In this chapter we will review how AIS data can be utilized for the commercial management of ships in two key areas: The real time tracking of trade flows for key commodities and freight trading analytics. The former can be interpreted as the tracking of demand for seaborne transportation, and the objective of the latter is to predict the direction of price movements for freight rates themselves or their financial derivatives.

11.2 Tracking of Trade Flows

A common application of AIS data for commercial analysis is the tracking of seaborne trade in key commodities [4, 10]. However, it is important here to remember what AIS can and cannot do. The system does a reasonably good job of tracking the position, speed and heading etc. of a vessel (see the chapter by Bereta et al. [7] for a detailed introduction). However, the data provided by the system does not, by itself, provide any information of the cargo type or cargo volume onboard. In order to be useful for commercial data analysis in this context, AIS data therefore has to be enriched with other datasets such as geospatial data on the location of cargo terminals and port agent or Bill of Lading[1] data containing further information on cargo sizes

---

[1] Bill of Lading is a document that confers ownership of the cargo. It contains information on the terms of carriage, such as the amount of cargo, and acknowledges that the cargo has been loaded on the ship.

and types (see the chapter by Santipantakis et al. [16] for AIS data enrichment techniques). Even with such information available, the analysis is generally limited to major commodities that are typically traded through single-purpose terminals as homogeneous loads: coal, iron ore, liquified natural gas (LNG) and crude oil.

The reason for this limitation becomes clearer when we consider the typical trading pattern for such commodities. Typically, a single shipment between a seller and a buyer occupies the full cargo carrying capacity of a ship (i.e. a full shipload), moving from clearly defined loading areas (say, the Arabian Gulf for crude oil) to separate consuming areas using purpose-made terminals. This means that a user with access to AIS data for the relevant ship type (e.g. crude oil tankers) can deduce the direction of trade and get a good handle on cargo size simply by observing the location of the loading/discharge ports and the ships' deadweight (DWT).[2] However, if cargoes are loaded from multi-use terminals that are not dedicated to a particular cargo type, then the mere observation of ship movements using AIS data alone cannot be used to infer anything about what the vessel is carrying, limiting the usefulness of such data for the aggregation of tradeflows. Similarly, for vessel types that by design can carry multiple cargo types or carry out both loading and discharge operations during the same port call (such as chemical carriers, product tankers), it is impossible to reliably infer the volumes carried or transferred of the various cargo types and, hence, to aggregate such information to reliable trade volumes.

11.2.1   Classification of Voyages

At the conceptual level, the algorithm for generating trade volume data from vessel observations can be described as follows:

1. Isolate the fleet of vessels capable of transporting the commodity (by IMO numbers).
2. Identify the location of ports/terminals serving this subset of the fleet and whether they are loading or discharge terminals.
3. Identify laden voyages by the relevant fleet between one or more loading ports and one or more discharge ports.
4. Estimate the cargo volume/weight onboard for each laden voyage.
5. Aggregate cargo sizes for individual voyages into a global trade matrix of exporting/importing ports or countries.

Some might argue that step 1 is not needed, and this would be correct in the case where the location of terminals and their properties have already been identified. For instance, if the analyst has already identified all terminals from which iron ore can be exported globally, it suffices to monitor the activity of the ships that call at

---

[2]Deadweight (DWT) refers to the carrying capacity of a ship in metric tonnes. The carrying capacity is used mainly for cargo (typically 90 – 95%) but also includes the weight of the crew, spare parts, fresh water, fuel oil and more.

these terminals. However, if such identification has not been done, the easiest way to locate the terminals used by a particular vessel type is to use AIS data (location and vessel speed) to identify the clusters in space where the ships are idle (at least approximately so, say vessel speed < 1 knot). We note here that such clusters may represent both anchorages where ships typically wait prior to cargo operations and the quay or offshore buoy where cargo operations take place. They may even represent random offshore locations where a ship has elected to wait for the next cargo contract. For the purpose of estimating trade flows such offshore locations must be excluded, though these data points can also represent valuable information as we shall see later. Similarly, it may not be necessary to distinguish between the various elements of a vessel's port call (anchorage, maneuvering, alongside terminal) when the objective is to monitor trade flows. However, the finer the resolution in time and space, the more statistics can ultimately be extracted from the data analysis, such as timeseries of port delays and port servicing times. The navigational status as reported in the AIS feed (e.g. underway, moored, at anchor) may also be useful in the classification of vessel activities, at least for cross-checking the results of clustering algorithms, yet we should keep in mind that the field is input by the crew and therefore may be unreliable.

Figure 11.1 illustrates the use of stationary vessels for identifying the polygons for where ships usually wait or perform loading operations, courtesy of Marinetraffic.com. The clusters of green dots in the vicinity of Port Hedland, Australia, outlines the location and extent of the anchorage and terminals. Refer also to the chapter by Tampakis et al. [19] for a detailed discussion on clustering techniques

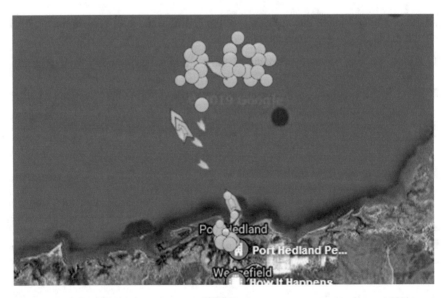

Fig. 11.1 Identification of anchorage and terminal polygons using AIS data. *Source* Marinetraffic.com/Google imagery

and the chapter by Andrienko and Andrienko [6] for visual analytics techniques for identifying anchorage areas.

When it comes to classifying single voyages as either laden or 'ballast', i.e. when the ship has no cargo onboard, the analyst can base the classification on changes in draught data, the location of terminals or, ideally, a combination of both. As the AIS-reported draught is manually input by the crew it is known to be notoriously inaccurate, particularly with regards to whether it is updated in a timely fashion during or after a port call. It follows that the classification of a vessel's loading condition (laden or ballast) based on draught alone can lead to the wrong outcome. A more reliable indicator is often the location and properties of the terminal itself. For certain commodities such as coal and iron ore, classification at the terminal level (or, at a higher resolution, quays) is usually straightforward as terminals are typically dedicated to either loading or discharge. For commodities such as crude oil, extensive use of temporary storage and transshipment hubs makes geographical location a less accurate predictor of whether a vessel has been loaded or discharged [4].

Once the identification of terminals and their properties (cargo type, loading/discharge) is done, the sequence of port calls can be isolated from a vessel's tracking history by extracting the occurrences in the data where a ship makes a stop (zero speed) inside the relevant polygon for a sufficient length of time. The threshold for the duration of what is likely to constitute a cargo operation will vary with ship type and size. Voyages from a loading port to a discharge port is then naturally classified as laden, while voyages from a discharge port to a loading port is classified as ballast. We note here that multi-porting, that is, the partial loading or discharging in a sequence of ports, is relatively common, particularly for large oil tankers, something the analyst must be aware of when identifying and classifying voyages by cargo status. Furthermore, care needs to be taken to ensure that all port calls are captured in the data, and that all port calls relate to cargo operations and not, for instance, maintenance in a shipyard or refueling. For instance, Singapore will often appear as a destination for bulk carriers even though there are no drybulk terminals in the city state. While such other vessel activities matter for the overall productivity of the fleet, the proper tracking of commodity trade flows requires that only port calls for cargo operations are included in the analysis.

11.2.2 Aggregating Cargo Volumes

Without access to Bill of Lading data stating the true cargo size loaded or discharge in a particular port, the analyst must rely on third-party estimates such as those provided by port agents in line-up reports or, more commonly, attempt to estimate cargo size based on the reported vessel draught from AIS messages. The underlying idea here is that for modern bulk carriers and tankers there is a roughly proportional

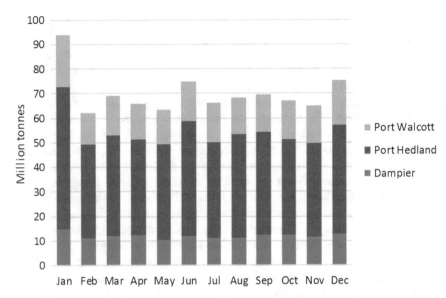

Fig. 11.2 Australian iron ore exports from main ports in 2018 derived from AIS data. *Source* Ocean Freight Exchange

relationship between the draught of the vessel and the cargo size.[3] For a detailed discussion on the various models for estimating cargo intake, the interested reader can refer to [11]. Given the well-known challenges with the accuracy of manually reported draught measures in AIS data, an equally useful approach may be to simply assume that the deadweight (DWT) utilization is a constant (say 90–95%). In other words, if the analyst has identified that the DWT of a bulk vessel calling a particular port is 100,000 tonnes, the cargo intake is estimated at 95,000 tonnes. Particularly where the purpose is to monitor changes in traded volumes at an aggregate level, such a simple approximation is likely to be satisfactory.

Once all port calls for the loading or discharge of the chosen commodity have been identified from the AIS data, it is possible to generate trade statistics at the chosen level of aggregation, for example, imports/exports at the port or country level, or a country-to-country trade matrix using the port pairs identified as laden trips. To the extent that terminals are dedicated to a single user, it may in some cases also be possible to generate exports by company, a useful input in financial analysis of listed companies such as the major iron ore miners (BHP, Rio Tinto, Vale, etc.).

As an example, Fig. 11.2 below shows monthly 2018 iron ore exports from Australian ports, as derived from AIS data.

---

[3]Assuming no ballast water and completely vertical ship sides in the relevant draught range, this holds exactly due to Archimedes' principle.

11.3 Analytics for Freight Trading

A major driver for the development of commercial applications based on AIS data in shipping has always been the quest to make more profitable trading decisions. Such decisions relate to either the commercial management of ships or trading in the freight derivative markets. Freight derivatives are financial contracts, such as options or futures, that derive their value from the development of spot rates in the physical freight market. An example is Forward Freight Agreements (FFA), which are effectively contracts that allow for bets on the market direction (see [5] for a detailed description and discussion of freight derivatives). For the purpose of derivatives trading, the objective of freight market analytics is therefore straightforward: How can we use AIS data and models to predict the direction of freight rates?

For the commercial management of ships, it is worth to first revisit how *shipowners* or *operators* can make money on trading their ships. By an owner we here mean a company that owns a fleet of ships and charters them out for single voyages in the spot market or longer periods of time under a so-called timecharter contract. An operator typically does not own the ships, but charters in vessels on short- or long-term contracts and attempts to re-let the vessels on different contracts at higher average rates in order to profit from the difference. Both types of freight market agents can additionally take on future obligations to carry cargoes at an agreed freight rate (so-called Contracts of Affreightments, COA).

Broadly speaking, the sources of profits in the physical freight market can then be classified as:

1. **Timing of cycles:** Chartering in vessels at low rates on longer-term contracts before an expected upturn and taking on COAs for cargoes or chartering out vessels at high rates on long-term contracts before an expected downturn. If trading in the spot market, a shipowner would want to take on short-duration voyages if rates are expected to increase, and long-duration voyages if rates are expected to decrease, thereby marginally increasing the average earnings of the ship.
2. **Geographical positioning:** While regional freight rates for different routes tend to move in sync in the long run (spatial co-integration of prices), there can be substantial differentials between them in the short run [1]. Because it takes time for vessels to physically move between regions to even out any local imbalances in supply and demand, such freight rate differentials can be large and persistent enough for smart operators to take advantage, as shown by [1, 13]. AIS data can have value here as an input to a model for the monitoring and prediction of short-term fleet and price movements.

We note that both sources of increased earnings involve an element of timing and positioning, albeit on two different scales. The first one covers the overall position of an owner or operator in the freight market: basically whether one should be "long" transportation capacity (i.e. benefit from an increase in freight rates, as is the natural position of a shipowner without any contractual obligations) or "short" transportation capacity (i.e. to have taken on more contractual obligations for the transportation of cargo than you have ship capacity available) and benefit from a decrease in freight rates. Even for companies which do not take on COAs, either as a policy or due to the lack of opportunities, the timing and contract mix is of crucial importance throughout the freight market cycle. The main question is then to manage the mix and duration of spot and longer-term timecharter contracts, typically over a period of many years. When it comes to the second source of increased earnings—short-term geographical positioning, the objective is to position the fleet around the globe such that its earnings are optimized, given the risk attitude of the owner/operator. Risk attitude here refers to how a company behaves under uncertainty. For instance, a highly risk-averse company would prefer to secure a low-paying contract for its vessel early while foregoing the opportunity of a potential future highly profitable but uncertain spike in freight rates (see [14] for a discussion on the relationship between risk attitude and behavior in the freight market). The risk appetite may, for instance, be a function of how much working capital the company has available and its financial leverage (i.e. the ratio of debt to equity capital). The risk aspect is important here, as without due attention to risk, an optimization algorithm might, for instance, suggest that the entire fleet be trading on a single route, which could turn out to be costly if the bet is incorrect. It is fair to say that until recently such chartering decisions have been made on the basis of human experience, perhaps complemented by some data on fleet and macroeconomic developments, but with little input from data-driven decision-support systems.

11.3.1 Monitoring Market Balance Using AIS Data

How can AIS data be used in this context? Principally, this type of data enables analysts to estimate and monitor the realized supply and demand for maritime transportation. We have already shown earlier in the chapter how it is possible to derive an estimate for the amount of cargo moving on a ship, whether a voyage is laden or ballast, and the duration or distance of every leg of a trip. This information can be converted to an estimate of the *tonnemile* demand that each cargo is generating by simply calculating the product of cargo size and distance sailed while laden. All else equal, an increase in the demand for transportation relative to the available supply of transportation will increase the freight rate [18]. As such, we now have a tool to monitor the demand for transportation that can be broken down or aggregated in a very flexible manner: by main trade route, by commodity (within the limits described initially, i.e. full loads of homogeneous cargo), by ship size range, and so on. This statement comes with one important caveat: When we observe a ship moving from loading port A to discharge port B using AIS data this actually does not represent current demand—it is instead a result of the matching (fixing) of a cargo and a ship in the past—perhaps as much as several weeks ago. What we are observing is *realized demand* and, therefore, we cannot rule out that there were cargoes that were in the market but were postponed or cancelled, even though they were part of the demand for transportation at some point. Nonetheless, the estimation of tonnemile demand on the basis of AIS data is a substantial improvement over what has been available to researchers and analysts in the past, not least because it can be done semi-automatically and in near real time as outlined in [10].

Similarly, AIS data cannot tell us what the true transportation capacity of the fleet is, i.e. the maximum tonnemiles that the fleet is capable of producing when it is sailing at full speed, using optimal routes and with the lowest port turnaround times. Again, what we are observing with ship movements is the response of the fleet in order to match realized demand, given constraints in trading patterns, weather and port conditions. Using AIS data to try to estimate the supply/demand balance, or what is often called fleet utilization, is therefore a precarious effort.

A more common approach is therefore to monitor changes in the supply/demand balance by using proxies for fleet productivity. We shall briefly review three possibilities here and how we can use AIS data to derive their estimates: average vessel speeds, port congestion and the number of idle vessels.

11.3.2   Monitoring Vessel Speeds

According to classical maritime economic theory [15, 18], ships should respond to higher earnings by speeding up to the point where the value of time gained equals the cost of the additional fuel burned. The availability of AIS data has allowed researchers to test this theory empirically [2, 3]. Given that a vessel's speed-over-ground (SOG) is reported in its AIS data feeds, we can aggregate the speed information for individual ships to some measure of average fleet speeds, keeping in mind that the reported speeds represent instantaneous snapshots reported at irregular intervals and not the average speed over the time interval. To adjust for this, and the prevalence of missing data, the analyst can alternatively estimate the average speed between two reported ship positions using the distance and reported timestamps. A further consideration is how to aggregate individual vessel speeds when analysing fleet behaviour. For instance, such aggregation can be done by ship type and vessel size, by route or as a global average, and by loading condition (i.e. ballast vs laden voyages). Academic research suggests that speed variations are smaller for laden ships, as this is when they are under contractual obligations to avoid delays [2, 3]. Accordingly, changes in ballast speeds are more likely to be reflecting the underlying changes in business conditions and what you want to monitor.

As an example, Fig. 11.3 shows the AIS-derived average daily vessel speed in the Capesize segment for the period 2015–2017. A Capesize bulk carrier is a vessel with a carrying capacity of 100,000 DWT or greater, and carries mainly iron ore and coal. Overall, higher average vessel speeds indicate that the fleet in question is responding by supplying more transportation (tonnemiles per time unit), all else equal. Hence, it is a short-term indicator of optimism and, possibly, increasing demand. In the medium term, higher supply will of course have a negative impact on freight rates if the increase in demand is not sustained [18].

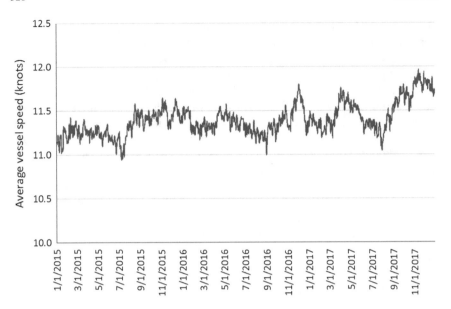

Fig. 11.3 Daily average Capesize speeds derived from AIS data. *Source* Cargometrics Technologies LLC. Includes all drybulk vessels with DWT > 150,000, both laden and ballast voyages, with average daily speeds above 7 knots

11.3.3 Monitoring Port Congestion

Another driver of fleet productivity which can be monitored using AIS data is port congestion, that is, the number of ships lying idle at anchorage awaiting cargo operations, and their waiting times. From a fundamental point of view, every ship that is lying idle due to port congestion contributes to a reduction in supply which, all else equal, will increase freight rates [18]. As an example, we note here that port congestion was a major driver of the strong drybulk freight market in the 2005–2007 period as China was ramping up its imports of iron ore, causing severe port congestion in both loading ports in Brazil and Australia and discharge ports in China.

In estimating port congestion we can rely on parts of the same pseudo-algorithm as outlined for the tracking of cargo flows. The key difference is that we here have to assess whether a vessel is at anchorage or alongside a terminal performing a cargo operations and, ideally, whether there is any remaining cargo onboard. Once again, the standard approach is to map the respective polygons where ships typically wait and undertake cargo operations based on observing vessel behavior in the AIS dataset. Figure 11.4 below (left panel) illustrates the separation of anchorage (black polygon) and port (red polygon) in the port of Santos, Brazil. At the most basic level, a measure of port congestion can then be based on counting unique ships (MMSI or IMO numbers) within a polygon across all ship types. This is illustrated for the port of Santos in Fig. 11.4 (right panel).

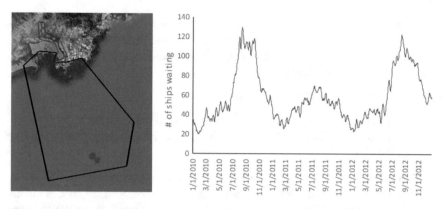

Fig. 11.4 Definition of anchorage (left) and number of ships at anchorage (right) in Santos, Brazil. Count of unique MMSI numbers within polygon, 7-day moving average, all vessel types. *Source* Cargometrics Technologies LLC

However, refinements to this approach may be needed (see, for example, the chapters by Etienne et al. [9], Tampakis et al. [19] and Andrienko and Andrienko [6]. Notably, for multi-commodity ports such as Santos, it may be necessary to distinguish by ship type to obtain a more accurate estimate of the port congestion relevant to a particular terminal, such as the loading of grain. Additionally, for the purpose of estimating the effect on the supply of transportation, it is necessary to track the waiting time at the individual vessel level, in which case the analyst also needs to capture the timestamps for when a vessel enters and leaves the anchorage area. It is the product of waiting times and the number of ships waiting (shipdays) which is the key driver in the overall fundamental market balance.

There is one important caveat in the above analysis: Ships may wait in the anchorage of a port, yet not be under contract or waiting for cargo operations to commence. Such ships are not part of the true port congestion, as they are merely waiting for a new contract or for other operational reasons (e.g. bunkering or waiting to enter a drydock). For instance, it is quite common for ships to lie idle near major exporting areas such as Southern Brazil for iron ore and the Arabian Gulf for crude oil. Unfortunately, it is not possible to assess on the basis of AIS data alone whether an empty ship at anchorage is under contract or unemployed. In this case the AIS data must be enriched with commercial data such as fixture data or information on open ships (i.e. ships looking for cargoes) from broker reports.

11.3.4   Monitoring Unemployed Ships

The number of idle ships not under contract can be a useful short-term indicator of changes in the supply/demand market balance. The reason is that these unemployed vessels represent the realization of oversupply in the market, and so a decrease in unemployment is a sign of improved prospects for the remaining idle fleet, and vice versa. While AIS data alone cannot track this variable perfectly due to the source of error described above (i.e. lack of contract information), it is possible to develop proxies. Consider the following:

- Count idle ships (speed < 1 knot) that

  - Are not located within port anchorage or terminal polygons, and
  - Have no cargo onboard (according to AIS-reported draught, which may be unreliable)

This variable, which represents a 'complex maritime pattern'(c.f. the chapter by Pitsikalis et al. [12]), will pick up ships that are idle in the open sea, say, along the coast of the South China Sea or off South Africa, where there is no reason for ships to normally wait. As they have no cargo onboard, we can be fairly certain that they are without contract. We note that this variable will not pick up unemployed ships that are stationary within normal port or anchorage areas for the vessel type (c.f. the discussion in Sect. 11.3.3 above) and therefore may represent an underestimate of the true level of vessel unemployment.

Figure 11.5 shows such an estimated proxy variable for monitoring the number of unemployed ships in the case of Capesize drybulk vessels. The inverse relationship between the average earnings of the fleet and the number of idle vessels is evident from the graph. When vessel unemployment increases, freight rates decrease, and vice versa. This is because the former indicator represents a measure of oversupply, or the difference between the demand for transportation and the available capacity. Lower demand relative to supply (i.e. an increasing number of idle vessels) will, thus, reduce the spot freight rate [18].

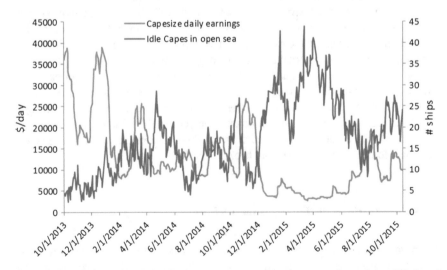

Fig. 11.5 Idle Capesize vessels waiting in open sea verses average spot earnings. *Source* Windward Ltd

11.4 Limitations and Summary

In this chapter we have discussed how AIS data can be utilized to generate new timeseries for variables that are suitable for the economic analysis of bulk freight markets, such as the tracking of commodity flows and transportation demand, port congestion and vessel speeds. There are numerous other potential applications that go beyond the scope of this chapter, for instance network analysis (i.e. how ports are connected within the various trades, see the chapter by Ducruet et al. [8]) and emission accounting [17].

We have also alluded to the fact that AIS data more often than not have to be enriched with other data sources to be useful. Mostly this is a matter of matching the vessel identity from AIS data with technical vessel information from fleet registers, such as the Clarksons World Fleet Register, to obtain, for instance, the vessel's DWT. This enables the matching of the observed vessels in the AIS data to the size and type categories used in commercial management (e.g. Capesize, Panamax, all crude oil tankers etc.), or the analysis of the fleet by ownership.

A more fundamental shortcoming of spatial data in the context of economic analysis is that the vessel position, or the past behavior of a ship as given by its historical track, does not reveal much if anything about its commercial status, i.e. whether it is under contract and, if so, what type of contract (spot voyage charter or longer-term period timecharter). Such information would be crucial to model the true regional supply of transportation. As an example, even if we can observe from AIS data that 30 Capesize drybulk vessels are sailing in ballast towards Brazil, the implication for future freight rates is dependent on how many of these that already are (or will be) fixed prior to arrival.

It is also worth keeping in mind that the observed movement of a ship is generally a result of a transaction in the chartering market that happened some time ago. For instance, the loading of oil cargoes for oil tankers can occur up to five weeks after the fixture was done in the freight market [14]. It follows that the present spatial distribution of ships contains limited information about the true supply of transportation. Perhaps more importantly, it tells us nothing about the demand side, as the arrival of cargoes and their destination is an exogenous process. Put differently, the timing of an 'arrival' of a cargo looking for a ship in the freight market can be thought of as independent of where ships are located. Instead, this process is determined by buyers and sellers in the commodity markets. The best we can do on the basis of AIS data is to derive the historical distribution of cargoes from a given terminal by counting the departure of laden vessels (e.g. establishing that between 50 and 80 cargoes for large oil tankers depart the Arabian Gulf every week). Getting a better handle on the demand side requires access to the flow of cargoes, most of which is private data and in the hands of shipbrokers and charterers. It is important when working with AIS data for the purpose of freight market analysis to keep these limitations in mind.

11.5 Bibliographical Notes

Readers that require a thorough introduction to the commercial shipping industry and the functioning of freight markets should read the book "Maritime Economics" by Dr. Martin Stopford. See Ref. [18].

Acknowledgements This research was supported in part by the Research Council of Norway as part of the project 280684 Smart digital contracts and commercial management. The author would also like to thank Dr. Vit Prochazka for assistance with formatting the manuscript, as well as Dr. Scott Borgerson of Cargometrics, John Hahn of Ocean Freight Exchange and Ami Daniel of Windward Ltd. for sharing their timeseries derived from AIS data.

References

1. Adland, R., Bjerknes, F., Herje, C.: Spatial efficiency in the bulk freight market. Marit. Policy Manag. **44**(4), 413–425 (2017). https://doi.org/10.1080/03088839.2017.1298864
2. Adland, R., Jia, H.: Vessel speed analytics using satellite-based ship position data. In: 2016 IEEE International Conference on Industrial Engineering and Engineering Management (IEEM), pp. 1299–1303 (2016). https://doi.org/10.1109/IEEM.2016.7798088
3. Adland, R., Jia, H.: Dynamic speed choice in bulk shipping. Marit. Econ. Logist. **20**(2), 253–266 (2018). https://doi.org/10.1057/s41278-016-0002-3
4. Adland, R., Jia, H., Strandenes, S.: Are ais-based trade volume estimates reliable? the case of crude oil exports. Marit. Policy Manag. 1–9 (2017). https://doi.org/10.1080/03088839.2017.1309470
5. Alizadeh, A., Nomikos, N.: Shipping Derivatives and Risk Management. Palgrave MacMillan (2009)
6. Andrienko, N., Andrienko, G.: Visual analytics of vessel movement. In: Artikis, A., Zissis, D. (eds.) Guide to Maritime Informatics, chap. 5. Springer (2021)
7. Bereta, K., Chatzikokolakis, K., Zissis, D.: Maritime reporting systems. In: Artikis, A., Zissis, D. (eds.) Guide to Maritime Informatics, chap. 1. Springer (2021)
8. Ducruet, C., Berli, J., Spiliopoulos, G., Zissis, D.: Maritime network analysis: Connectivity and spatial distribution. In: Artikis, A., Zissis, D. (eds.) Guide to Maritime Informatics, chap. 10. Springer (2021)
9. Etienne, L., Ray, C., Camossi, E., Iphar, C.: Maritime data processing in relational databases. In: Artikis, A., Zissis, D. (eds.) Guide to Maritime Informatics, chap. 3. Springer (2021)
10. Jia, H., Lampe, O., Solteszova, V., Strandenes, S.: An automatic algorithm for generating seaborne transport pattern maps based on ais. Marit. Econ. Logist. (2017). https://doi.org/10.1057/s41278-017-0075-7
11. Jia, H., Prakash, V., Smith, T.: Estimating vessel payloads in bulk shipping using ais data. Int. J. Shipp. Transp. Logist. **11**, 25 (2019). https://doi.org/10.1504/IJSTL.2019.096864
12. Pitsikalis, M., Artikis, A.: Composite maritime event recognition. In: Artikis, A., Zissis, D. (eds.) Guide to Maritime Informatics, chap. 8. Springer (2021)
13. Prochazka, V., Adland, R., Wallace, S.W.: The value of foresight in the drybulk freight market. Transp. Res. Part A: Policy Pract. **129**, 232–245 (2019). https://doi.org/10.1016/j.tra.2019.07.003. http://www.sciencedirect.com/science/article/pii/S0965856418312497
14. Prochazka, V., Adland, R., Wolff, F.C.: Contracting decisions in the crude oil transportation market: evidence from fixtures matched with ais data. Transp. Res. Part A: Policy Pract. **130**, 37–53 (2019). https://doi.org/10.1016/j.tra.2019.09.009. http://www.sciencedirect.com/science/article/pii/S0965856418311650

15. Ronen, D.: The effect of oil price on the optimal speed of ships. J. Oper. Res. Soc. **33**(11), 1035–1040 (1982). http://www.jstor.org/stable/2581518
16. Santipantakis, G.M., Doulkeridis, C., Vouros, G.A.: Link discovery for maritime monitoring. In: Artikis, A., Zissis, D. (eds.) Guide to Maritime Informatics, chap. 7. Springer (2021)
17. Smith, T., Jalkanen, J., Anderson, B., Corbett, J., Faber, J., Hanayama, S., O'Keeffe, E., Parker, S., Johansson, L., Aldous, L., Raucci, C., Traut, M., Ettinger, S., Nelissen, D., Lee, D., Ng, S., Agrawal, A., Winebrake, J., Hoen, M., Chesworth, S., Pandey, A.: Third IMO Greenhouse Gas Study 2014. International Maritime Organization, United Kingdom (2015)
18. Stopford, M.: Maritime Economics, 3rd edn. Routledge, London (2009)
19. Tampakis, P., Sideridis, S., Nikitopoulos, P., Pelekis, N., Theodoridis, Y.: Maritime data analytics. In: Artikis, A., Zissis , D. (eds.) Guide to Maritime Informatics, chap. 4. Springer (2021)